GENE ENVIRONMENT INTERACTIONS

GENE ENVIRONMENT INTERACTIONS

Nature and Nurture in the Twenty-first Century

Moyra Smith

Professor Emerita, Department of Pediatrics and Human Genetics, UCI Institute for Clinical and Translational Science, UCI Campus & Medical Center, University of California, Irvine, Irvine, CA, United States

ACADEMIC PRESS

An imprint of Elsevier

Academic Press is an imprint of Elsevier
125 London Wall, London EC2Y 5AS, United Kingdom
525 B Street, Suite 1650, San Diego, CA 92101, United States
50 Hampshire Street, 5th Floor, Cambridge, MA 02139, United States
The Boulevard, Langford Lane, Kidlington, Oxford OX5 1GB, United Kingdom

Notices
Knowledge and best practice in this field are constantly changing. As new research and experience broaden our understanding, changes in research methods, professional practices, or medical treatment may become necessary.

Practitioners and researchers must always rely on their own experience and knowledge in evaluating and using any information, methods, compounds, or experiments described herein. In using such information or methods they should be mindful of their own safety and the safety of others, including parties for whom they have a professional responsibility.

To the fullest extent of the law, neither the Publisher nor the authors, contributors, or editors, assume any liability for any injury and/or damage to persons or property as a matter of products liability, negligence or otherwise, or from any use or operation of any methods, products, instructions, or ideas contained in the material herein.

Library of Congress Cataloging-in-Publication Data
A catalog record for this book is available from the Library of Congress

British Library Cataloguing-in-Publication Data
A catalogue record for this book is available from the British Library

ISBN 978-0-12-819613-7

For information on all Academic Press publications
visit our website at https://www.elsevier.com/books-and-journals

Publisher: Andre Gerhard Wolff
Acquisition Editor: Peter B. Linsley
Editorial Project Manager: Samantha Allard
Production Project Manager: Swapna Srinivasan
Designer: Mark Rogers

Typeset by SPi Global, India

Working together to grow libraries in developing countries

www.elsevier.com • www.bookaid.org

Contents

Acknowledgments

Going far back in time, I am grateful to my Grandfather, who in his later years became a farmer who demonstrated harnessing wind for energy and who husbanded earth's natural resources with forethought and care.

I am grateful to our professors at Medical Schools in South Africa who inspired us to consider evolution and who took us to sites where remains of archaic Homo species were discovered.

I acknowledge the inspiration and excitement inspired by my mentors at University College London, Harry Harris and David Hopkinson who taught all the world so much about individual genetic variation in humans and about differences in population frequencies of specific genetic variants.

I am very grateful for the encouragement and advise I received from Peter Linsley at Elsevier in planning and preparing the contents of this book.

I am very grateful for access to the extensive resources available through the University of California Library System, and for insights provided by patients, faculty and students at the University of California, Irvine.

Throughout the stages of preparation and production of this book I received outstanding help from production editors at Elsevier, Samantha Allard and Swapna Srinivasan, and I sincerely thank them both.

My thanks are also due to Dr Simon Prinsloo for his encouragement throughout this process.

Epigraph

Science has a simple faith which transcends utility. It is the faith that it is a privilege of man to learn to understand and that this is his mission.

Vannevar Bush in "Searching for Understanding" published 1967

INTERACTING WITH THE ENVIRONMENT RECEIVING AND INTERPRETING SIGNALS

Sensory receptors

General factors

In considering sensory systems it is important to take into account initiating stimuli, cell membranes, cellular receptors, ion channels, ion pumps and intra-cellular signaling systems, particularly G- protein coupled systems. Julius and Nathans[1] categorized stimuli of sensory systems as small molecules, mechanical changes or radiation changes, e.g. heat or light energy radiation.

G-protein coupled receptors are activated when a specific ligand couples to the receptor. The G protein then activates intra-cellular second messengers. The passage of ions into cells can be accomplished through specific ionotropic receptors or through specific ion channels that only conduct passage of ions. The latter include calcium and sodium channels, chloride channels, potassium channels. Passage of ions into cells can result in changes in electrical charge.

Touch sensation

Touch sensation is enabled by mechanoreceptors in the skin. Mechanoreceptors are sometimes referred to as encapsulated mechanoreceptors. Purves et al.[2] described four types of encapsulated mechanoreceptors: Meissner corpuscles, Pacinian corpuscles, Merkel's discs and Ruffini corpuscles.

Meissner corpuscles occur beneath the epidermis and their capsular components include connective tissue and myelin producing Schwann cells. They detect low frequency stimulation. Pacinian corpuscles occur in subcutaneous tissue. They are also present in other locations, such as in the connective tissue in the skeletal system and in the gut mesentery. Purves et al. described Pacinian corpuscles as having onion like layers and their outmost layer surrounds a fluid filled section. Pacnian corpuscles detect high frequency stimulation.

Gene Environment Interactions. https://doi.org/10.1016/B978-0-12-819613-7.00001-3

Merkel's disks occur in the epidermis and they respond to light pressure. The disks form a saucer like structure that accommodates nerve endings. They also provide information on contours. Merkel's disks contain vesicles that can release neurotransmitters. Ruffini corpuscles form spindle shaped structures that occur deep in the skin and in ligaments. These structures are sensitive to stretching.

Touch and pressure on the skin lead to opening of mechanosensitive channels located within the sensory receptors. Hao et al.[3] noted that influx of cations through these channels generated an electric potential that can be further amplified by voltage gated channels. They documented the following excitatory channels and voltage gated channels:

TRPA1	transient receptor potential
PIEZO2	a cation channel with 30 transmembrane domains
Nav 1–8 (SCN4A)	voltage gated sodium channel
Cav3.2 (CACNA1H)	voltage gated calcium channel

Hao et al. also documented signals that inhibited mechanosensitive channels and specific molecules that amplified inhibitory signals, these included:

TREK1 (KCNK2)	two pore potassium channel
TRAAK (KCN4	potassium channel

The signal generated in sensory nerve terminals can be transmitted through the connected axon to neuronal cell bodies in dorsal root ganglia and then subsequently transmitted through secondary axons to the central nervous system.

Jenkins and Lumpkin[4] noted that low threshold mechanoreceptors arose from neural crest cells and that development of somatosensory neurons requires expression of the transcription factor neurogenin. Additional factors involved in specification of mechanoreceptors include the transcription factor MAF and the transmembrane receptor RET that interacts with the ligand GDNF (glial derived neurotrophic factor).

Jenkins and Lumpkin drew attention to the altered sensory perception that has been reported in cases of autism. Some children with autism have been report to have tactile hypersensitivity while other children with this disorder have tactile hyposensitivity.

Nociceptors

These are sensory receptors that detect extreme change in temperature and pressure and can detect the application of harmful chemicals. They may also be activated by chemicals released as a result of inflammatory processes. The stimulation of these receptors then triggers the pain pathway. Inability to detect painful stimuli that occurs in consequence of specific mutations, is a dangerous condition.

Sherrington in 1903[5] first reported the existence of pain receptors and referred to these as nociceptors. In a 2007 review Woolf and Ma[6] noted that nociceptor associated neurons are frequently unmyelinated C fibers or in some cases may be associated with thinly myelinated fibers A delta fibers.

The cell bodies of nociceptors are located in dorsal root ganglia within spinal nerves. They are also located in the trigeminal ganglia. The axons that arise from dorsal root and trigeminal ganglia cell bodies give rise to peripheral branches. In addition, cell bodies give rise to central axons that enter the central nervous system and end at specific central terminals.

Woolf and Ma noted that studies by a number of investigators have revealed that nociceptors are derived late in neurogenesis from the neural crest stem cells in the dorsal neural tube. There is also evidence that the cells that give rise to nociceptors express receptors for the TRKA nerve growth receptor (also known as NTRK1 neurotrophic receptor tyrosine kinase 1). Neurogenin 1 is important for their differentiation and maintenance also requires expression of transcription factor Brna 3A (POUAF1).

The sequence of events following activation of sensitizer with nociceptors may involve direct interaction with specific ion channels or phosphorylation of specific small G-protein followed by ion channel activation.

Activation of ion channels leads to generation of electrical current. The ion channels therefore act as transducers and transmit electrical signaling along the nerve axons.

Di Mario[7] described the pain receptors (nociceptors) as unmyelinated or small diameter myelinated axons with distal ends located in end-organs such as the skin.

Sodium ion channels

The nomenclature of these genes was changed from Nav 1 to SCN. Dib-Hajj and Waxman[8] reported that 9 different genes encode sodium channels alpha subunits. Each gene encoded channel is composed of 4 domains. Collectively the domains give rise to 24 transmembrane segments. The 4th domain forms the voltage sensor of the channel.

The N terminal and C terminal domain of the protein are intracellular. In addition, the transmembrane segments of the proteins are linked to each other by means of loops.

The SCN1A genes encode subunits that form the Nav 1 channels. Dib-Hajj and Waxman[9] reported that within peripheral neurons only a subset of sodium channels occur. These include channels Nav1.7, Nav1.8 and Nav 1.9 that are expressed in peripheral sensory neurons and in dorsal root ganglia. They noted that Nav1.7 channel is also expressed in sympathetic ganglion neurons.

Sodium ion channels play important roles in amplifying signals received on excitation of sensory receptors and nociceptors.

Dib-Hajj and Waxman[9] noted that many pain syndromes are due to defect in activity of Nav1 type channels caused by mutations in sodium channel genes, SCNA genes.

In 2006 Cox et al.[10] reported 3 consanguineous families from Pakistan that each reported individual who manifested congenital insensitivity to pain. They mapped the locus for this recessive condition to chromosome 2q24.3. This chromosome region was found to harbor the locus for a voltage gated sodium channel Nav1.7 (SCN9A). Each of the three families harbored a different homozygous nonsense mutation. This finding led them to conclude that SCN9A sodium channel was essential for pain sensitivity. Key genes and channels associated with pain syndromes, that may include hypersensitivity to pain or diminished sensitivity include SCN9A gene (NAV1.7 channel, SCN10A gene (Nav1.8 channels, and SCN11A gene (Nav1.9 channels).

Steven and Stephens[11] noted that specific calcium channels, including Cav2.2 (CACNA1B) played key roles in neurotransmitter and neuropeptide regulation and release in the dorsal root ganglia. They reported that 5 s order ascending neuronal pathways carry nociceptive information from the dorsal root ganglia to the thalamus and the cerebral cortex. They noted further that thalamic nuclei express high levels of T type low voltage calcium channels. Steven and Stephens reported that there are descending inhibitory pathways from specific brain regions to the dorsal root ganglia. The neurotransmitters in these descending inhibitory pathways include 5-hydroxytryptamine and nor-adrenaline. They reported that increased production of specific proteins, including ion channels could lead to increase sensitivity to pain or to a condition referred to as allodynia where non-painful stimuli are perceived as painful.

Therapeutic agents to treat pain specific that act on ion channels.

Skerratt and West[12] reported that 55 of the 215 ion channels described in humans are linked to pain pathways. Carbamazepine that impacts sodium channels was first approved in 1963 for treatment of epilepsy and has also been used for treatment of pain. Other sodium channel impacting medications used in pain management include Lidocaine, used as local anesthetic. Specific therapeutic agents that target calcium channels, particularly Cav2.2 (CACNA1B) have been developed for pain reduction, e.g. Gabapentin.

Various toxins from plants, insects, mollusks and fish are being investigated for their capacity to inhibit sodium channels and to reduce pain

Erickson et al.[13] reviewed the role of sodium channels in generation of chronic visceral pain resulting from disorders of the lower gastro-intestinal tract and the bladder. They reported that sodium channels Nav1.1 (SCN1A), Nav1.6 (SCN8A, Nav 1.8 (SCN10A) and Nav1.9 (SCN11A) contribute to generation of pain from these sites.

Neuropathic pain

This form of pain results from nerve injury, Cardosa and Lewis[14] reported that following nerve injury sodium channels Nav1.3 through Nav1.9 accumulate at different sites in the axon.

Transient receptor potential (TRP) channels

Veldhuis et al.[15] reviewed the transient receptor potential channel axis and coupling to the intracellular G protein signaling pathway. They reported that 28 different TRP channel proteins occur. The G-protein coupled TRP axis is involved in detecting and transmitting signals related to pain and itch. The TRP channel forms particularly involved in detection of pain and inflammatory signals and the chromosomal location of genes that encode them include:

TRPV1 17p13.2
TRPV2 17p11.2
TRPV3 17p13.2
TRPV4 12q24.1
TRPA1 8q21.11
TRPM2 21q22.3
TRPM8 2q37.1
TRPC3 4q27
TRPC5 X23
TRPC6 11q22.1

High intensity noxious stimuli open TRP channels. Direct stimulation of TRP channels leads downstream of phosphatidyl inositol signaling pathways.

Opioids reduce activity of TRP channels and K channel activity in presynaptic location. Analgesics that impacts peripheral TRP channels and peripheral G coupled receptors are under intense investigations.

Potassium channels

The opening of potassium channels permits influx of potassium (K) ions that counteract the conduction of signal Tsantoulas and McMahon[16] reviewed the relevance of potassium channels to the treatment of pain. They noted that C type nerve fibers are unmyelinated and that unmyelinated and thinly myelinated AS fibers are primarily involved in pain conduction. K channels are also important in neurotransmission and in cardiac function. These investigators noted that voltage gated potassium channels particularly relevant to pain included the following 7 channels.

Kv1.2 KCNA2 1p13.3
Kv2.2 KCNB2 8q21.11
Kv2.1 KCNB1 20q13.3
Kv3.4 KCNC4 1p13.3
Kv4.3 KCND3 1p13.2
Kv7.2 KCNQ2 20q13.33
Kv9.1 KCNS1 20q13.12

Hearing

In considering hearing it is useful to briefly review aspects of the anatomy of the ear, particularly the middle ear and the inner ear. The outer ear leads into the auditory canal. The middle ear is separated from the outer ear by the tympanic membrane. The key structures of the middle ear include the moveable bones, malleus, incus and stapes. One part of the malleus is connected to the tympanic membrane, another part of the malleus connects to the incus. The incus also connects to the stapes. The stapes also connects with the fenestra ovalis in a membranous structure that separates the middle ear and the inner ear. Through this structure the movements of bones in the middle ear are conveyed to the inner ear, Gray's Anatomy.[17]

The inner ear is lined with membranous tissue. The inner ear is divided into 3 regions, the vestibule that is contiguous with the fenestra ovalis, the semicircular canals and the cochlea. The key components of the cochlea include the basilar epithelium, the organ of Corti, sensory epithelium with outer and inner hair cells, and the stria vascularis, Gray's Anatomy.[17]

Fluid and ions are produced by the membranous tissue and ions particularly potassium (K+) are provided from the stria vascularis. Calcium ions also regulate processes in the inner ear.[18] These authors noted that hearing is dependent not only on mature functional hair cells. It is also dependent on non-sensory cell networks and on the transfer of ions and nutrient molecules through the gap junctions between cells.

The three different compartments of the inner ear, the scala vestibuli, scala media and the scala tympani are filled with fluids. The fluid that fills the scala vestibuli and the scala tympani is defined as perilymph. The scala media is filled with endolymph and has a higher concentration of Potassium (K+) and a higher electron potential. The stria vascularis is responsible for supplying a high concentration of K to the scala media endolymph.

Gap junctions

In 2009 Martinez et al.[19] reviewed the role of gap junctions in hearing. The gap junctions connect connexin proteins. They reported that mutations in five different genes that encode connexins had been reported in cases of deafness, the implicated genes encoded connexins 26, 31, 30, 32 and 43. Mutations in connexin 26 encoded by the gene GJB2 constituted a relatively common cause of deafness.

Some connexin 26 mutations lead not only to deafness but also to corneal lesions and skin lesion (keratoderma).

In describing gap junctions Martinez et al. noted that each cell forms a hemichannel and hemichannels on one cell dock with compatible hemichannels on an adjacent cell to form a channel that connects the two cells. A particular gap junction is not necessarily composed of identical forms of connexins. Channel function is influenced by cation concentrations and by pH. Some connexin 26 mutations lead to reduced channel function while other lead to hyperactivity of channels.

In 2009 Martinez reported that 90 different connexin mutations were known to lead to non-syndromic deafness and the GJB2 mutations accounted for almost half of all cases of hereditary deafness. Connexin muttions and gap junction defects particularly disrupt the transfer of potassium (K+) between cells. However, calcium transfer is likely also disrupted.

Fluid filled cavities of the inner ear

Three cavities in the inner ear include the scala vestibuli, the scala media that accommodates the cochlea and the scala tympani that lies beneath the cochlea. The fluid that fills the scala vestibuli and the scala tympani is referred to as perilymph. Martinez et al. noted that the perilymph has ion concentrations similar to those of extra-cellular fluid. The endolymph has higher concentrations of potassium (K+) and higher electron positive charge than the perilymph.

The cochlea is accommodated in a canal described as the membranous canal of the cochlea or the scala media. The roof of this canal is the membrane of Reissner and the floor of the canal is formed by the basilar membrane. The cochlea canal is then separated from the scala vestibuli above and from the scala tympani below. The stria vascularis is located on a wall of the cochlea canal and is rich in capillaries and blood vessels that produce the endolymph.

A particular vascular structure on the wall of the scala media, the stria vascularis is responsible for supplying high concentrations of K+ to the endolymph. Appropriate function of the hair cells in the cochlear is dependent on the presence of adequate concentrations of K+.

There is evidence that aquaporins, that form water channels also play critical roles on the inner ear.[20]

Solute carriers such as SLC26A4 that transports iodide are mutated in Pendred syndrome This disorder is associated with cochlear abnormalities, sensorineural deafness, diffuse thyroid enlargement. SLC26A4 transports a number of different ions and solutes, particularly chloride and bicarbonate, solutes.

Hair cells

Hair cells occur in both the auditory systems and in the vestibular system. They act as sensory receptors that detect movement. Key elements of the hair cells include the stereocilia at the apex of the hair cells. Hair cells occur in two regions of the cochlea. The outer hair cells respond to low level sound. The inner hair cells respond to different sound. Cochlea hair cells detect movement of fluids in the cochlea and transform these into electrical signals that can be conveyed to nerves.

Sound waves and the inner ear

Sound waves that pass through the auditory canal are amplified in the middle ear through movement of the tiny bones. Movement of the stapes that is attached to the fenestra ovalis leads to movement of fluid in the scala media. This fluid movement stimulates the hair cells of the cochlea.

In a 2011 review Appler and Goodrich[21] noted that each hair cell detects a narrow range of sound frequencies based on its position in the cochlea.

Disruption of hearing can arise from defects in hair cell function, from impaired function of the stria vascularis leading to impaired ion homeostasis and from impaired neuronal function.

The inner hair cells connect to specialized synapses, the ribbon synapses. Moser et al.[22] reported that these synapses have distinct molecular components. A key component of the ribbon synapse was initially referred to as Ribeye it is now designated terminal binding protein (TBP2). Ribbon synapses utilize calcium channels for signaling downstream to neurons. Importantly the ribbon synapses accommodate synaptic vesicles that are subsequently released to the downstream neurons in the spiral ganglion. A key protein was discovered that is involved in exostosis of ribbon synapse vesicles. This protein is designated as otoferlin. In 2017 Michalski et al.[23] reported that otoferlin acts as a calcium sensor and binds to membranes of synaptic vesicles and impacts fusion of synaptic vesicles to synapses. Otoferlin mutations have been found to cause deafness.

The spiral ganglion has a cell body a peripheral process that connects to the organ of Corti and a separate process that projects into the auditory nerve. The spiral ganglion receives signals from the hair cells via the ribbon synapses, axons from the spiral ganglion pass to the auditory nerve.

The vestibular system

Components of the vestibular system include the semicircular canals and two ampulae the utricula and the saccule. The saccule is connected to the ductus endolymphahticus and that ductus connects to the vestibular system and to the auditory system. The semicircular canals are bony structures lined with membranes. The utricle saccule and semicircular canals are lined with three-layered membranes. Regulation of fluid is essential for the functioning of the neurosensory cells in the vestibular system. Through positioning of the semicircular canals with the posterior and superior canals oriented vertically and the lateral canal oriented at 30 degrees from the horizontal, head movements can be detected, Gray's Anatomy.[17]

Vestibular impulses originate from movement of the fluids and stimuli to the sensory ciliated cells in the semicircular canals, and sensory cells in the utricle and saccule. Nerve fibers pass signal from these sensory cells to the vestibular ganglion. Fibers from the vestibular ganglion then join the vestibulo-cochlear nerve

Benoudiba et al.[24] reviewed the paths and processes of the 8th cranial nerve. The auditory branch and the vestibular branch join together in the auditory meatus to form the vestibulocochlear nerve (the 8th cranial nerve). Fibers from this nerve reach two nuclei in the brain stem the anterior nucleus and the dorsal nucleus.

Acoustic fibers from the dorsal nucleus in the brain stem them pass to the transverse temporal gyri (Heschl area).

Vestibular fibers follow two separate paths. The sub-conscious balance control system connects to the cerebellum. The conscious balance control system connects through the corpus striatum and the thalamus to the post-central gyrus in the cortex.

There is evidence that the thalamo-cortical connections are essential in general for sensory processing. Harris and Mrsic-Flogel[25] noted that sensory stimuli trigger cascades of electrical activity through thalamo-cortical connection. Signal pass primarily to principal neurons in the cortex in layers L4 and also in L5 and L6. There is also evidence for multiple connection between the principal cells in these areas. Principal neurons utilize the excitatory neurotransmitter glutamate.

Vestibular disorders

Different forms of vestibular disorders arise. These disorders are generally associated with vertigo (dizziness). Some disorders are associated with vertigo and migraine headaches, other disorders are associated with vertigo and tinnitus (ringing in the ears) and hearing loss. The latter disorder is referred to as Meniere disease and may be due to genetic defect it may also be due to environmental factors, including infections.[26]

Congenital hearing loss

In 2011 Richardson et al.[27] published a review that described insights gained into the physiology of hearing through discovery and analysis of gene defects that lead to hearing loss. By 2011 135 loci for monogenic forms of hearing loss had been mapped to the human genome and 55 genes responsible for these disorders had been identified (http://hereditaryhearingloss.org). Mouse models for deafness have been particularly useful in discovery and analysis of deafness genes. Audiometric testing can be carried out in mice.

Connexin 26 (GJB2) mutations are the most common genetic cause of deafness. Connexin mutations disrupt assembly of the macromolecular complex that is essential for gap junction function.[28]

Early onset deafness may also arise due to mutations that impact the cochlear hair cells. Richardson et al. described the sterocilia at the tips of hair cells as actin filled rods that contain the kinocilium, a microtubule-based structure. The sterocilia and kinocilia are interconnected. There are 3 different type of connectors.

Richardson et al.[27] noted that mechanoelectrical transduction occurs at the hair bundles that connect hair cells to overlying membrane. The current then flows from the stereocilia into the cells via mechanicoelectrical transduction (MT) channels. Following transmission, the MT channels close. There is evidence that the links between stereocilia act as springs. Fettiplace and Kim[29] described these channels as cation channels with high sensitivity for calcium Ca^{2+}.

By 2011, 50 different hair bundle proteins had been identified. Richardson emphasized that most of these had been identified through investigations of the cause of deafness. They classified these hair bundle proteins into 4 sub-groups, membrane proteins, sub-membrane-protein, motor proteins and actin and actin binding proteins. It is interesting to note that 3 of the membrane proteins had been identified in the form of deafness known as Usher syndrome.

These proteins include Cadherin 23, protocadherin and a protein referred to as Usherin.

Other interesting proteins identified through analyses in deafness include the transmembrane protein Clarin and channel proteins including TMC1 (transmembrane channel like), CLCC1 a chloride channel protein and PMP2 a calcium pump component. Sub-membrane scaffold proteins defined with mutations in specific forms of deafness include Harmonin, Whirlin and a calcium dependent kinase CASK.

Motor proteins important in hair bundle function are encoded by 6 different myosins and actin binding proteins are also important in hair bundle functions.

Usher syndrome is defined as a sensory disorder characterized by deafness and blindness due to retinal defects. This condition will be discussed further below. Mechanical support cells for hair cells are sometimes referred to as cochlear non-sensory cells.

Korver et al.[30] reviewed congenital hearing loss. They noted that specific prenatal factors could contribute to this. Congenital infections including cytomegalovirus infection, rubella or toxoplasmosis, can lead to deafness. Low birthweight and prematurity are also risk factors for hearing loss.

Korver et al. noted that in the majority of children hearing loss is due to genetic factors and that autosomal recessive hearing loss occurred in 80% of cases. In approximately half of these GJB2 mutations were present. In their studies only 1.4% of children had a family history of deafness.

Potassium ion channel KCNQ1

The flow of K+ from the stria vascularis through channels to the endolymph in the scala media is essential for the health and proper functioning of the cochlea. Mutations in particular ion channel genes KCNQ1 and KCNE1 have been found in two divergent pathologies, congenital deafness and cardiac arrythmias.

Specific KCNQ1 mutations lead to a form of cardiac arrythmia, long QT syndrome characterized by episodes of syncope and risk for sudden death. This condition is most commonly due to autosomal dominant mutations (heterozygous) in KCNQ1 and is sometimes referred to as Romano-Ward syndrome, long QT syndrome type 1.

Jervell Lange Nielsen types of deafness are most commonly thought to be due to autosomal recessive mutation homozygous recessive or compound heterozygous mutations in KCNQ1 gene on chromosome 11p15. This syndrome is also associated with cardiac arrythmia.[31,32]

Jervell Lange Nielsen syndrome may result from recessive or compound heterozygous mutations in KCNE1 potassium channel gene. Heterozygous mutations in KCNE1 may lead to long QT syndrome type 5.[33]

Chang et al.[34] reported that virally mediated gene replacement of KCNQ1 into the scala media of young mouse models of Jervell Lange Nielsen restored hearing.

Hearing impairment and deafness

Epidemiology

In 2018 Sheffield and Smith[35] reviewed the epidemiology of deafness. They reported that deafness impacts 5% of the world's population. Analyses across different countries revealed that the highest degrees of deafness in children occurred in South Asia and in the Pacific island nations. The highest degrees of deafness in adults occurred in Eastern Europe and Central Asia.

Sheffield and Smith emphasized that deafness in young infants could be due to genetic hearing loss or due to specific harmful factors and conditions present around the time of birth or during the early prenatal period. These harmful factors will be discussed in a subsequent section.

Genetic etiology of deafness

Sheffield and Smith included in this category, syndromic and non-syndromic deafness due to Mendelian factors and deafness due to complex factors. In syndromic deafness individuals have defects in other systems. However, it is interesting to note that defects in specific single genes can lead to specific syndromes in which deafness occurs. They reported that more than 75% of cases of genetically determined deafness in children are due to autosomal recessively inherited defects. However, in more than approximately 20% of cases, deafness is an autosomal dominant trait; in about 2% of cases deafness is an X linked traits and a low percentage of cases are reported to be due to defects in the mitochondrial genome.

Sheffield and Smith confirmed that the most common gene defect in autosomal non-syndromic hearing loss occurs in the *GJB2* gene. The protein product of this gene is connexin 26, Connexins form channels and connexin channels in the inner ear are essential for recycling ions, particularly potassium ions that are required to ensure homeostasis within the cochlea.[36] Sheffield and Smith noted that more than 100 different pathogenic variants have been identified in *GJB2*. The most common variant in *GJB2* is a nucleotide deletion, 35delG. This variant was reported to have a carrier frequency of approximately 2.5% in European and American populations A different *GJB2* variant 235delC was reported to be common in the Japanese population. Several investigators have reported that GJB2 mutations do not occur in African individuals with deafness.

An Important gene defect that leads to autosomal dominant deafness occurs in KCNQ4 a potassium channel protein.

Smith and Sheffield identified 3 other genes as harboring mutations that frequently cause autosomal recessive deafness. The proteins encoded by these genes and their functions include STRC sterocilin that forms a ciliary structure in the hair cells that responds to sound waves, SLC26A4 a solute carrier, and TECTA tectorin alpha that is present in the tectorial membrane of the inner ear.

With respect to X linked deafness Sheffield and Smith stressed the importance of POUF4, a transcription factor. Defects in the function of POUF4 were reported to lead to fixation of the stapes bone and to cochlea hypoplasia.

Sheffield and Smith reported that syndromic form of hearing loss occurred in 30% of cases with congenital deafness. It is interesting to note that solute carrier defects and collagen defects are among the causes of syndromic deafness. Solute carrier (transporter) defects include SLC26A4 mutations that lead to Pendred syndrome. SLC52A2 and SLC25A3 mutations occur in a condition Brown Vialetto-Laere syndrome where deafness is associated with neurological impairments. Importantly SLC25A2 and SLC25A3 are carriers of riboflavin and symptom of the syndrome can be relieved by administration of high doses of Riboflavin.

Stickler syndrome, where deafness occurs in individuals who also manifest joint hypermobility and ocular defects can arise due to defects in specific collagens including COL9A1, COLA2, COL2A1, COL11A1.

Deafness associated with a specific bone disorder otospondylomegaepiphyseal dysplasia can occur due to defects in COL11A2.

It is important to note that some cases of sensorineural deafness may be associated with defects in functions of other cranial nerves.

Syndromic forms of congenital deafness are sometimes associated with facial abnormalities, e.g. in Treacher Collins syndrome. The Treacher Collins syndrome can result from defects in any one of three genes, TCOF1 a ribosome biogenesis factor, POLR1D and POLR1C that encode RNA polymerase subunits that form ribosomal RNAs and other small RNAs. Some forms of congenital deafness are associated with renal defects, e.g. Alport syndrome and Brancho-oto renal syndrome BOR syndrome, Alport syndrome can result from autosomal dominant or autosomal recessive mutations in COL4A3, An X linked form of Alport syndrome can result from mutation in COL4A5. BOR1 syndrome, hearing loss, renal defects a with or without cataracts results from mutation in the EYA1 gene that encodes a protein tyrosine phosphatase. BOR2 syndrome has manifestation similar to those in BOR1 syndrome and is caused by mutations in the SIX5 protein that interacts with the EYA1 proteins.

Particularly important are associations of syndrome forms of deafness with visual defects, e.g. in Usher syndrome. This will be discussed further below.

Distefano et al. and the ClinGen Hearing Loss Curation Expert Panel[37] examined genetic variant curated information on 153 genes that had been implicated in deafness. They classified genetic variants into different categories based on the strength of evidence for their pathogenicity and for their association with deafness. Categories with definitive and strong evidence of association with deafness, included 94 genes. Other categories defined as having moderate limited, disputed or refuted evidence of associated with deafness included 70 genes.

Deafness due to environmental factors in the perinatal period

These factors particularly operate in premature infants and can include exposure due to unusual noise levels through artificial ventilation systems and other factors in neonatal intensive care units (NICU). Premature infants are also at increased risk for hemolytic disease of the newborn and hyperbilirubinemia that can cause nerve damage. In addition, these infants may require medications some of which are damaging to hearing. Particularly damaging are aminoglycosides these antibiotics can cause damage when administered over long periods. However, the presence of particular genetic variants can lead individuals to incur hearing damage even after a single dose. Aminoglycosides include Kanamycin, Gentamycin, Streptomycin, Tobramycin, Amikacin. The particular variant in the mitochondrial genome A1555G that occurs in the sequence that encodes the mitochondrial 12 s Ribosomal gene, increases risk for hearing loss in individuals medicated with aminoglycosides. This variant leads to aminoglycoside sensitivity throughout life.[38]

Newborn screening for deafness

In a 2017 review Wroblewska-Seniuk et al.[39] reported that the incidence of sensori-neural deafness in healthy newborns was 2–3 per 1000, however the incidence was 2–4 per 100 high risk infants, especially infants who were in neonatal intensive care units.

Wroblewska-Seniuk et al. reported that failure to intervene therapeutically in the first 6 months of life in cases with sensori-neural hearing loss led to impaired speech development, learning and psychological disorders. They noted that newborn screening programs were being implemented throughout the world. They also noted that there are now recommendations to check for hearing loss in all infants with methods that do not require participation of the individuals being tested.

Electrophysiological exploration of hearing

Bakhos et al.[40] reviewed objective electrophysiologic audiometry, defined as objective since the methods did not require active participation of the individual being examined. They defined three types of such studies, otoacoustic emission electrocochleography (OAE), auditory brain stem responses (ABR) and auditory steady state responses (ASSR).

Otoacoustic emissions derive from contraction of the outer hair cells in response to acoustic stimulation. Bakhos noted that specific acoustic stimuli used in this testing include clicks and pure tones. The miniaturized probe used for testing includes a transmitter the emits signal and a microphone receiver that records response. The miniaturized probe is inserted into the external auditory canal in close approximation to the tympanic membrane.

Specific testing in newborns involves transient stimuli. The test is best administered when the infant is asleep. Testing can be negatively impacted by ambient noise.

Auditory brain response test (ABR)

Bakhos et al. noted that this involves assessment of transmission of signal activity along the auditory nerve to the brain stem. Head Phones or ear canal inserts deliver sound and electrodes are placed at specific position on the scalp forehead and mastoid. Testing requires a sound proof chamber and a relaxed test subject.

Five specific waves are generated and together they document the passage of signal from the distal cochlear nerve, proximal cochlear nerve, cochlear nucleus and superior olivary complex and inferior colliculus in the brain stem.

The test used in neonates is defined as AABR automated auditory brainstem responses and in this test click sounds are emitted.

Benefits of newborn screening and early detection of hearing loss

Kral and O'Donoghue[41] reported evidence that profound hearing loss in early childhood leads to loss of spoken language development that restricts learning and education and later employment opportunities. They emphasized that children who had hearing restored before 2 years of age benefitted significantly. They emphasized the benefits of universal neonatal hearing screening.

Wroblewska-Seniuk et al. in a 2016[39] review emphasized outcome analyses that revealed early detection of hearing loss and follow-up treatment significantly improved development of language skills. They noted that detection and initiation of treatment before 6 months of age is particularly important. There is evidence that prior to initiation of newborn assessment of hearing, impairment was first assessed when children had significant speech delay or compromised speech and development.

Treatment options for hearing impairment

Hearing Aids are amplification devices designed to intensify sound to fall into the range at which the patient can hear. Wroblewska-Seniuk et al. emphasized that cochlear implants are used in cases of profound hearing loss where amplification of sound does not result in ability to hear.

White et al.[42] reviewed recommendations for use of implantable devices in special populations. They noted that in the USA implantable cochlear devices were approved for children in the 12 to 23-month age range who had hearing loss greater than 90 decibels (dB). Children older than 2 years with hearing loss greater than 70 dB were also approved for cochlear implants.

White et al. noted that pre-implantation imaging was essential in the pediatric population since young children with profound hearing loss frequently had anatomic abnormalities and were at risk for misplacement of devices.

Osseointegrated bone conduction devices are approved for use in children older than 5 years. In cases with absent cochlea who were 12 years of age or older, implanted prosthetic devices in the brain stem constituted approved treatment.

Public health and pediatric hearing impairment

Kaspar et al.[43] reported that children living in the Pacific Islands have the greatest rates of deafness in the world. They noted that otitis media and meningitis were the leading causes of infection related hearing loss.

The Pacific Islands are included in the Oceania the region of the Globe and include Micronesia, Melanesia, and Polynesia.

Kaspar et al. reported that examinations revealed that in some cases the children had compromised hearing due to impacted cerumen in the external auditory canal, however, in the majority of cases conductive hearing loss was associated with chronic otitis media. Causes of sensorineural hearing impairment included consequences of meningitis, measles or rubella infections.

The authors concluded that the WHO Global School Health Initiative could provide a platform for school-based hearing screening and facilitate medical intervention.

In 2015 le Roux et al.[44] reported information on 264 South African children with congenital and early hearing impairment. They documented risk factors and reported that the most significant risk factor was admission to the Neonatal Intensive Care Unit, 28.1% of cases had NICU admissions; prematurity was noted in 15.1% of cases. Another important post-natal risk factor was the presence of hyperbilirubinemia, this occurred in 10% of cases. The most significant prenatal risk factor was family history of deafness. This was present in 19.6% of cases.

Hearing loss in adults

The frequency of hearing loss increases with age. In their (2018) review, Sheffield and Smith[35] documented a steep rise in hearing loss in each 10-year period between 40 and 80 years of age. They noted that with increasing age, there was a particularly marked decline in hearing ability in the range of frequencies into which speech falls.

Adult onset loss of speech was attributed to the interplay of genetic and environmental factors. Noise exposure represents an important risk factor in adult onset hearing impairment.

Cochlear implants, vestibular implants

Georg von Békésy[45] was awarded the 1961 Nobel Prize in Physiology or Medicine for his work on cochlea stimulation to correct hearing deficiency.

It is also important to note that vestibular implants have been developed to treat balance related disorders.[46]

Advances in the treatment of hearing defects

In a 2017 report Dabdoub and Nishimura[47] noted that despite benefits to hearing, cochlear implant users had difficulties in processing of complex sound. They specifically noted that music represents complex sound.

They also noted that the degree of benefit of cochlear implants in a particular individual is dependent on the extent of degeneration or loss of primary auditory neurons (spiral ganglion neurons). However, survival of 10% of primary auditory neurons was sufficient for benefit from cochlear implants.

Spiral ganglion neurons also referred to as primary afferent auditory neurons, send projections to the cochlear hair cells and they then receive signals from hair cell movements. These signals can be transmitted along projections from the spiral ganglion to the brain stem.

Dabdoub and Nichimura noted that primary afferent auditory neurons can degenerate in aging or as a results of noise exposure.

Suzuki et al.[48] have reported advantages of local delivery of neutrophin-3 to stimulate regeneration of primary auditory neurons. Other approaches to stimulate regeneration include use of implants to deliver plasmids expressing brain derived neurotrophic factor (BDNF) to primary afferent auditory neurons that were degenerating.

Animal studies have revealed that transfer of optogenetic channel rhodopsins to neurons permits neuronal stimulation by photoelectric signals.

Auditory neuropathy Spectrum disorder

Yawn et al.[49] defined auditory neuropathy spectrum disorder (ANSD) as a disorder characterized by "altered neural synchrony in response to auditory stimuli". They noted that children with this disorder presented with poor speech perception. They noted that in this disorder the outer hair cells of the cochlea were not impacted and that auditory otoacoustic emission responses (AOAE) were not impacted, however abnormalities were detected on ABR testing.

Yawn et al. noted that risk factors for auditory neuropathy syndrome included, hypoxia, mechanical ventilation, exposure to ototoxic drugs. In a few cases genetic defects were postulated to cause this disorder.

They noted that development of a treatment plan was challenging and that it was important to rule out structural defects or absence of the cochlear nerve.

Usher syndrome

This syndrome is characterized by blindness and deafness. It was first described by Charles Usher in 1914. Different types of Usher syndrome are distinguished primarily by age of onset and rate of progression of manifestations.

Mathur and Yang[50] reported that defects in different gene loci had been found to lead to Usher syndrome, gene encoded proteins that were often present in multiprotein complexes. Many of the genes impacted in Usher syndrome encoded components of the inner ear hair cells. In the retina the Usher gene encoded proteins occurred in photoreceptors. Usher syndrome is one of the causes of syndromic retinitis pigmentosa. In Retinitis pigmentosa (RP) progressive degeneration of cells in the retina occurs. Most types of Usher syndrome manifest autosomal recessive inheritance. However there are forms that are digenic in inheritance, indicating that defects in two genes lead to the disorder. Different types of Usher syndrome and specific proteins that manifest defective functions are listed below.

Type	Defective protein and function
USH1/USH1B	MYO7A Myosin7A, mechanosensitive protein
USH1C/PDZ73	Harmonin scaffold protein involved in assembly of protein complexes
USH1D/USH1F	Digenic mutations in PCDH15 protocadherin and CDH23 cad-herin, cell adhesion
USH1G	SANS scaffold protein interacts with harmonin
USH1J	CIB2 calcium and integrin binding family member 2, cell maintenance
USH2A	Usherin basement membrane protein important in cell homeostasis
USH2D	WHRN whirlin, involved in actin cytoskeletal assembly, cilia stabilization
USH3A	CLRN1 Clarin transmembrane protein involved in cellular homeostasis
USH3B	HARS histidyl-transfer RNA, incorporation of histidine into proteins
USH4	ARSG, hydrolyzes sulfate esters, essential lysosome function

Vision and the retina

The choroidal region of the eye is a vascular membrane the occupies the most posterior region of the eye and is penetrated by the optic nerve. The choroidal region had 4 layers. These include a layer of large blood vessels, a layer of medium sized blood vessels, a layer of choroidal capillaries and Bruch's membrane. This membrane, sometimes referred to as the vitreous lamina, is the innermost layer of the choroidal region.

The retina is composed of layers of neurons, Jacob's layer of rods and cones and the pigmentary layer. The 10 layers of the retina are numbered in order from the most interior to the outer retinal regions, Gray's Anatomy.[17]

1. Membrana limitans interna (internal limiting membrane)
2. Stratum opticum, nerve fiber layer
3. Ganglionic layer
4. Inner molecular plexiform layer)
5. Inner nuclear layer
6. Outer molecular plexiform layer
7. Outer nuclear layer
8. External limiting membrane
9. Jacob's membrane of rods and cones
10. Pigmentary layer

The internal limiting membrane (1) forms part of the supporting network of the retina. The nerve fiber layer (2) is composed of optic nerve radiations; these fibers are most dense in the vicinity of the optic nerve. The ganglionic layer (3) accommodated the nuclei of nerve fibers.

The inner plexiform layer (4) contains the dendrites of ganglion cells. The inner molecular layer (5) contains diverse cell types including bipolar cells, amacrine cells and horizontal cells. The outer molecular layer (6) has projections from the rods and cones. The outer nuclear layer (7) contains cell bodies that are connected to the rods and cones. The external limiting membrane (8) is a layer with fibers lying between the rods and cones and the cell bodies in layer 7.

Jacob's membrane (layer 9) contains the rods and cones. Layer 10 is the pigmentary layer.

Regions of the retina

The optic disc overlies the site where the optic nerve leaves the retina. This region of the retina lacks photoreceptors. The relative distribution of rods and cones differs in different regions of the retina. Cones predominate in the central retina and in the peripheral retina rods predominate. Retinal rods and cones are involved in the processes whereby light is converted to neural action potential.

Retinal pigment epithelium

The retinal pigment layer lies outside the neurosensory layer. It lies interior to the to the choroid region that is rich in blood vessels. A specialized membrane, Bruch's membrane lies between the choroid and the retinal epithelium. The retinal epithelial layer lies interior to the photoreceptor layer.

In a comprehensive review of the retinal epithelium Strauss[50] emphasized that functional vision is dependent on the interaction between the retinal epithelial pigmentary cells and the photoreceptor cells. These two regions are co-dependent and gene defects that impact one of the region frequently have secondary impact on the other region.

The retinal pigmentary epithelium (RPE) is described as a single cell layer with cells rich in pigment granules composed of melanin. Tight junctions connect the cells. Nutrients are supplied from blood vessels in the choroid region to the RPE. These nutrients include glucose, fatty acids, and phospholipids. The presence of glucose transporters GLUT, GLUT3 facilitate transport of glucose. In addition, ions and H_2O are transported. Ions and water and metabolites such as lactic acid can also be transported in reverse direction from the RPE to the choroid. Special chloride channels were also important. Strauss noted that the transport of ions involved the activity of $Na^+ K^+$ ATPases.

Strauss noted that the RPE is also responsible for the transfer of specific growth factors from the blood vessels in the choroid to the retinal tissues. These include vascular endothelial growth factor (VEGF), fibroblast growth factor (FGF) and insulin like growth factor (IGF).

Melanin in the retinal pigment epithelium

Melanin in the retinal pigment epithelium has a photoprotective role. Defects in melanin synthesis, such as occur in albinism lead to significant visual impairment. Seven different genes have been implicated in oculocutaneous albinism.[51] The different forms of oculocutaneous albinism and the genes that are defective in each are listed below

OCA1	TYR tyrosinase that is involved in the first step of melanin synthesis from tyrosine
OCA2	Due to defect in a gene that transfers tyrosine into the melanosome
OCA3	TYRP1 involved in melanin biosynthesis
OCA4	SLC45A2 transporter protein that facilitates melanin synthesis
OCA5	gene defect not yet identified
OCA6	SLC24A5 calcium potassium dependent solute carrier
OCA7	LRMDA leucine rich melanocyte differentiation associated

The visual cycle, retinal pigmentary epithelium and photoreceptor interactions

A specific type of opsin is present in rods, this is rhodopsin. Different types of cones are present and each type has different forms of opsin. The wavelength of light detected by a specific cone is dependent on the type of opsin present. Light waves detected by cones include long medium or short waves. These include blue 420 nm (short), green 530 nm (medium) and red 560 nm (long), Strauss.[50]

Visual pigments are chromophores that contain 11-cis retinol, an aldehyde of Vitamin A. 11-cis retinal, also known as retinaldehyde is derived from Vitamin A through activity of retinol dehydrogenase. On activation, the 11-cis retinol changes configuration and activates transducin that is encoded by the GNAT1 gene. Activation of transducin causes it to bind to GTP and to activate the phosphodiesterase PDE6 that then hydrolyzes cyclic guanosine monophosphate (cGMP). This results in lowering of the intracellular concentration of cGMP. This in turn leads to hyperpolarization in the cells and closing of the sodium channels. This hyperpolarization ultimately leads to activation of a neuronal response. PDE6 is composed of three subunits each encoded by a different gene.[52]

The impact of phosphodiesterase is subsequently terminated through activity of specific kinases.

In order to be able to react to a subsequent photon, the all transretinal need to be converted back to 11 cis retinal. This reconversion can only be carried out in the retinal pigmentary epithelium since this is where the appropriate enzyme complex exists. A specific protein ATP binding cassette protein (ABCA4) can transfer all trans retinal. The transformation of all transretinal to 11-cis retinal requires three enzyme, lecithin retinal transferase (LRAT), RPE65, and retinal isomerase.

Visual cycle. In the retinal pigment epithelium trans-retinol is converted to cis-retinal and captures photons that then enter the rod outer segment to activate rhodopsin in G-coupled receptor, Transducin and phosphodiesterase are then activated. This leads to breakdown of guanosine monophosphate followed by ion channel closing, intra-cellular hyperpolarization and neuronal activity.

The region adjacent to the photo conducting rods and cones includes bipolar cells, horizontal cells amacrine cells and ganglion cells. Signals from the rod and cone photoreceptors pass through specialized synapses, the photoreceptor ribbon synapses to post-synaptic contact.[53]

Strauss[50] noted that another important function of the retinal pigment epithelium is to participate in phagocytosis of damaged photoreceptor membranes. These membranes can be damaged during the process of photo-oxidation.

Retinal dystrophies

In 2015 Nash et al.[54] reviewed retinal functions that are impaired in retinal dystrophies. These may arise through defects in specific cells in the retina or through

1. Defects in the visual cycle processes which require generation of light sensitive pigments and their recycling.
2. Defects in photoreceptors, rods and cones.
3. Defects in phototransduction that involves propagation of signal. These may arise through defects in specific cells or to defects in genetic elements that control gene expression of through mutations that impact functions of specific gene products.

Retinitis pigmentosa

Clinical manifestations

This disorder may present in childhood or later. Initially patients manifest night blindness and a loss of peripheral vision. Early manifestations include difficulty seeing in dim light, loss of peripheral vision that gradually increases to limit central vision. Complete blindness may subsequently result. Retinitis pigmentosa can be a progressive disease. In some cases, it may be stationary and is then sometimes referred to as stationary night blindness. Retinitis Pigmentosa is associated with progressive degeneration of cells in the retina.

Daiger et al.[55] noted that retinitis pigmentosa accounted for approximately half of the cases of retinal dystrophy. They note further that both syndromic and non-syndromic causes of retinitis pigmentosa (RP) occur. Syndromic forms of RP occur in Usher syndrome; this syndrome may be caused by defects in any one of 12 different genes, and Bardet-Biedl syndrome associated with defects in any one of 17 different genes.

Daiger et al. reported that the age of onset of RP is variable. RP may be present at birth. There is genetic heterogeneity in RP, with different genes leading to the disorders. In addition, there is clinical heterogeneity associated even with the same gene mutation and specific pathogenic gene mutations may even be non-penetrant so that there are individuals where a specific pathogenic mutation does not lead to disease symptoms.

Nash et al.[54] reported that more than 60 different genes had been implicated in Retinitis pigmentosa. These genes fall into different categories and include defects in phototransduction, in retinal metabolism, in tissue development and maintenance and in RNA splicing. Damaged photo-receptors undergo apoptosis.

In rod dystrophies or rod-cone dystrophies those receptors are first affected. Autosomal recessive, autosomal dominant and X linked forms occur. In addition, digenic mutations may lead to photoreceptor disease. In digenic mutation deleterious mutations in two different genes act together to cause disease. Nash et al. noted however that molecular diagnoses were only made in approximately 50% of cases of rod-cone dystrophies. The most commonly impacted gene was the RHO gene that encodes rhodopsin. Rhodopsin defects occur in 20–30% of cases with retinitis pigmentosa. Rhodopsin is a protein that binds to 11-cis retinal, it is activated by light, Rhodopsin is essential for vision in dim light.

Defects in nine other different genes lead to phototransduction defects. Defects in 8 different genes impact retinal metabolism. Important genes in this category include RPE65 and ABCA4. RPE64 encodes an enzyme retinoid isomerohydrolase a component of the Vitamin A visual cycle. It is involved in carotenoid cleavage. ATP binding cassette transporter that transports ATP adenosine triphosphate across membranes.

Genes involved in retinal tissue development and maintenance were reported to be relatively common causes of retinal dystrophy and Retinitis pigmentosa. The RP1 gene encodes a protein involved in determining the structure of microtubules in the rod photoreceptors. The RP2 gene is involved in beta tubulin folding and is essential to the photoreceptor and defects in this gene were reported in 10–20% of cases of RP. The CRB1 gene encoded an essential component in the photoreceptors and defects in this gene occur in 6–7% of cases of RP. The USH2A gene encodes a protein important in determining membrane structure. Defects in this gene were reported in 10–15% of cases of RP.

The EYS gene is involved in signaling in the photoreceptors and was found to be defective in a significant number of cases of RP in China and was defective in 10–30% of cases of RP in Spain.

Leber's congenital amaurosis

Kumaran et al.[56] described Leber's hereditary amaurosis as a severe early onset retinal dystrophy that manifested both genetic and phenotypic heterogeneity. Clinical manifestations include vision loss, nystagmus, reduced or absent signal on retinogram.

Theodor Leber first described this condition in 1869 and it was designated Leber's Hereditary amaurosis. Later he referred to a milder form of the disease that presented after infancy but before the age of 5 years, as Early onset severe retinal dystrophy.

Kumaran et al. noted that by 2017 mutations in 25 different genes had been implicated in Leber's Hereditary amaurosis and together these explained 70–80% of cases. Detailed studies on animal models have led the way to understanding the functions of the gene products of the different genes found to be mutated in Leber's hereditary amaurosis. These studies are also serving as a basis for design of therapies. An example of this initiation of clinical trials of gene therapy in cases of congenital blindness due to mutations in RPE65.

Kumaran et al. reported that some gene defects lead primarily to Early Onset Severe Retinal Dystrophies. However other genes have been shown to be defective both in cases of Leber's congenital amaurosis and in Early Onset Severe Retinal Dystrophies.

The genes that are defective in these disorders impact different functions these include the retinoid cycle and metabolic processes, photoreceptor transport, photoreceptor morphogenesis, photoreceptor structural stabilization. Genes most frequently impacted:

GUCY2D	Guanylate cyclase phototransduction
CEP290	Centrosomal protein, photoreceptor ciliary transport
RFH12	Retinol dehydrogenase 12 retinoid-cycle
RPE65	Retinoid isomerase, retinoid cycle

Defects in certain genes that impact the macula may also lead to congenital or early onset blindness.

These include:

TULP1	Tubby-like1, photoreceptor ciliary transport
AIPL1	Arylhydrocarbon interacting protein phototransduction
NMNAT1	Nicotinamide nucleotide adenyltransferase, important in neural function

Kumaran et al. emphasized that a child presenting with congenital or early onset blindness should be investigated for syndromic disorders.

Mitochondrial functional defects leading to visual impairment

Yu-Wai-Man et al.[57] reviewed mitochondrial DNA defects and nuclear gene defects that impact mitochondrial function and lead to visual defects. They noted that vision is particularly vulnerable when mitochondrial function is compromised. Visual impairment may occur in disorders of mitochondrial function that lead to widespread systemic defects. There are however examples of mitochondrial functional disorders where visual impairment is the key manifestation; Leber's Hereditary optic atrophy and autosomal dominant optic atrophy represent two of these conditions.

Yu-Wai-Man et al. reported that Leber's Hereditary Optic Atrophy (LHON) is most frequently due to mutations at any one of three specific sites in mitochondrial DNA. Each of these mutations impacts the function of mitochondrial electron transfer complex 1, resulting in impaired oxidative phosphorylation and also raised levels of reactive oxygen species. They also noted the gene-environment interaction also play roles in this disorder. In individuals at increased genetic risk, smoking further increases risk. There is also evidence that estrogens have a protective effect.

Autosomal dominant optic atrophy was reported to be most frequently due to mutations in the OPA1. The product of this gene is a component of the inner mitochondrial membrane. Yu-Wai-Man et al. noted that more than 250 different pathogenic mutations had been reported in OPA1 these mutations included deletions insertions, splice-site mutations and missense mutations. A specific OPA1 mutation p. Arg455His was reported to be associated with visual defects and sensorineural deafness. Other OPA1 mutations have been described in patients with neurological and or muscle defects in a condition referred to as OPA plus syndrome.

In Leber's hereditary optic atrophy and in autosomal dominant optic atrophy Yu-Wai-Man et al. reported that there was loss of retinal ganglion cells.

It is interesting to note that optic atrophy also results from mutations in the nuclear gene TMEM126A that encodes a different mitochondrial inner membrane protein.

Age related macular dystrophy (AMD)

Fritsche et al.[58] described AMD as a multi-factorial late onset disorder Patients often presented first with loss of central vision. Major risk factors for AMD included advanced age, family history and smoking. Fritsche et al. noted that identification of genetic factors in AMD causation had led to increased understanding of AMD pathology and biological processes implicated in AMD.

Pathological features of AMD in the early stages include accumulation of lipid rich deposits and inflammation in the region of the macula. Later changes include neurodegeneration of the macula. Pathologic features can include a "dry form" with geographic atrophy and "wet form" with vascularization and infiltration of choroidal vessels. The wet form of macular degeneration can be treated with injections into the eye of anti-angiogenic compounds.

Fritsche et al. noted that in AMD the photoreceptor support system was primarily impacted. The photoreceptor support system included the retinal pigment epithelium, Bruch's membrane and choroidal vasculature. Defects in the support system led to progressive damage to photoreceptors.

Newborn eye screening and retinopathy of prematurity

Various tests have been devised to screen newborns for ocular diseases. The Red reflex test is a test that can be used to test for cataracts. However retinal screening is important for detection of retinal hemorrhages and retinopathy in infants who were born prematurely. Simkin et al.[59] reported that wide angle retinopathy is important for screening in such cases.

Blindness in children

In a review in 2017 Solebo et al.[60] reported that 14 million children in the world are reported to have severe visual impairment or to be blind. They noted that key causes of child blindness differ in different parts of the world.

In high income countries key causes were reported to include retinopathy of prematurity, cerebral visual impairment and optic nerve anomalies. In lower income countries key causes include infection damage, and nutrition damage that led to corneal opacities and congenital anomalies.

Solebo et al. noted that a blind child is more likely than a seeing child, to have impaired development. Furthermore, the death rate of children with blindness is higher than that of seeing children. Severe visual impairment in adults is associated with lower socio-economic status.

Solebo et al. noted that in low income setting programs of Vitamin A supplementation, vaccination and sanitation improvements are leading to decreases in blindness in children.

Retinopathy of prematurity

Hellström et al.[61] described several phases of retinopathy of prematurity. In early phases following birth, administration of high levels of oxygen, leads to vessel constriction and suppression of oxygen regulated growth leading to cessation of retinal development and vessel development.

A second phase of retinopathy of prematurity is characterized by proliferation of blood vessels as level of erythropoietin and vascular endothelial growth factor increase. The newly formed blood vessels were noted to be leaky and exudates occur. These exudates lead to formation of fibrous scars that can lead to retinal detachment.

Important factors in retinopathy of prematurity include degree of prematurity, low gestational age and low birthweight. Hellström et al. noted that raised neonatal glucose concentrations also increased risk for prematurity. Neonatal infections, especially fungal infections also increased risk for retinopathy of prematurity.

In phase 2 factors that suppressed vessel proliferation, including anti- vascular endothelial growth factor, were reported to be helpful.

Adequate neonatal nutrition aimed at optimization of essential fatty acids and levels of omega poly unsaturated fatty acids was reported to be important in avoiding complications.

Solebo et al.[27] emphasized that lowering levels of oxygen content of respired air in premature infants has been found to reduce the incidence of retinopathy of prematurity. Recommended reduction in levels of oxygen were from previous levels used 91–95% of oxygen to 85–89%.

They noted that in the second phase of retinopathy of prematurity anti-vascular epithelial growth factor agents were recommended. In specialized facilities ablation of excess vessels was sometimes performed.

References

1. Julius D, Nathans J. Signaling by sensory receptors. *Cold Spring Harb Perspect Biol* 2012;**4**(1):a005991. https://doi.org/10.1101/cshperspect.a005991. 22110046.

2. Purves D. Mechanoreceptors Specialized to Receive Tactile Information. In: *Neuroscience*. 2nd ed. Sinauer Associates; 2001. https://www.ncbi.nlm.nih.gov/books/NBK10895/.

3. Hao J, Bonnet C, Amsalem M, Ruel J. Transduction and encoding sensory information by skin mechanoreceptors. *Pflugers Arch* 2015;**467**(1):109–19. https://doi.org/10.1007/s00424-014-1651-7. 25416542.

4. Jenkins BA, Lumpkin EA. Developing a sense of touch. *Development* 2017;**144**(22):4078–90. https://doi.org/10.1242/dev.120402. 29138290.

5. Sherrrington CS. Qualitative differences of spinal reflex corresponding with qualitative difference of cutaneous stimulus. *J Physiol* 1903;**30**:39–46.

6. Woolf CJ, Ma Q. Nociceptors–noxious stimulus detectors. *Neuron* 2007;**55**(3):353–64. https://doi.org/10.1016/j.neuron.2007.07.016. 17678850.

7. DiMario Jr FJ. Inherited pain syndromes and ion channels. *Semin Pediatr Neurol* 2016;**23**(3):248–53. https://doi.org/10.1016/j.spen.2016.10.009. 27989333.

8. Dib-Hajj SD, Black JA, Waxman SG. NaV1.9: a sodium channel linked to human pain. *Nat Rev Neurosci* 2015;**16**(9):511–9. https://doi.org/10.1038/nrn3977. Review, 26243570.

9. Dib-Hajj SD, Waxman SG. Sodium channels in human pain disorders: genetics and pharmacogenomics. *Annu Rev Neurosci* 2019. https://doi.org/10.1146/annurev-neuro-070918-050144. 30702961.

10. Cox JJ, Reimann F, Nicholas AK, Thornton G, Roberts E, et al. An SCN9A channelopathy causes congenital inability to experience pain. *Nature* 2006;**444**(7121):894–8. https://doi.org/10.1038/nature05413. 17167479.

11. Stevens EB, Stephens GJ. Recent advances in targeting ion channels to treat chronic pain. *Br J Pharmacol* 2018;**175**(12):2133–7. https://doi.org/10.1111/bph.14215. 29878335.

12. Skerratt SE, West CW. Ion channel therapeutics for pain. *Channels (Austin)* 2015;**9**(6):344–51. https://doi.org/10.1080/19336950.2015.1075105. Review. 26218246.

13. Erickson A, Deiteren A, Harrington AM, Garcia-Caraballo S, et al. Voltage-gated sodium channels: (Na$_V$)igating the field to determine their contribution to visceral nociception. *J Physiol* 2018;**596**(5):785–807. https://doi.org/10.1113/JP273461. 29318638.

14. Cardoso FC, Lewis RJ. Sodium channels and pain: from toxins to therapies. *Br J Pharmacol* 2018;**175**(12):2138–57. https://doi.org/10.1111/bph.13962. 28749537.

15. Veldhuis NA, Poole DP, Grace M, McIntyre P, Bunnett NW. The G protein-coupled receptor-transient receptor potential channel axis: molecular insights for targeting disorders of sensation and inflammation. *Pharmacol Rev* 2015;**67**(1):36–73. https://doi.org/10.1124/pr.114.009555. Review, 25361914.

16. Tsantoulas C, McMahon SB. Opening paths to novel analgesics: the role of potassium channels in chronic pain. *Trends Neurosci* 2014;**37**(3):146–58. https://doi.org/10.1016/j.tins.2013.12.002. Review. 24461875.

17. Gray's Anatomy. In: Pick TP, Howden R, editors. *Organs of Special Senses*. 15th ed. Barnes and Noble; 1995.

18. Ceriani F, Mammano F. Calcium signaling in the cochlea—molecular mechanisms and physiopathological implications. *Cell Commun Signal* 2012;**10**(1):20. https://doi.org/10.1186/1478-811X-10-20. 22788415.

19. Martínez AD, Acuña R, Figueroa V, Maripillan J, Nicholson B. Gap-junction channels dysfunction in deafness and hearing loss. *Antioxid Redox Signal* 2009;**11**(2):309–22. https://doi.org/10.1089/ars.2008.2138. 18837651.

20. Eckhard A, Gleiser C, Arnold H, Rask-Andersen H, et al. Water channel proteins in the inner ear and their link to hearing impairment and deafness. *Mol Aspects Med* 2012;**33**(5–6):612–37. https://doi.org/10.1016/j.mam.2012.06.004. Review, 22732097.

21. Appler JM, Goodrich LV. Connecting the ear to the brain: molecular mechanisms of auditory circuit assembly. *Prog Neurobiol* 2011;**93**(4):488–508. https://doi.org/10.1016/j.pneurobio.2011.01.004. Review. 21232575.

22. Moser T, Predoehl F, Starr A. Review of hair cell synapse defects in sensorineural hearing impairment. *Otol Neurotol* 2013;**34**(6):995–1004. https://doi.org/10.1097/MAO.0b013e3182814d4a. Review, PMID23628789.

23. Michalski N, Goutman JD, Auclair SM, Boutet de Monvel J, et al. Otoferlin acts as a Ca^{2+} sensor for vesicle fusion and vesicle pool replenishment at auditory hair cell ribbon synapses. *Elife* 2017;**6**. pii: e31013 https://doi.org/10.7554/eLife.31013. 29111973.

24. Benoudiba F, Toulgoat F, Sarrazin JL. The vestibulocochlear nerve (VIII). *Diagn Interv Imaging* 2013;**94**(10):1043–50. https://doi.org/10.1016/j.diii.2013.08.015. Review. 24095603.

25. Harris KD, Mrsic-Flogel TD. Cortical connectivity and sensory coding. *Nature* 2013;**503**(7474):51–8. https://doi.org/10.1038/nature12654. 24201278.

26. Harcourt J, Barraclough K, Bronstein AM. Meniere's disease. *BMJ* 2014;**349**:g6544. https://doi.org/10.1136/bmj.g6544. 25391837.

27. Richardson GP, de Monvel JB, Petit C. How the genetics of deafness illuminates auditory physiology. *Annu Rev Physiol* 2011;**73**:311–34. https://doi.org/10.1146/annurev-physiol-012110-142,228. Review, 21073336.

28. Kamiya K, Yum SW, Kurebayashi N, Muraki M, et al. Assembly of the cochlear gap junction macromolecular complex requires connexin 26. *J Clin Invest* 2014;**124**(4):1598–607. https://doi.org/10.1172/JCI67621. 24590285.

29. Fettiplace R, Kim KX. The physiology of mechanoelectrical transduction channels in hearing. *Physiol Rev* 2014;**94**(3):951–86. https://doi.org/10.1152/physrev.00038.2013. 24987009.

30. Korver AM, Smith RJ, Van Camp G, Schleiss MR, et al. Congenital hearing loss. *Nat Rev Dis Primers* 2017;(3): 16094. https://doi.org/10.1038/nrdp.2016.94. Review, 28079113.

31. Neyroud N, Tesson F, Denjoy I, Leibovici M, et al. A novel mutation in the potassium channel gene KVLQT1 causes the Jervell and Lange-Nielsen cardioauditory syndrome. *Nat Genet* 1997;**15**(2):186–9.

32. Nishimura M, Ueda M, Ebata R, Utsuno E, et al. A novel KCNQ1 nonsense variant in the isoform-specific first exon causes both jervell and Lange-Nielsen syndrome 1 and long QT syndrome 1: a case report. *BMC Med Genet* 2017;**18**(1):66. https://doi.org/10.1186/s12881-017-0430-7. 28595573.

33. Schulze-Bahr E, Wang Q, Wedekind H, Haverkamp W, et al. KCNE1 mutations cause Jervell and Lange-Nielsen syndrome. *Nat Genet* 1997;**17**(3):267–8. https://doi.org/10.1038/ng1197-267. 9354783.

34. Chang Q, Wang J, Li Q, Kim Y. Virally mediated Kcnq1 gene replacement therapy in the immature scala media restores hearing in a mouse model of human Jervell and Lange-Nielsen deafness syndrome. *EMBO Mol Med* 2015;**7**(8):1077–86. https://doi.org/10.15252/emmm.201404929. 26084842.

35. Sheffield AM, Smith RJH. The epidemiology of deafness. *Cold Spring Harb Perspect Med* 2018. pii: a033258 https://doi.org/10.1101/cshperspect.a033258. 30249598.

36. Mammano F. Inner ear connexin channels: roles in development and maintenance of Cochlear function. *Cold Spring Harb Perspect Med* 2018. pii: a033233 https://doi.org/10.1101/cshperspect.a033233. 30181354.

37. DiStefano MT, Hemphill SE, Oza AM, Siegert RK, Grant AR, et al. ClinGen expert clinical validity curation of 164 hearing loss gene-disease pairs. *Genet Med* 2019. https://doi.org/10.1038/s41436-019-0487-0. 30894701.

38. Selimoglu E. Aminoglycoside-induced ototoxicity. *Curr Pharm Des* 2007;**13**(1):119–26. Review, 17266591.

39. Wroblewska-Seniuk K, Dabrowski P, Greczka G, Szabatowska K, Glowacka A, et al. Sensorineural and conductive hearing loss in infants diagnosed in the program of universal newborn hearing screening. *Int J Pediatr Otorhinolaryngol* 2018;**105**:181-6. https://doi.org/10.1016/j.ijporl.2017.12.007. 29447811.

40. Bakhos D, Marx M, Villeneuve A, Lescanne E, Kim S, Robier A. Electrophysiological exploration of hearing. *Eur Ann Otorhinolaryngol Head Neck Dis* 2017;**134**(5):325–31. https://doi.org/10.1016/j.anorl.2017.02.011. 28330595.

41. Kral A, O'Donoghue GM. Profound deafness in childhood. *N Engl J Med* 2010;**363**(15):1438-50. https://doi.org/10.1056/NEJMra0911225. 20925546.

42. White JR, Preciado DA, Reilly BK. Special populations in implantable auditory devices: pediatric. *Otolaryngol Clin North Am* 2019;**52**(2):323-30. https://doi.org/10.1016/j.otc.2018.11.015. Review, 30827361.

43. Kaspar A, Kei J, Driscoll C, Swanepoel de W, Goulios H. Overview of a public health approach to pediatric hearing impairment in the Pacific Islands. *Int J Pediatr Otorhinolaryngol* 2016;**86**:43-52. https://doi.org/10.1016/j.ijporl.2016.04.018. 27260578.

44. le Roux T, Vinck B, Butler I, Cass N, Louw L, et al. Predictors of pediatric cochlear implantation outcomes in South Africa. *Int J Pediatr Otorhinolaryngol* 2016;**84**:61–70. https://doi.org/10.1016/j.ijporl.2016.02.025. Epub 2016 Mar 3, 27063755.

45. von Békésy G. Concerning the Pleasures of Observing, and the Mechanics of the Inner Ear. In: *The Nobel Prize in Physiology or Medicine*; 1961. https://www.nobelprize.org/prizes/medicine/1961/bekesy/lecture/.

46. Guinand N, van de Berg R, Cavuscens S, Stokroos RJ, Ranieri M, et al. Vestibular implants: 8 years of experience with electrical stimulation of the vestibular nerve in 11 patients with bilateral vestibular loss. *ORL J Otorhinolaryngol Relat Spec* 2015;**77**(4):227-40. 26367113.

47. Dabdoub A, Nishimura K. Cochlear Implants Meet Regenerative Biology: State of the Science and Future Research Directions. *Otol Neurotol* 2017;**38**(8):e232-6. https://doi.org/10.1097/MAO.0000000000001407. Review, 28806331.

48. Suzuki J, Corfas G, Liberman MC. Round-window delivery of neurotrophin 3 regenerates cochlear synapses after acoustic overexposure. *Sci Rep* 2016;**6**:24907. https://doi.org/10.1038/srep24907. 27108594.

49. Yawn RJ, Nassiri AM, Rivas A. Auditory neuropathy: bridging the gap between hearing aids and cochlear implants. *Otolaryngol Clin North Am* 2019;**52**(2):349-55. https://doi.org/10.1016/j.otc.2018.11.016. 30765091.

50. Strauss O. The retinal pigment epithelium. In: Kolb H, Fernandez E, Nelson R, editors. *Webvision: The Organization of the Retina and Visual System [Internet]*. Salt Lake City (UT): University of Utah Health Sciences Center; 1995-2011. 21563333.

51. Kamaraj B, Purohit R. Mutational analysis of oculocutaneous albinism: a compact review. *Biomed Res Int* 2014;**2014**:905472. https://doi.org/10.1155/2014/905472. 25093188.

52. Koch KW, Dell'Orco D. Protein and signaling networks in vertebrate photoreceptor cells. *Front Mol Neurosci* 2015;**8**:67. https://doi.org/10.3389/fnmol.2015.00067. eCollection 2015. Review, 26635520.

53. Mathur P, Yang J. Usher syndrome: hearing loss, retinal degeneration and associated abnormalities. *Biochim Biophys Acta* 2015;**1852**(3):406-20. https://doi.org/10.1016/j.bbadis.2014.11.020. Review. 25481835.

54. Nash BM, Wright DC, Grigg JR, Bennetts B, Jamieson RV. Retinal dystrophies, genomic applications in diagnosis and prospects for therapy. *Transl Pediatr* 2015;**4**(2):139-63. https://doi.org/10.3978/j.issn.2224-4336.2015.04.03. 26835369.

55. Daiger SP, Sullivan LS, Bowne SJ. Genes and mutations causing retinitis pigmentosa. *Clin Genet* 2013;**84**(2):132-41. https://doi.org/10.1111/cge.12203. Review, 23701314.

56. Kumaran N, Moore AT, Weleber RG, Michaelides M. Leber congenital amaurosis/early-onset severe retinal dystrophy: clinical features, molecular genetics and

therapeutic interventions. *Br J Ophthalmol* 2017;**101**(9):1147–54. https://doi.org/10.1136/bjophthalmol-2016-309,975. Review. 2868916.

57. Yu-Wai-Man P, Votruba M, Burté F, La Morgia C, et al. A neurodegenerative perspective on mitochondrial optic neuropathies. *Acta Neuropathol* 2016;**132**(6):789–806. Review, 27696015.

58. Fritsche LG, Fariss RN, Stambolian D, Abecasis GR, et al. Age-related macular degeneration: genetics and biology coming together. *Annu Rev Genomics Hum Genet* 2014;**15**:151–71. https://doi.org/10.1146/annurev-genom-090413-025610. Review, 24773320.

59. Simkin SK, Misra SL, Battin M, McGhee CNJ, Dai S. Prospective observational study of universal newborn eye screening in a hospital and community setting in New Zealand. *BMJ Paediatr Open* 2019;**3**(1). pii: bmjpo-2018-000376. https://doi.org/10.1136/bmjpo-2018-000376 eCollection 2019.

60. Solebo AL, Teoh L, Rahi J. Epidemiology of blindness in children. *Arch Dis Child* 2017;**102**(9):853–7. https://doi.org/10.1136/archdischild-2016-310,532. Review. 28465303.

61. Hellström A, Smith LE, Dammann O. Retinopathy of prematurity. *Lancet* 2013;**382**(9902):1445–57. https://doi.org/10.1016/S0140-6736(13)60178-6. Review, 23782686.

ENVIRONMENT AS PROVIDER

Introduction

Earth and its biosphere provide air water, nutrients including food and minerals for human survival and flourishing. In this chapter each of these essentials will be considered in light of their functions and current status on planet earth. Insights into functions of these essential components have been in part gathered through clinical and research studies in cases of deficiencies and also in cases where genetic mutations impact functions (Fig. 1).

Nutrients

Nutrient production food security and environmental sustainability

In 2007 Borlaug[1] reported that each year 800 million people in the world experienced chronic or frequent hunger. He emphasized that continued agricultural development would be necessary to alleviate poverty and improve human health. He noted that two-thirds of the people who experienced food insecurity lived in Asia or Africa. Borlaug also emphasized the importance of linking efforts to improve food production with environmental sustainability and consideration of population growth.

Important points to take into account when considering Earth as Provider include soil quality and production potential. In a review on soils and food sufficiency, Lal[2] emphasized the poor crop yield that occurred in areas where soil was eroded and areas where soils were degraded and depleted of organic matter.

Ecosystems and food security

Bommarco et al.[3] noted that industrial farming methods used to increase food production included crop breeding experimentation, improved irrigation, addition of inorganic fertilizer, use of pesticides for pest and weed control. In addition, mechanical loosening of soil was implemented. They emphasized that although these practices

Gene Environment Interactions. https://doi.org/10.1016/B978-0-12-819613-7.00002-5

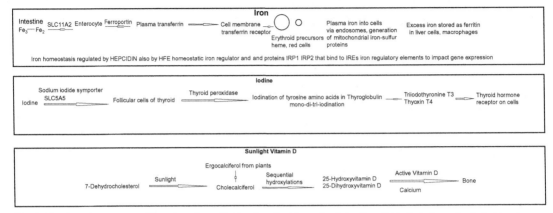

Fig. 1 *Earth and sun as providers.* Iron from diet is absorbed from the intestine. It binds to transferrin in blood and is then transported into cells to facilitate metabolic processes or it may be stored. Body iron homeostasis is highly regulated. Absorbed iodine is taken up into the thyroid gland where it is converted to the hormone Thyroxin. Sunlight activates and converts 7 dehydrocholesterol in skin and ergocalciferol in plants to Vitamin D that is essential for numerous processes especially bone health.

increased food production, they had significant negative environmental effects and impacted biodiversity.

Bommarco et al. proposed ecological intensification approaches to minimize the negative impacts of increased agricultural productivity. They defined four categories of ecosystem services that should be promoted in support of agricultural productivity.

They noted that soil ecology is an important factor to consider, but emphasized that there were difficulties in analyzing soil microbial organisms and in defining the degree of species diversity and function of different species present in soil.

Important aspects of maintaining sustainable soil include use of organic material to increase fertility. These include mulch, and biochar derived from burning plant materials. In addition, crop rotations are recommended. Specific plowing methods have also been found to have deleterious effects on soil and can promote soil erosion.

Another important feature of the proposed ecological intensification includes the intensification of biological methods of pest control. Bommarco et al. noted that this requires knowledge concerning the natural enemies of particular pests. Increased attention to pollinators is also required if food production is to be increased. Environmental factors that can increase pollinator population include increasing floral populations near fields.

Pollinator services

Potts et al.[4] reported that 75% of all crops used in human food production are dependent on pollinator services. Pollinators are key contributors to biodiversity and agricultural productivity. Bees constitute the key pollinator species.

In 2010 there was evidence that bee population in the USA declined 59% between 1947 and 2005. In Central Europe between 1985 and 2005 bee population were reported to have declined by 25%. Potts et al. noted that the wild bee population declined in part due to infection with an ectoparasitic mite that originated in Asia. In addition bee-keeping as an industry declined substantially in the USA and in Europe.

Important drivers of pollinator loss include fragmentation of habitats, decreased plant density pesticide application, and environmental pollution.

Agroecology

A new discipline, Agroecology the ecology of food systems was founded at several institutions in the early years of the twenty-first century.[5] Founders of the discipline encouraged faculty and students "to embrace the wholeness and connectivity of systems".

Key underlying principles of agroecology are to apply ecological principles to agriculture and to foster sustainable ecosystems. Francis et al. emphasized that within natural ecological systems plans and animals were sustained.

Agroecologists stress the importance of taking into account negative environmental impacts of specific agricultural practices so that loss of soil and entry of damaging chemical into ground water can be reduced.

Nutrition and ecology

In 2005 McMichael[6] noted that nutrition sciences traditionally focused on the roles of specific nutrients in physiology and metabolism, growth, human health and deficiency states. He urged that there be greater focus on food production and consumption and on inequities between and within regions and countries. He stressed the importance of integrating human health and biosphere health.

McMichael noted that as population grew and food production escalated, the food producing industries were doing damage to natural resources. He noted further that while the is evidence that hunger and malnourishment is declining slightly, based on reports from the Food and Agriculture Organization (FAO) millions of peoples remained undernourished. In ironic contrast the prevalence of obesity was increasing in middle class populations particularly in urban areas.

Food production trends

McMichael[6] reported that food production, particularly production of cereals rice, wheat and maize increased in the decades that coincided with the green revolution during the period from the 1960s until the latter years of the twentieth century. In his 2005 report MacNeil noted that there was evidence of a downturn in food production and that in some areas this was due to decreasing soil fertility. In some areas there was evidence that irrigation had led to depletion of underground aquifers. In addition, the chemicalization of soils and water increases and levels of nitrates in particular increased due to use of chemical fertilizers.

Nutritional deficiencies

Nutritional deficiency can include insufficient intake of macronutrients or micronutrients. In a 2012 report Ahmed et al.[7] reported that undernutrition and micronutrient deficiency affected significant percentage of the population particular in Africa and Asia. Protein and energy deficiency malnutrition led to stunting of growth of children. Micronutrient deficiencies included particularly deficiency of iron, iodine, vitamin A and zinc.

In 2015 Christian et al.[8] reviewed the central importance of nutrition in advancing health of women, newborn and child health. They emphasized that optimal nutrition was particularly important during the 1000- day period between conception and the 2nd year of life.

Vitamin A

Wiseman et al.[9] reported that information on functions of Vitamin A has greatly expanded. Studies have revealed that vitamin A deficiency in women and children has been shown to be a significant problem in women and children. A number of countries and organization have implemented Vitamin A supplementation programs.

Vitamin A deficiency can be clinically established through measurement of plasma retinol levels.

Wiseman et al. stressed that vitamin A deficiency can also occur as a consequence of decreased nutrient absorption in cases of intestinal diseases. They noted that vitamin A deficiency can interact with iron deficiency since vitamin regulates expression of genes involved in iron metabolism. Vitamin A deficiency has also been reported to affect the composition of the intestinal microbiome.

Vitamin A is particularly abundant in fish oil and in liver and it is also present in many plants particularly sweet potatoes. Vitamin A consists of a mixture or retinol, retinal, retinoic acid and provitamin carotenoids.

The key function of vitamin in the eye include binding of a specific molecule 11Cis retinol to opsin and the formation of the visual pigment in rods, rhodopsin. Tragically vitamin A deficiency in infants and children can lead to blindness.

Nutrition and growth

Undergrowth

Stunting is the term applied to undergrowth in children. Stunting is defined as child height two standard deviations below the average population height for individuals of the same age and sex. The most prevalent cause of stunting is under-nutrition.

In a review published in 2017, Campisi et al.[10] reported that comprehensive studies indicated that 165 million children in the world manifested sub-optimal growth. They noted however data gathered until 2015 revealed that in 2015 there were 98.5 million fewer children with stunted growth then were recorded in 1990. Countries with the most significant decreases in stunting were in East Asia, South Asia and Pacific islands. Stunting rates increased in Sub-Saharan Africa.

Campisi et al. stressed that under-nutrition from pre-conception to beyond the first 1000 days of life plays key roles in childhood stunting. Factors leading to stunting during this period are particularly prevalent in low and middle-income countries. Reduced availability of food, availability of only poor-quality food and infections increase the rate of stunting. In some countries staple food may have low nutrient density and poor mineral availability.

Campisi et al. presented data on stunting in the 2015 World Bank defined regions.

Regions	Numbers of children with stunting in 2015	Numbers of children with stunting in 1990
South Asia	62.7 million	100.1 million
Sub-Saharan Africa	57.2	44.8
East Asia and Pacific	22.2	79.1
Middle-East, North Africa	7.6	16.7
Latin America and Caribbean	5.4	11.65
Europe and Central Asia	5.1	11.6

Genetic factors and height

in times of plenty, height is highly inheritable; in a famine much less so.
RC Lewontin 1970[11]

Genome-wide studies by the Hirschhorn group (see Guo et al.[12]) have revealed that height is determined by thousands of genes across the genome. Their studies have revealed that variants in genes expressed in bone growth plates have important roles in height determination. Of particular interest are genes in the Fibroblast Growth Factor (FGF) signaling pathway.

Guo et al. also reported discovery of a new pathway important in height determination. This pathway involves proteins encoded by 3 genes Stanniocalcin 2 (STC2) a secreted glycoprotein, Pappalysin 1 (PAPPA1) a metalloprotein that cleaves insulin like growth factor, and insulin like growth factor binding protein 4 (IGFBP4). Cleavage of IGFBP releases insulin like growth factor IGF that plays an important role in bone formation.

Overall increase in population height

The NCD Risk Factor Collaboration[13] reported data of world-wide population height in 1896 and 1996 in 200 countries. They reported that adult population height increased in many populations, however, in Sub-Saharan Africa adult height of men remained the same or decreased.

In 1896 the tallest men in the world were in the USA and Sweden. In 1996 the tallest men in the world were in the Netherlands. Significant additions to height were observed in Mexico, South America, in China, Japan. Iran and in Southern and Eastern Europe. Changes in height in the USA and Canada, were much less marked. Importantly in the USA height increased only minimally but Body Mass Index increased significantly. BMI weight in kilogram divided by height in meters squared (kg/m^2).

Requirement for specific minerals

A number of different metals are essential to humans. Much attention has been paid to calcium and to iron. However other metals that are essential factors for enzyme and coenzyme functions these include zinc (ZN), magnesium (Mg), manganese (Mn) other metals also occur within enzymes and proteins and are essential for their function they include copper (Cu), molybdenum (Mb), chromium, selenium. Several of these metals are required only in trace amounts.

Zinc and zinc deficiency

Zinc is present in many fruits and vegetables and also in fish and meat. Krebs et al.[14] reported that zinc deficiency contributed to poor health and could even lead to death in infants and children younger than 5 years of age. In addition to limited intake in pregnancy mothers and young children zinc deficiency resulted from poor intestinal absorption due to enteric diseases and diarrhea.

Krebs et al. noted that more the 20 different zinc transporters are active in humans indicating the importance of maintenance of zinc homeostasis.

Manifestations of zinc deficiency include impaired growth, altered immune response and behavioral changes. In addition, children may develop erosive lesions around the mouth (cheilosis), dermatitis and hair loss. Zinc levels can be determined through measurement of blood zinc levels.

Ackland and Michalczyk[15] reported that zinc is necessary for structural conformation and catalytic functions of many different enzymes. They noted that zinc is a component of transcription factors and plays roles in signal transduction and neural transmission.

Iron

Lane et al.[16] reported that in adult humans 3–5 g of iron are present. Erythrocyte hemoglobin contains 80% of the body iron, the remaining 20% is contained in cytochromes, in iron-sulfur proteins, and iron is also present bound to other proteins and to enzymes.

Iron is an essential nutrient for life forms. Iron is obtained primarily from consumed foods or supplements. In considering the role of iron in biological processes, Gerhard et al.[17] noted that since iron exists in range of reduction and oxidation states, −2 to +6, it plays key roles in electron transfer reactions as an oxygen carrier. They noted also that iron functions as a cofactor for many proteins including iron-sulfur proteins. Gerhard et al. emphasized that iron is taken up from the intestine not only as elemental iron, it is also taken up from sources where it is bound to heme. They noted however that the processes utilized in the processing of dietary heme and iron uptake from that source have not been clearly delineated.

In a 2012 review Fleming and Ponka[18] defined four major cell types involved in iron metabolism, duodenal enterocytes that absorb iron, erythroid precursors that utilize iron, reticulo-endothelial macrophages that are involved in iron storage and hepatocytes that are involved in iron storage and in regulating iron homeostasis.

The first step in iron absorption involves reduction of heminic iron FEIII to ferrous iron FEII, by means of the enzyme DCYTB (CYPBRD1) cytochrome B reductase 1. The reduced iron is then taken up into the enterocyte by means of a specific transporter DMT1 (SLC11A2) that transport divalent metals. Ferroportin (SLC40A1) is involved in the transfer of iron from the enterocyte into the plasma.

Fleming and Ponka noted that iron absorption through the intestine is regulated by iron regulatory proteins IRP1 and IRP2. In addition, the amount of iron transferred from the enterocyte into the systemic circulation is impacted by hepcidin.

Within the circulation iron is bound to transferrin and the degree of iron saturation is normally 30%. The amount of iron absorbed from transferrin by cells is dependent on cell surface transferrin receptors, TFR1 and TFR2 (TFRC1 and TFRC2). The entry of iron into erythroid precursors is facilitated by transferrin receptors on those cells. Iron is then taken up from the surface into endosomes.

Fleming and Ponka noted that when transferrin is oversaturated non-transferrin iron may be present in the circulation and can be taken up by cells including liver cells and cardiomyocytes, leading to cell damage.

Storage levels of iron in cells including reticulo-endothelial cells is regulated by hepcidin. Reticuloendothelial cells also derive iron from degrading erythrocytes. Iron can also be exported from reticulo-endothelial cells and ferroportin and hepcidin regulate this export.

Hepatocytes can store iron as ferritin. When transferrin levels of iron saturation are very high, non-transferrin bound iron can enter into hepatocytes.

Uptake of iron into cells

After iron is taken up into the endosomes and lysosomes compartments in cells it is released by acidic pH, and it undergoes reduction. Lane et al.[16] noted that transporters move ferrous iron out of the endosome into the cytoplasm of the cell. These transporters include DMT1 and ZIP14. now designated as solute carriers SLC11A2 and SLC39A14.

Iron in the cytoplasm is designated as the Labile iron pool. Iron from this pool can be imported into heme and other proteins including enzymes and can be incorporated in mitochondria it can also be released back into the circulation. Export of iron is expedited by the protein ferroportin.

In storage cells ferritin is the major iron storage proteins. Ferritin is a multimeric protein composed of light chain and heavy chain subunits. Light chains are encoded by *FTL* a gene on chromosome 19 and heavy chains are encoded by a gene *FTH1* on chromosome 11. Together the subunits of ferritin form a hollow structure and iron is stored in the hollow cavity. Release of iron from ferritin occurs through an autophagic process.

Mechanisms exist to tightly control cellular homeostasis. Hepcidin is a key controller of iron homeostasis, it functions by controlling ferroportin the key molecule involved in controlling cellular efflux of iron out of cells into the blood stream. When iron levels in blood are low ferroportin transfers iron that is stored in liver cells or macrophages out of these cells into the bloodstream. When blood levels of iron are adequate levels of hepcidin increase and ferroportin is suppressed so that iron remains inside cells. However, hepcidin levels and functions are also influenced by infections.

Other key proteins in control of iron metabolism are RNA binding proteins IRP1 and IRP2 that bind to iron regulatory elements (IREs) in the untranslated regions of RNA and moderate mRNA stability and translation of key mRNAs that encode proteins related to iron metabolism.

Iron homeostasis

Fleming and Ponka[18] emphasized the important role of Hepcidin that is produced by liver cells, in controlling iron homeostasis. Hepcidin can down-regulate ferroportin release of iron into the circulation through a complex signaling mechanisms pathway that involves BMP6 and hemojuvelin (a BMP coreceptor) and the transferrin receptors and the HFE proteins (homeostatic iron regulators).

In situations where erythropoiesis is increasing the production of hepcidin by the liver is decreased. Also, under hypoxic condition there is reduced signaling through the BMP6 pathway and reduced hepcidin inhibition of iron uptake.

Mechanisms exist to tightly control cellular homeostasis. Hepcidin is a key controller of iron homeostasis, it functions by controlling ferroportin the key molecule involved in controlling cellular efflux of iron out of cells into the blood stream. When iron levels in blood are low ferroportin transfers iron that is stored in liver cells or macrophages out of these cells into the bloodstream. When blood levels of iron are adequate levels of hepcidin increase and ferroportin is suppressed so that iron remain insides cells. However, hepcidin levels and functions are also influenced by infections.

Other key proteins in control of iron metabolism are RNA binding proteins IRP1 and IRP2 that bind to iron regulatory elements (IREs) in the untranslated regions of RNA and moderate mRNA stability and translation of key mRNAs that encode proteins related to iron metabolism.

Hemochromatosis

Detailed studies on Hereditary Hemochomatosis syndromes have expanded insights into factors involved in iron homeostasis. It is however important to noted that hemochromatosis can also result from high dietary iron intake that leads to over-saturation of transferrin and to increases levels of circulating non-transferrin iron.

Manifestations of hereditary hemochromatosis (HH) include elevated transferrin saturation, increased levels of serum ferritin and increased tissue iron deposit. Particularly hazardous are increase iron deposits in liver, cardiomyocytes and in the islet cells of the pancreas.

Juvenile hemochromatosis (Type2a) is a severe form of hemochromatosis that presents in childhood and is associated with liver enlargements and pain and joint defects.[19]

In the 2012 review by Fleming and Ponka[18] and in a 2018 review by Gerhard et al.[17] four main types of Hereditary Hemochromatosis are defined.

In Type 1 Hereditary Hemochromatosis, mutations occur in in *HFE* gene, that encodes a homeostatic iron regulator; three different forms have been described:

Type 1a	HFE mutation Cys282Tyr homozygosity
Type 1b	HFE mutation Cys282Tyr/His/63Asp compound heterozygosity
Type 1c	Other pathogenic HFE genotypes.

Other forms of hereditary hemochromatosis are due to mutations in other genes:

Type 2a	Hemojuvelin (HJV) (HFE2) mutations
Type 2b	Hepcidin (HAMP) mutations
Type 3	Transferrin receptor mutations TFR1 (TFTC) and TFR2
Type 4	Ferroportin (SLC40A1) mutations

Gene discoveries and identification of key functional proteins in iron metabolism

Gerhard et al.[17] traced the history of discovery of the genes defects that lead to Hereditary hemochromatosis. The key discovery relevant to hemochromatosis type 1 was that the locus for this phenotype mapped to human chromosome 6 in the vicinity of the HLA gene. Intense cloning and sequencing efforts led to identification of the *HFE* gene (HFE1). The Cys282Tyr (C282Y) mutation occurs with high frequency in populations of Northern European origin. Fleming and Ponka reported that 36–76% of males homozygous for this mutation have disease manifestation; 2 to 38% of women primarily post-menopausal women, manifest symptoms. They noted that penetrance of the pathogenic mutation is influenced by modifier genes and by environmental factors.

The type 2a hemochromatosis phenotype, a severe disorder that manifestations in infants and children, was mapped to chromosome 1q21.1 through genetic linkage studies. Following this mapping, positional cloning was carried out and a specific gene, designated hemojuvelin was identified in which a G320V substitution occurred in affected individuals in more than one third of families.[19] Gerhard et al. reported that subsequently additional disease-causing mutations were found in the hemojuvelin gene, HJV in families with hereditary hemochromatosis type 2a (HH2a).

The gene on chromosome 19q13.2 that encodes hepcidin, also known as hepcidin antimicrobial peptide (HAMP), was investigated in cases of juvenile hemochromatosis that did not have Hemojuvelin defects. This led to discovery of several pathogenic variants in HAMP and the associated form of hemochromatosis was referred to as HH2b.

Gerhard et al. reported that the transferrin receptor gene was investigated as a possible candidate gene for hereditary hemochromatosis in Sicilian families. These studies led to identification of mutations in the transferrin receptor TRF2 encoding gene on chromosome 7q22,1. The transferrin 1 receptor TRF encoding gene has also been shown to carry mutations in certain families with hemochromatosis.

Mapping studies of Hereditary hemochromatosis type 4, were carried out by researchers in the Netherlands and this led to assignment of the condition to a locus on chromosome 2q.[20] Positional cloning and sequencing studies led to identification of the gene that encodes a metal transporter, Ferroportin. This transporter gene is now known as SLC40A1. HH4 can be inherited as an autosomal dominant or as an autosomal recessive condition.

Dietary iron overload

This condition has been described in populations in South Africa and is apparently due to high consumption of alcoholic beverages brewed in iron containers. Kew[21] presented data on the role of iron overload in causation of liver cirrhosis and hepato-cellular carcinoma.

Troadec et al.[22] reported evidence for genotoxicity and DNA damage of excess iron and roles of genotoxicty in hepatocellular carcinoma.

Iron deficiency

Troadec et al.[22] reported that iron deficiency is the most common nutritional deficiency and that it occurs throughout the world and that it is particularly common in children and women. They emphasized that iron is essential for appropriate cell cycle progression. In addition, iron dependent enzymes and protein are required for DNA synthesis, DNA proof reading and for telomere replication.

Stelle et al.[23] reported that iron deficiency impacts 1.24 billion people. People in resource poor countries are particularly impacted. They emphasized the concomitant infections frequently increased the burden of iron deficiency. They noted that this degree of iron deficiency persists despite efforts aimed at nutritional improvements.

Calcium and vitamin D

In 2014 Pettifor[24] reported that in many developing countries, including those in tropical and sub-tropical regions, nutritional rickets remained a significant health problem. He emphasized that although sunlight was abundant in those regions air pollution and other problems led to vitamin D insufficiency. Importantly he emphasized that adequate calcium intake was also essential for prevention of rickets.

Important sources of calcium include milk- based products, certain vegetables beans, lentils certain green vegetables, nuts (almonds), certain seeds.

Vitamin D can be provided by certain foods e.g. cheese, eggs, fish. Vitamin D can also be synthesized in the body. Exposure of skin to sunlight can convert 7-dehydrocholesterol to cholecalciferol. Cholecalciferol is converted in the liver to 25-hydroxycholecalciferol (also known as calcifediol), through activity of an enzyme CYP2R1. The 25-hydroxycholecalciferol binds to a transporter protein and enters the circulation. A further hydroxylation reaction occurs in tissues, especially in the kidney, leading to the production of 1alpha 25dihydrocholecalciferol (also known as calcitriol), through activity of an enzyme CYP27B1. This is the reaction that leads to the formation of the biologically active form of vitamin D also known as vitamin D3 or as calcitriol.

It is important to note that vitamin D deficiency can occur as a manifestation of kidney disease.

Synthesis of active Vitamin D3 is regulated by parathyroid hormone. In recent years it has also been shown to be regulated by a hormone FGF23 produced by osteoclasts and osteoblasts.

In 2016 Munns et al.[25] published information on the incidence of calcium and Vitamin deficiencies in different countries and nutritional guidelines to counteract nutritional rickets. Their analyses revealed that rickets, osteomalacia and calcium deficiency represent significant health problems particularly in infants, children adolescents. Nutritional rickets was defined as a disorder that impacted differentiation and survival of chondrocytes in the growth plate and was associated with defects in mineralization of bones and may lead to death of chondrocytes in the bone plate.

In situations where minerals and vitamin D are deficient and in the presence of disease, especially kidney disease, bone mineralization is inadequate. Inadequate mineralization of the bone matrix leads to osteomalacia and rickets.

In children inadequate calcium and vitamin D intake leads to clinical features of rickets. These include, swelling at wrists and ankles, limb deformities, enlarged costochondral junctions of ribs, pelvis deformities and bone pain.

It is important to note that if calcium intake is greatly reduced and calcium blood levels fall parathyroid hormone acts to release calcium from bone. Parathyroid hormone also acts in the kidney to promote resorptions of calcium in the renal tubules and parathyroid hormone increases calcitriol absorption in the intestine, when it is provided in supplements.

Insights into processes underlying calcium, phosphate and Vitamin D roles and factors involved in bone mineralization have been advanced through detailed genetic studies in genetic conditions associated with familial forms of rickets, occurring despite adequate intake of calcium and vitamin D and through studies associated with defective bone formation, e.g. osteogenesis imperfecta.

Insights gained into bone formation through studies in osteogenesis imperfecta

In a 2018 review Bacon and Crowley[26] reported that 16 different types of osteogenesis imperfecta were known Five forms manifest as autosomal dominant genetic diseases and 11 forms are autosomal recessive genetic diseases. In many of the forms of Osteogenesis imperfecta (OI)) the genes that carried the disease -causing mutations had been identified and their protein functions were analyzed. In 8 forms of OI the mutations impacted collagen protein that forms the bone matrix. The mutations led to defective collagen that failed to undergo appropriate folding or mutations led to defects in post-translational modification of collagen. Three forms of OI were found to lead to defects in osteoblasts that secrete the bone matrix and in three forms of OI there were defects in bone mineralization processes.

Hypophosphatasia and alkaline phosphatase

Extensive studies were carried out over several decades by Harry Harris' research group on human alkaline phosphatase loci, their products and genetic variations. In 1990 Harris[27] published a review on human alkaline phosphatases. He noted that in humans at least 3 main gene loci encoded these enzymes.

The L/B/K locus encodes an enzyme referred to as liver, bone, kidney alkaline phosphatase. This locus is now known as the *ALPL* gene locus and the enzyme it encodes is sometimes referred to as non-specific alkaline phosphatase TNAP or TNSALP since it has been found in additional tissues and cell types. This locus maps to human chromosome 1p36.12.

A second locus was found to encoded an alkaline phosphatase expressed primarily in the intestine. This locus is now referred to as the *ALPI* locus. It maps to chromosome 2q37.1.

A third locus encodes an alkaline phosphatase that is expressed primarily in placenta (*ALPP*). The placental alkaline phosphatase was found to manifest many polymorphic variants. This locus maps on chromosome 2q37.1 close to *ALPI.*

In 1988 Weiss et al.[28] reported discovery of a pathogenic missense mutation in the liver/bone/kidney form of alkaline phosphatase in a case of hypophosphatasia. Millán and Whyte[29] reviewed alkaline phosphatase defects and hypophosphatasia. They reported that patients with this disorder manifest a broad range of clinical severity but all patients had pathogenic ALPL mutations. One form of classification of hypophosphatasia is based on the age of the patient with clinical manifestation, there are perinatal, infantile childhood forms of hypophosphatasia.

Millán and White reported that hypophosphatasia manifests with clinical signs, rickets or osteomalacia, however the circulating levels of calcium, inorganic phosphate and vitamin D are not decreased. Some patients may manifest with dental abnormalities and tooth loss. The main functional defect is apparently a block in the entry of mineral into the skeleton. The *ALPL* gene product (ALPL) is present in osteoblasts and chondrocytes.

By 2016, 300 different pathogenic ALPL mutations had been identified in hypophosphatasia patient. Millán and White noted that 70% of these mutations were reported to be missense mutations. It is interesting to note that both autosomal dominant and autosomal recessive forms of hypophosphatasia occur. In infantile and perinatal forms of the disorder, pathogenic homozygous mutations or compound heterozygous mutations were found.

Millán and White reported that enzyme replacement therapy has been developed for this disorder. Asfotase alpha, a recombinant form of human alkaline phosphatase has been approved in several countries, including the USA, for treatment of infantile and juvenile hypophosphatasia.

Rickets/osteomalacia due to genetic defects

Autosomal dominant hypophosphatemic rickets (ADHR) and X-linked hypophosphatemic rickets

In 2000 the ADHR consortium[30] published the results of genetic linkage studies in several families with autosomal dominant hypophosphatemic rickets. Results of their studies revealed that the disease phenotype was linked to a region on chromosome 12p13.3. They carried our extensive gene cloning and DNA sequencing in this region and established that the disease in the families they studied was due to mutations in a gene designated *FGF23*.

Extensive studies have subsequently been carried out on FGF23 protein. FGF23 is produced by osteoclasts and osteoblasts in bone. However, its function occurs in the kidney. Erben[31] emphasized that under physiological conditions the endocrine activity of FGF23 is depends on activity of a cofactor Klotha that facilitates its binding to FGF (fibroblast growth factor) receptors. In that disorder there are pathologic mutations in a gene that encodes an endopeptidase PHEX. It turned out that PHEX functions to cleave and inactivated FGF23.

Another important aspect of FGF23 that emerged as a result of genetic studies on X linked hypophosphatasia. X-linked hypophosphatasia occurs due to defective PEX and continued FGF23 activity.

One specific mutation found in ADHP autosomal dominant hypophosphatasia rendered the protein unresponsive to cleavage by PHEX, so that FGF23 was not inactivated by cleavage.

The normal function of FGF23 in the kidney is to reduce reabsorption of phosphate in the renal tubules. In cases where FGF is overexpressed (e.g., when it is not inactivated), this results in a condition associated with elevated levels of FGF23 and low levels of circulating phosphate result.

Shimada et al.[32] reported that certain tumors, e.g. bone tumors produce excess FGF23 and that this leads to hypophosphatemic and abnormal bone resorption. In some such cases there are sometimes abnormal calcium deposits in soft tissues.

FGF23 has been found to suppress Vitamin D synthesis in the kidney through suppression of hydroxylase CYP27B1. FGF23 also to increase calcium resorption.

Erben[31] reported that there is evidence for a feedback loop whereby synthesis of normal FGF23 in bone cells is inhibited by high circulating levels of phosphate or by high levels of phosphate in bone.

Iodine and thyroid function

Between the twelfth and the fourteenth century writings emerged that described goiters and cretinism and the occurrence of these in regions in great distances from the ocean.[33] The documentation in 1825 by Boussinggault that salts with iodide cured goiters and the promotion by Lombroso in Switzerland, starting in 1873,[34] of use of iodine to cure goiters are among the key medical breakthroughs of the nineteenth century.

Stathatos[35] in reviewing thyroid physiology emphasized that the functional unit of the thyroid gland is thyroid follicle, a cystic structure lined by epithelial cells, the thyrocytes. Thyroglobulin is present within the follicle cavities.

The key regulator of thyroid function is thyroid stimulating (TSH), produced in the anterior pituitary gland. It binds to TSH receptor, a 7 transmembrane domain G-protein coupled receptor, is located in the basal membrane of the thyrocytes. Binding of TSH to this receptor leads to activation of signaling through adenylate cyclase. Signal is ultimately transmitted from thyrocyte cytoplasm to nucleus leading to activation of expression of specific genes that will ultimately lead to production of active thyroid hormone.

Mutations in the gene that encodes the thyroid receptor gene TSHR, were reported to play critical roles is hyper-proliferation and hyperfunctioning of thyroid nodules in in some cases of follicular cancer, Mon et al.[36] Inactivating mutations of this receptor leading to hypothyroidism have also been described.[37]

Congenital hypothyroidism

In a 2011 review of congenital hypothyroidism Grasberger and Refetoff[38] noted that previously congenital hypothyroidism was frequently attributed to defects in the development formation and migration of the thyroid gland. They noted that more recent studies determined that congenital hypothyroidism frequently results from defects in specific steps involved in distinct steps in the synthesis of thyroid hormone.

A review by Grasberger and Refetoff[38] provided detailed insights into the processes involved in the synthesis of thyroid hormone. Thyroid gland cells, thyrocytes, lie in close contact with a network of vascular capillaries. Iodine is transferred from the circulating blood in the capillaries through activity of a transporter a sodium iodide symporter (NIS). The this a solute carrier and is currently referred to as SLC5A5.

The next key step involves linking of iodide to tyrosine amino acids in thyroglobulin. This process the function of thyroid peroxidase (TPO). Since TPO is a peroxidase it requires activation by hydrogen peroxidase. This reaction requires the activity of hydrogen peroxidase generating oxidase DUOX2 and its maturation factor DUOXA2. Following its oxidation TPO transfers iodide to tyrosines in thyroglobulin.

Iodinated thyroglobulin is then taken up into the interior of the thyrocyte through processes of endocytosis and pinocytosis. It is taken up into endosomes and lysosomes where it undergoes proteolysis. The leads to release into the cell interior of mono-di, tri and tetra iodothyronines. Mono and diiodothyronines serve as a reservoir of iodide, Iodine can be released through activity of a dehalogenase enzyme DEHAL1 (IYD) iodotyrosine deiodinase.

The solute carrier SLC26A4 (also known as Pendrin) and the membrane protein Anoctamin (TMEM16A) mediate the efflux of mono- di, and tri- iodothyronines into circulation through the apical membrane of thyrocytes, Silviera and Kopp 2015. The name Pendrin was assigned to a protein found to be defective in a specific syndrome Pendred syndrome, characterized by congenital hypothyroidism and deafness.

The transporter SLC16A2 facilitates the cellular import of thyroxine (tetraiodothyronine T4) and triiodothyronine T3, in cells in different parts of the body.

Iodine deficiency goiter, hypothyroidism

In 2012 Andersson et al.[39] stressed the effects of iodine deficiency and consequent thyroid hormone deficiency on growth and development including cognitive development. They noted that iodine deficiency remained a global threat since it leads to preventable mental impairment.

Major efforts have been made to ensure iodine sufficiency, particularly in countries where the soil is poor in iodine and cultivated plants therefore have insufficient iodine.

Andersson et al.[39] reported that data from the WHO indicated that between 2003 and 2011 major progress had been made. During that period the number of countries where iodine deficiency was detected decreased from 54 to 32. Distribution of iodized salt had led to this progress. Andersson et al. note however that regional deficiencies still occurred in S.E Asia. Furthermore, little progress in reducing iodine deficiency had been made in Africa. In 2015 Zimmerman et al.[40] reported that iodine deficiency occurred in Africa and SE Asia.

In 2013 Büyükgebiz[41] wrote, "Newborn screening for congenital hypothyroidism is one of the major achievements of preventive medicine."

Manousou et al.[42] emphasized the importance of adequate iodine intake during pregnancy. Even as efforts continue to improve distribution of iodized salt some investigators have cautioned that iodine concentrations in salt should be appropriately monitored to avoid iodine excess.

Requirement of other minerals present in soil

Copper is an essential mineral for eukaryotes since it serves as a cofactor for many enzyme reactions. However free copper is damaging therefore in the body copper is bound to specific proteins. Baker et al.[43] reported that copper bound proteins (cuproproteins) play important roles in oxidation reduction reactions, in energy metabolism in collagen deposition and in neurotransmitter biosynthesis. Excess copper intake is damaging.

Zinc is another important metal used in the body as a cofactor for enzymes. Zinc deficiency can be precipitated by intake of food high in the plant component phytate that binds zinc, Sandstead and Freeland-Graves.[44]

Magnesium is another earth derived mineral required by the body and is used as a co-factor in enzyme reactions. Specific investigators have postulated that magnesium deficiency can promote vascular disease.[45]

Water

Statistics published by WHO in 2017 revealed that world-wide 2.1 billion people in the world lacked regular access to safe drinking water.

Managed sanitation services were not available to 4.5 billion people and untreated waste water flowed into the ecosystem. https://www.who.int/water_sanitation_health/publications/jmp-2017/en/.

UNICEF published evidence in 2015 that 340,000 children died each year of diarrhea related complications. https://www.unicefusa.org/mission/survival/water.

WHO stressed that water is at the core of sustainable development and healthy ecosystems. Precise assessments of water availability will continue to fluctuate in this age of climate change.

Mekonnen and Hoekstra[46] reviewed freshwater security. They noted that in developed countries water scarcity prevailed in some areas of high population density and in areas where large quantities of water were used for agricultural irrigation. In some regions water shortage was cyclical. They also drew attention to disappearing lakes and lack of water in downstream regions of rivers due to heavy upstream use. In several countries there is evidence of ground water depletion.

Water pollution

In 2010 Schwarzenbach et al.[47] reviewed water pollution and its impact on human health. They also addressed improved methods to cope with water pollutants. They noted that lack of appropriate sanitation and lack of safe drinking water impacted more than one third of the planet population. The also noted problems related to the bioaccumulation of toxic chemicals in the food chain as a result of water contamination.

Agriculture industry and domestic activities were reported to consume more than one-third of the accessible fresh water on earth.

The Schwarzenbach et al. review concentrated primarily on chemical water, nitrogen, phosphorus and organic constituents derived from biomass including toxic algal blooms. They noted that particular pollutants can be toxic even at relatively low concentrations; high salt concentration can also be problematic Other inorganic pollutants that pose problems include heavy metals mercury, chrome copper, cadmium and nickel.

Mining operations can generate contaminants that make their way into water, these include cyanide mercury and copper.

Persistent organic pollutants include highly chlorinated compounds DDT, polychlorinated biphenyls (PCBs), polycyclic aromatic hydrocarbons and dioxins. Other emerging organic pollutants include flame retardant polybrominated diphenyl esters. Schwarzenbach emphasized that persistent organic pollutants present a particular problem because they accumulate if the food web.

Pesticides

Data indicate that between 3 and 7 million tons of pesticides are produced annually. Schwarzenbach et al. noted that in agricultural areas water catchment contaminants constitute threats to human life forms and other life forms.

In aquifers and ground water hazardous contaminating elements identified include arsenic fluoride and chromium.

Air and oxygen

Humans and most life forms require oxygen. The optimal oxygen concentration is breather air for humans is reported to be between 19.5% and 23.5%. Components of clean atmospheric air include: Oxygen 21%, Nitrogen 78%, CO_2 and other gases 1% and water vapor 1%.

Holgate and Stokes-Lampard[48] considered the most damaging components in polluted air to be the small particulates pm 2.5 and pm 0.1. The small particulates are oxides of nitrogen, organic chemicals and ozone O_3. Emission from diesel trucks are particularly abundant in NO_2. Coal fired plants emit sulfur dioxide, nitrogen oxides and carbon dioxide. In certain environments lead levels in air may be high particularly due to use of leaded fuels.

Ozone is defined by the WHO as a secondary pollutant produced by oxidation of other pollutants, including organic compounds carbon monoxide and nitrous oxides.

Household air pollution is an important problem in certain regions of the world. Pollution results from burning of solid fuels wood, animal dung coal and is a problem particularly when ventilation is limited in the space where fuel is burned. Household pollution is reported to be responsible for pneumonia in children and chronic obstructive lung disease in adults.

Central nervous system and air pollution

Babdjouni et al.[49] reviewed the effects of air pollution on the central nervous system. They noted evidence that air pollution with particulate matter are sources of neuro-inflammation and lead to generation of reactive oxygen species. They also noted that there are studies indicating that prolonged exposure to particulate matter and pollutants can lead to white matter degeneration.

Studies on residents living close to major roadways in New York and New Jersey revealed that chronic exposure to particulate pollution PM2.5 was associated with increased carotid intima media thickening and carotid artery stiffness that can compromise cerebral blood flow. High levels of black carbon in air was also reported to increase carotid artery stiffness.

Black carbon is a component of fine particular matter emitted from coal fired power plants and engines that burn fossil fuels (EPA). https://www.epa.gov/air-research/black-carbon-research.

References

1. Borlaug N. Feeding a hungry world. *Science* 2007;**318**(5849):359. https://doi.org/10.1126/science.1151062. 17947551.

2. Lal R. Soils and food sufficiency. A review. In: *Agronomy for Sustainable Development*. 29 (1). Springer Verlag/EDP Sciences/INRA; 2009. p. 113-33. ffhal-00886505.

3. Bommarco R, Kleijn D, Potts SG. Ecological intensification: harnessing ecosystem services for food security. *Trends Ecol Evol* 2013;**28**(4):230-8. https://doi.org/10.1016/j.tree.2012.10.012. 23153724.

4. Potts SG, Biesmeijer JC, Kremen C, Neumann P, Schweiger O, Kunin WE. Global pollinator declines: trends, impacts and drivers. *Trends Ecol Evol* 2010;**25**(6):345-53. https://doi.org/10.1016/j.tree.2010.01.007. 20188434.

5. Francis C, Lieblein G, Gliessman S, Breland TA, Creamer N, Harwood R, et al. Agroecology: the ecology of food systems. *J Sustain Agric* 2003;**22**(3):99-118. Received 27 Sep 2001, Accepted 11 Jul 2002, Published online: 17 Oct 2008.

6. McMichael AJ. Integrating nutrition with ecology: balancing the health of humans and biosphere. *Public Health Nutr* 2005;**8**(6A):706-15. 16236205.

7. Ahmed T, Hossain M, Sanin KI. Global burden of maternal and child undernutrition and micronutrient deficiencies. *Ann Nutr Metab* 2012;**61**(suppl 1):8-17. https://doi.org/10.1159/000345165. 23343943.

8. Christian P, Mullany LC, Hurley KM, Katz J, Black RE. Nutrition and maternal, neonatal, and child health. *Semin Perinatol* 2015;**39**(5):361-72. https://doi.org/10.1053/j.semperi.2015.06.009. 26166560.

9. Wiseman EM, Bar-El Dadon S, Reifen R. The vicious cycle of vitamin a deficiency: a review. *Crit Rev Food Sci Nutr* 2017;**57**(17):3703-14. https://doi.org/10.1080/10408398.2016.1160362.

10. Campisi SC, Cherian AM, Bhutta ZA. World perspective on the epidemiology of stunting between 1990 and 2015. *Horm Res Paediatr* 2017;**88**(1):70-8. https://doi.org/10.1159/000462972 Review. 28285312.

11. Lewontin RC. The Heritability fallacy. *Bull At Sci* 1970;**26**:2-8.

12. Guo MH, Hirschhorn JN, Dauber A. Insights and implications of genome wide association studies of height. *J Clin Endocrinol Metab* 2018. https://doi.org/10.1210/jc.2018-01126. 29982553.

13. NCD Risk Factor Collaboration (NCD-RisC). A century of trends in adult human height. *elife* 2016;**5**. pii: e13410. https://doi.org/10.7554/eLife.13410. 27458798.

14. Krebs NF, Miller LV, Hambidge KM. Zinc deficiency in infants and children: a review of its complex and synergistic interactions. *Paediatr Int Child Health* 2014;**34**(4):279-88. https://doi.org/10.1179/2046905514Y.0000000151. Review, 25203844.

15. Ackland ML, Michalczyk AA. Zinc and infant nutrition. *Arch Biochem Biophys* 2016;**611**:51-7. https://doi.org/10.1016/j.abb.2016.06.011. Review, 27317042.

16. Lane DJ, Merlot AM, Huang ML, Bae DH, Jansson PJ, et al. Cellular iron uptake, trafficking and metabolism: key molecules and mechanisms and their roles in disease. *Biochim Biophys Acta* 2015;**1853**(5):1130-44. https://doi.org/10.1016/j.bbamcr.2015.01.021. Review, 25661197.

17. Gerhard GS, Paynton BV, DiStefano JK. Identification of genes for hereditary hemochromatosis. *Methods Mol Biol* 2018;**1706**:353-65. https://doi.org/10.1007/978-1-4939-7471-9_19.

18. Fleming RE, Ponka P. Iron overload in human disease. *N Engl J Med* 2012;**366**(4):348-59. https://doi.org/10.1056/NEJMra1004967. 22276824.

19. Lee PL, Beutler E, Rao SV, Barton JC. Genetic abnormalities and juvenile hemochromatosis: mutations of the HJV gene encoding hemojuvelin. *Blood* 2004;**103**(12):4669-71. 14982867.

20. Njajou OT, Vaessen N, Joosse M, Berghuis B, et al. A mutation in SLC11A3 is associated with autosomal dominant hemochromatosis. *Nat Genet* 2001;**28**(3):213-4. https://doi.org/10.1038/90038. 11431687.

21. Kew MC. Hepatic iron overload and hepatocellular carcinoma. *Liver Cancer* 2014;**3**(1):31-40. https://doi.org/10.1159/000343856. Review, 24804175.

22. Troadec MB, Loréal O, Brissot P. The interaction of iron and the genome: for better and for worse. *Mutat Res* 2017;**774**:25-32. https://doi.org/10.1016/j.mrrev.2017.09.002. Review, 29173496.

23. Stelle I, Kalea AZ, Pereira DIA. Iron deficiency anaemia: experiences and challenges. *Proc Nutr Soc* 2018;1-8. https://doi.org/10.1017/S0029665118000460.

24. Pettifor JM. Calcium and vitamin D metabolism in children in developing countries. *Ann Nutr Metab* 2014;**64**(suppl 2):15-22. https://doi.org/10.1159/000365124. Review, 25341870.

25. Munns CF, Shaw N, Kiely M, Specker BL, et al. Global consensus recommendations on prevention and management of nutritional rickets. *J Clin Endocrinol Metab* 2016;**101**(2):394-415. https://doi.org/10.1210/jc.2015-2175Review. 26745253.

26. Bacon S, Crowley R. Developments in rare bone diseases and mineral disorders. *Ther Adv Chronic Dis* 2018;**9**(1):51-60. https://doi.org/10.1177/2040622317739538. Review, 29344330.

27. Harris H. The human alkaline phosphatases: What we know and what we don't know. *Clin Chim Acta* 1990;**186**(2):133-50. Review, 2178806.

28. Weiss MJ, Cole DE, Ray K, Whyte MP, et al. A missense mutation in the human liver/bone/kidney alkaline phosphatase gene causing a lethal form of hypophosphatasia. *Proc Natl Acad Sci U S A* 1988;**85**(20):7666-9. 3174660.

29. Millán JL, Whyte MP. Alkaline phosphatase and Hypophosphatasia. *Calcif Tissue Int* 2016;**98**(4):398-416. https://doi.org/10.1007/s00223-015-0079-1. 26590809.

30. ADHR Consortium. Autosomal dominant hypophosphataemic rickets is associated with mutations in FGF23. *Nat Genet* 2000;**26**(3):345-8.

31. Erben RG. Physiological actions of fibroblast growth factor-23. *Front Endocrinol (Lausanne)* 2018;**9**:267. https://doi.org/10.3389/fendo.2018.00267. eCollection 2018. Review, 29892265.

32. Shimada, T., Muto, T., Urakawa, I., Yoneya, T et al. Mutant FGF-23 responsible for autosomal dominant hypophosphatemic rickets is resistant to proteolytic cleavage and causes hypophosphatemia in vivo. Endocrinology 2002 143: 3179-3182. PubMed: 12130585.

33. Merke F. *History and iconography of endemic goiter and cretinism* [translated by S.D. Stephenson]. Berne Switzerland: H. Huber; 1984. first published 1971.

34. Lombroso C. *Sulla microcephalia e sul cretinismo. Revista Clinica di Bologna fasc. 7 July*; 1873.

35. Stathatos N. Thyroid physiology. *Med Clin North Am* 2012;**96**(2):165-73. https://doi.org/10.1016/j.mcna.2012.01.007. Review, 22443969.

36. Mon SY, Riedlinger G, Abbott CE, Seethala R, et al. Cancer risk and clinicopathological characteristics of thyroid nodules harboring thyroid-stimulating hormone receptor gene mutations. *Diagn Cytopathol* 2018;**46**(5):369-77. https://doi.org/10.1002/dc.23915. 2951668.

37. Briet C, Suteau-Courant V, Munier M, Rodien P. Thyrotropin receptor, still much to be learned from the patients. *Best Pract Res Clin Endocrinol Metab* 2018;**32**(2):155-64. https://doi.org/10.1016/j.beem.2018.03.002. activating mutations; receptor; signalization; small molecule agonists; specificity; thyrotropin, 29678283.

38. Grasberger H, Refetoff S. Genetic causes of congenital hypothyroidism due to dyshormonogenesis. *Curr Opin Pediatr* 2011;**23**(4):421-8. https://doi.org/10.1097/MOP.0b013e32834726a4. Review, 21543982.

39. Andersson M, Karumbunathan V, Zimmermann MB. Global iodine status in 2011 and trends over the past decade. *J Nutr* 2012;**142**(4):744–50. https://doi.org/10.3945/jn.111.149393. 22378324.

40. Zimmermann MB, Boelaert K. Iodine deficiency and thyroid disorders. *Lancet Diabetes Endocrinol* 2015;**3**(4):286–95. https://doi.org/10.1016/S2213-8587(14)70225-6. 25591468.

41. Büyükgebiz A. Newborn screening for congenital hypothyroidism. *J Clin Res Pediatr Endocrinol* 2013;**5**(Suppl 1):8–12. https://doi.org/10.4274/jcrpe.845. 23154158.

42. Manousou S, Johansson B, Chmielewska A, Eriksson J, et al. Role of iodine-containing multivitamins during pregnancy for children's brain function: protocol of an ongoing randomised controlled trial: the SWIDDICH study. *BMJ Open* 2018;**8**(4):e019945. https://doi.org/10.1136/bmjopen-2017-019945. 29643159.

43. Baker ZN, Cobine PA, Leary SC. The mitochondrion: a central architect of copper homeostasis. *Metallomics* 2017;**9**(11):1501–12. https://doi.org/10.1039/c7mt00221a. Review, 28952650.

44. Sandstead HH, Freeland-Graves JH. Dietary phytate, zinc and hidden zinc deficiency. *J Trace Elem Med Biol* 2014;**28**(4):414–7. https://doi.org/10.1016/j.jtemb.2014.08.011. 25439135.

45. Kostov K, Halacheva L. Role of magnesium deficiency in promoting atherosclerosis, endothelial dysfunction, and arterial stiffening as risk factors for hypertension. *Int J Mol Sci* 2018;**19**(6). pii: E1724. https://doi.org/10.3390/ijms19061724. 29891771.

46. Mekonnen MM, Hoekstra AY. Four billion people facing severe water scarcity. *Sci Adv* 2016;**2**(2):e1500323. https://doi.org/10.1126/sciadv.1500323. eCollection 2016 Feb, 26933676.

47. Schwarzenbach RP, Egli T, Hofstetter TB, von Gunten U, Wehrli B. Global water pollutionand human health. *Annu Rev Environ Resour* 2010;**35**:109–36.

48. Holgate S, Stokes-Lampard H. Air pollution--a wicked problem. *BMJ* 2017;**357**:j2814. https://doi.org/10.1136/bmj.j2814. 28615175.

49. Babadjouni RM, Hodis DM, Radwanski R, Durazo R, Patel A, et al. Clinical effects of air pollution on the central nervous system, a review. *J Clin Neurosci* 2017;**43**:16–24. https://doi.org/10.1016/j.jocn.2017.04.028. Review, 28528896.

EVOLUTION

Connections between paleoclimate and evolution

Darwin[1] first proposed what is now known as the Savanna hypothesis that suggested that the transition from ape-like to human-like species occurred as life in the Savanna regions of Africa became more challenging through climate change. More recent studies have revealed that oscillation in paleoclimate occurred in Africa leading to changes between moist and dry conditions.

In 2011 de Menocal[2] proposed that climate changes with aridity arose in Africa between 2.8 and 1.7 million years ago. Campisano[3] noted that geologic evidence indicates fluctuations in lake sizes in Africa occurred between these periods.

Several scientists, including Forster[4] have also pointed out that analyses of ancient population migration patterns must take into account not only wet dry cycles, but also tectonic movements and differences in sea levels.

In a 2018 review James et al.[5] emphasized that early primates occurred not only in Africa but also in Eurasia. They cited evidence that pre-homo lineages diverged between 7 and 9 million years ago and that human like lineages emerged at least 2 million years ago and that *Homo sapiens* arose more than 300, 000 years ago (Fig. 1).

Modern humans and their relationship to archaic humans

As new discoveries are reported, concepts of human evolution are constantly modified.

In 2015 Paabo[6] reviewed work carried out over several decades on DNA derived from remains of ancient humans. Analyses of DNA from the originally excavated Neanderthal skeleton were carried out in 1990s. Subsequently Neanderthal skeletons recovered elsewhere in Europe also yielded DNA that could be sequenced. Their initial sequencing studies involved analyses of mitochondrial DNA. Those studies revealed that Neanderthal mitochondrial DNA was different than human DNA. In 2008 they began analyses of DNA from a Denisovan bone.

Gene Environment Interactions. https://doi.org/10.1016/B978-0-12-819613-7.00003-7

Evolution: Gene Duplications and Dispersal

New genes may become widely dispersed in the genome **NBPF (DUF1220) domain genes**

Chromosome 1 (29 dispersed locations), chromosomes 3, 4, and 5.

New genes may disperse to a few chromosomes e.g. **NOTCH1: 9q34.3, NOTCH2: 1p12; NOTCH3 1913.2**

New genes may remain close to original: **ADH class 1 ADH1A, ADH1B, ADH1C: chromosome 4q23**

ADH class 1 genes

ADH1A 99,276,369-88,290,985 kbADH1C vivgvppdsq
ADH1B 99,304,971-99,321,401 kb
ADH1C 99,336,492-99353,045 kb

Protein sequence 92-101, 102-11

ADH1A iplaipqcgk cricknpesn

ADH1B iplftpqcgk crvcknpesn

ADH1C iplftpqcgk cricknpesn

Protein sequence 152-161

ADH1A flgistfsqy

ADH1B flgtsfsqy

ADH1C f**V**gvstfsqy

Protein sequence 292-301

ADH1A vivgvppdsq

ADH1B vivgvppasq

Protein sequence 362-371, 372-376

ADH1A dllhsgksir til**l**mf

ADH1B dllhsgksir tv**l**tf

ADH1C dllhsgksir tv**l**tf

Fig. 1 (A) *Evolution gene duplications and dispersals.* New genes that evolve by duplications may widely disperse e.g. NBPF (DUF1220 domain genes) disperse from chromosome 1 to 29 different locations. New genes may disperse to a few chromosomes e.g. NOTCH genes. New genes may remain close to the original gene e.g. ADH class 1 genes (B) *Class I ADH genes protein homologies and differences.* Note close similarity of amino acid sequences in different regions of the ADH gene. Specific amino acids that differ are in bold text.

Based on study data, Paabo postulated that around half a million (500,000) years ago common ancestors for Neanderthals and Denisovans migrated out of Africa. They gave rise to Neanderthals in Western Eurasia and to Denisovans in Eastern Eurasia. Paabo postulated that other "human forms" were likely present in Eastern and Western Eurasia at the same time.

Paabo in 2015 noted that Neanderthals and Denisovans subsequently became extinct; however, they live on in modern humans in the form of segments of DNA in the human genome, estimated to be between 1% and 2% in non-African modern humans and up to 7% in individuals of Asians and Melanesian descent.

In 2018 Vernot and Paabo[7] noted that archeological studies and dating have led to estimations that Neanderthals were present in Europe and Asia for half a million years and that their disappearance apparently coincided with the migration of modern humans into Europe and Asia around 40,000 years ago.

Remains from a different Archaic population of humans were discovered in Southern Siberia and this population was defined as Denisovan. The Denisovans were apparently descended from the Neanderthals. There is evidence for Denisovan related sequences in current populations of modern humans living in certain parts of Asia and in Oceania. Current individuals from Asia also have segments of Neanderthal sequences in their genomes.

In 2018 Browning et al.[8] reported data from comparative DNA sequencing analyses in Denisovan and Neanderthal genomes and genomes of 18 European, Asian and American populations. They reported that Papuans harbored the most significant amounts of Neanderthal and Denisovan ancestry. Very low amounts of Denisovan sequences were described in population of north-eastern European descent.

Browning et al. also examined sequence data for evidence of positive selection of introgressed Neanderthal or Denisovan sequences in current humans. Their studies and results of previous studies, revealed evidence of positive selection of introgressed Neanderthal sequence variants related to skin pigmentation and hair traits in genes that encode BNC2 basonuclein 2, POU2F3 a homeodomain transcription factor, and KRT71 keratin 71. Previous studies and those of Browning et al.[8] also highlighted introgression of a region on chromosome 3p21.21 that is related to immune response. Variants in this region that were positively selected included genes CCR9 and CXCR6 that encode chemokine receptors. A second region that showed evidence of introgression and selected variants, involved the immune response genes on chromosome 14q33.3. The region includes IGHA1, immunoglobulin heavy chain alpha 1, IGHG1, immunoglobulin heavy chain gamma 1 and IGHG3 immunoglobulin heavy chain gamma 3.

In a 2018 review Wolf and Akey[9]emphasized that archaic humans, Neanderthals and Denisovans, shared more ancestry with the current non-African populations than with current African populations They noted that apparently introgressed sequences from Archaic humans in non- African modern humans include variants in genes STAT2, signal transducer and activator of transcription, OAS1 oligoadenylate synthetase that inhibits viral replication, and TLR1, 5 and 10 genes that encode Toll receptors involved in pathogen receptors.

Dannemann and Racimer[10] emphasized that there is clear evidence for negative selection of Archaic DNA sequences in current populations since the retention in archaic genome sequences in these populations is very low. They concluded that we still have along way to go to understand the genetic legacy bequeathed from archaic humans to modern humans.

Archaic genomes

Discoveries of additional forms of archaic humans continue to be made. These include Homo floresienis discovered in Indonesia[11] and the Homo Luzonensis discovered in the Philippines.[12]

Genomic sequencing

Segments of Neanderthal genome sequence were published by Green et al. in 2010[13] and in 2012 Mayer et al.[14] published segments of Denisovan sequence. In 2014 Prüfer et al.[15] published complete genome sequences from Neanderthals excavated at two different sites. Availability of genomic sequence has facilitated detailed comparisons between archaic human genome and the genomes of anatomically modern humans (AMH).

Reher et al.[16] reported that in Neanderthal genomes and in Denisovan sequences the levels of heterozygosity are decreased relative to levels of heterozygosity found in modern humans. The degree of heterozygosity in genes that encode products involved in immune function provide some evidence of effectiveness of individuals in dealing with pathogens. Reher et al. reported that the degree of heterozygosity in immune function genes was greatly lower in archaic DNA relative to DNA of modern humans. However, the major histocompatibility loci genes showed higher degrees of diversity than did their other immune function genes. Based on analysis of introgression of archaic genome sequence variants in human genomes investigators have proposed that interbreeding between Neanderthals and anatomically modern human likely occurred 47 to 65 thousand years before the present.

Sankararaman et al.[17] reported that evidence of introgression of Denisovan DNA into modern human genomes is highest in Austro-Melanesians where levels of introgression may reach 6%. Questions arise regarding the evolutionary and medical consequences of the genomic sequences that were introgressed from archaic genomes into the genomes of modern humans and that have been retained in the human genome over many, many generations. Dolgova and Lao[18] reported that examples of introgressed and retained sequences include genes that encode products related to skin pigmentation and hair morphology BNC2 basonuclin 2 a protein reported to function in skin color saturation, and MC1R, melanocortin receptor 1, a receptor protein for the melanocortin hormone. In addition, specific gene related to immune function were introgressed into the human genome from Neanderthal.

Quach et al.[19] carried out RNA isolated from human monocytes to identify transcripts produced in response to bacterial and viral stimuli. They specifically analyzed transcripts of genes in the toll receptor pathway. In addition, they searched for regulatory elements that enhanced transcription, expression quantitative trait loci. They determined that a number of important regulatory loci that impacted the immune response were located in genomic sequences previously identified as sequences introgressed into the human genome from Neanderthal genome.

Introgression from of archaic sequences into genomes of modern humans

Genes that impact immune function are among the genes derived from Neanderthal that are introgressed into the modern human genome. They were reported by Deschamps et al.[20] and Quach and Quintana-Murci.[21] The products encoded by these genes include:

STAT2	signal transducer and activator of transcription 2
OAS12'-5'	oligoadenylate synthetase 1, induced by interferon
NLRC5	NLR family CARD domain containing 5, plays a role in cytokine response
IRF6	Interferon regulatory factor 6
IFITM1–3	Interferon induced transmembrane protein 1 and 3
IL17A, IL17F	Interleukin 17A and 17F, proinflammatory cytokines activated T cells
TLR6–1-10	Toll like receptor cluster, recognizes pathogen-associated molecular patterns
SIRT1	Sirtuin 1, intra-cellular regulatory protein

There are also other genetic variants present in modern humans that encode proteins that apparently impact susceptibility or response to specific pathogens, Deschamps et al.[20] documented several of these. Variation in MERTK (MER tyrosine kinase) involved in inflammatory response was reported impact propensity to develop fibrosis following hepatitis C infection. Specific variants in LARGE (that encodes an *N*-acetylglucosaminyltransferase and IL21 (interleukin 21) were reported to increase resistance to Lassa fever in West African population.[22]

Population migrations

The ice-age and its impact on populations

In 2016 Fu et al.[23] reviewed studies designed to address movements of peoples relative to the ice aged period. It is estimated that modern humans arrived in Europe approximately 45,000 years before the present. The ice-age occurred between 19,000 and 25,000 years before the present.

Fu et al., analyzed genomic DNA from 51 modern humans who had lived between 45,000 and 7000 years before the present. Their DNA analysis system involved the use of prior designed oligo-nucleotide probes to target between 390,000 to 37 million single nucleotide polymorphisms. The isolated DNA segments were then sequenced.

The data generated in the Fu et al. study and comparison with data from previous studies of pre-Neolithic populations indicated that following the ice-age, the European population derived from primarily one population. Fu et al. proposed that at the end of the ice-age this population migrated back into Europe from South-Eastern Europe or Western Asia.

Nielsen et al.[24] reviewed peopling of the world based on genomic studies. They noted that a population of farmers migrated from Anatolia to the Iberian Peninsula approximately 7000 years ago. This population assimilated local hunter gatherers. They reached Britain and Scandinavia at around 6000 years ago.

Nielsen et al. cited evidence that during the late Neolithic Early Bronze age period 4500 years before the present, herders from the Caspian Steppe migrated to Central Europe. This population is thought to have been originators of the Indo-European languages to Europe.

Nielsen et al. noted that modern day European populations are derived from three components and migrations, initial hunter-gatherers, then recolonization of Europe from south-eastern Europe-western Asia after the maximum glacial period and the migration of Neolithic farmers (from Anatolia). They noted that the origins of these three components explain the modern diversity in Europe with the Neolithic genetic component most common in Southern Europe and decreasing toward the Northern latitudes.

Minoans and Myceneans

Lazaridis et al.[25] reported data they assembled from 19 ancient individuals. These individuals included Minoans from Crete and Myceneans from the Greek mainland and individuals from south-western Anatolia. Based on their analyses both the Myceneans and the Minoans had significant proportions (at least three-quarters) of their ancestry from Anatolia. They also shared some ancestry with populations from the Caucasus region.

Africa

Skoglund et al.[26] reported results of genome wide analyses from 15 pre-historic African individuals. Three individuals were recovered from the western Cape region of South Africa, and 12 were recovered from eastern and south eastern Africa.

Based on their analyses they proposed that the San people in South-Africa shared ancestry with the Hazda group in Tanzania and with the Sandawe group and that these groups originated in Ethiopia. All three of these population groups use clic languages.

The predominant South African Bantu speaking population apparently originated from farmers in West-Africa. Herders from East Africa also contributes to this population.

It is important to note that variation between individuals exist at the single nucleotide level and at genomic segment levels.

Brain evolution: Primate and human divergence

The DUF1220/NBPF domain also known as the Olduvai domain

Dumas and Sikela[27] reported that a specific amino acid domain known as DUF1220 has undergone duplication and translocation during primate evolution and the greatest increase in copy number of this domain has occurred in human evolution. Repeats of DUF1200 domains are particularly abundant on human chromosome 1q21.1–21.2. However domain repeats are also present on human chromosomes 3, 4, and 5.

Dumas et al.[28] reported that the DUF1200 domain had undergone increases in copy number in parallel with evolution from primates to human. The DUF1220 copy number in humans was reported to be 272, in gorillas there are 99 DUF repeats, in orangutan 92 repeats in macaques 35 repeats.

Specific variations with the DUF1200 DNA sequence indicate that there are 2 subgroups, each with 3 members: HLS1, HLS2, HLS3 and CON1, CON2 and CON3. The HLS segments have particularly increased in number in humans.

The DUF1220 encoding region has undergone specific name changes, they were often designated NBPF. In 2017 Sikela and Roy[29] suggested that the DUF1220 domain be renamed the Olduvai domain in light of its evolutionary significance.

Documenting gene, metabolite differences and brain differences between primates and humans

Somel et al.[30] reviewed evidence for potential roles of genes, their transcripts and regulators and possible metabolite differences that potentially play roles in differences in brain and cognition between humans and primates. They noted that key abilities that distinguish human and primates include the ability of humans to "create, accumulate and transmit cultural knowledge between generations".

One measure that has been examined in many studies involves brain volumes and brain/body ratios. Somel et al., noted that there is evidence that during the course of evolution cranial volumes in hominid species have progressively increased. Reports by paleontologists and measurement of skull indicate that Homo erectus had a brain volume trice as large as that of chimpanzees.

It is interesting to note that the cranial size in *Homo sapiens* neanderthalis was greater than the average cranial size in modern human *Homo sapiens* sapiens.

Interesting data has emerged from comparisons on modern human and Neanderthal genome sequences and primate sequences. Modern humans, Neanderthals and Denisovans share two specific amino acid changes in the FOXP2 gene that are absent in primates. FOXP2 plays and important role in language development.

Somel et al. noted that an early change that apparently developed during differentiation of later *Homo sapiens* involves duplication of SLIT-ROBO Rho GTPase activator (SRGAP2) that is involved in determination of density of dendrites at neuronal synapses.

Several studies have been carried out so search for genes that have undergone sequence changes in evolution of humans. Somel et al., listed the following genes that encode products with significant changes in humans and the functions of the proteins encoded by those genes.

FOXP2	Forkhead Box P2 transcription factor
PCDH11X	Protocadherin 11X (defects lead to speech inability)
PCDH11Y	Protocadherin 11Y central nervous system development
MCPH1	Microcephalin; functions at cell cycle checkpoint
ASPM	Abnormal spindle microtubule, mitotic spindle function
CDK5	Cyclin dependent kinase 5, phosphorylation of proteins involved in neuronal migration
CDKRAP2	CDK5 regulatory subunit
SLC2A1 SLC2A4	Solute carriers, glucose transporters
NBPF	Neuroblastoma breakpoint family, neurodevelopment
GADD45G	Growth arrest damage control, neuronal development
RFPL1,2,3	Ret finger-like proteins, brain expressed
DRD5	Dopamine receptor D5
GRIN2A	GRIN2B Glutamate ionotropic receptors at synapses
SLIT-ROBO	Cortical neurodevelopment
SRGAP2	Slit-Robo Rho-GTPase, SLIT-ROBO activator

Somel et al. stressed that in addition to changes in amino-acid coding sequences, it is important to consider that species specific differences may emerge in transcript abundance due to differences that arise at regulatory elements.

It is however import to note that it may be important to assess transcript levels not only in tissue from specific brain regions but also to analyze transcript levels within particular cells in a brain region.

Somel et al. noted that there is interesting evidence that indicates difference in timing of maximal gene expression when human and primates are compared. Expression of genes that encode products expressed in synapses peaks later in humans than in chimpanzee. In addition, the time frames during which synaptogenesis occurs is later in humans relative to chimpanzees. In humans, maximal synaptic density in the prefrontal cortex occurs between 3.5 and 10 years of age. In rhesus and macaque peaks synaptic density is present shortly after birth.

Somel et al., emphasized that alterations in regulation of gene expression may particularly impact timing of developmental changes.

Key to altered expression could include alterations in transcription factor activity. They noted evidence for accelerated evolution of transcription factor sequences in human evolution.

FOXP2

There is evidence that FOXP2 plays roles not only in human speech, but also in sound production by other life-forms. Recent evidence indicates that FOXP2 impacts the plasticity of specific brain circuits. FOXP2 as a transcription factor, has a number of downstream targets genes and was shown to be expressed in specific cell types within the cortex, thalamus, basal ganglia and cerebellum.[31]

Trajectory of brain evolution

Somel et al., proposed two distinct phases. The first involved primarily an increase in brain size. The second phase involved region specific remodeling. Heide et al.[32] emphasized the importance of prolonged proliferation of neural progenitor cells during development. They noted that specific genomic factors that promoted this prolonged proliferation included altered gene regulation, development of novel genes, e.g. through increases of gene number through emergence of gene paralogs, and the possibility that new gene paralogs evolved to have new functions.

Sousa et al.[33] reviewed brain-expressed genes that show significant differences in humans and chimpanzee brains. Some genes listed are expressed only in excitatory neurons, other gene are expressed only in inhibitory neurons; some genes are expressed in both.

Note gene products marked * are expressed in both excitatory and inhibitory neurons; other gene products occur primarily in excitatory neurons. Note the first 11 genes have also been found to have disease associated variants.

Products expressed primarily in excitatory neurons

MET	MET receptor tyrosine kinase
NRGN*	Neurogranin, post-synaptic protein kinase substrate
NGLN3*	Neuroligin 3, neuronal cell membrane protein
AHI1*	Abelson helper integration site, required for development of cortex and cerebellum
TSC2*	Tuberous sclerosis 2, stimulates GTPase activity
UBE3A*	Ubiquitin protein ligase 3A
AGAP1*	ADP-ribosylation factor GTPase-activating protein
GLRA3	Glycine receptor alpha-3
HTH3*	Histamine receptor H3 (can regulate neurotransmitter release)
GRIN1*	N-methyl-D-aspartate (NMDA) receptors, glutamate-regulated ion channel
GRM3	Metabotropic glutamate receptor, major excitatory neurotransmitter
UBE2H*	ubiquitin conjugating enzyme E2H
CLCN4*	voltage-dependent chloride channel gene
KCNN1/2	potassium calcium-activated channel subfamily N members 1 and 2
KCNJ1/2	potassium voltage-gated channel subfamily J member 1
KCNG1	potassium voltage-gated channel modifier subfamily G member 1
KCND3	potassium voltage-gated channel subfamily D member 3
KCNJ11	potassium voltage-gated channel subfamily J member 11
PKD2L1	polycystin 2 like 1, transient receptor potential cation channel

Products expressed primarily in inhibitory neurons

Note the first 3 genes encode products listed have been found to have variants in specific diseases.

COMT*	catechol-*O*-methyltransferase
HTR5A	5-hydroxytryptamine (serotonin) receptor
GAD1	glutamic acid decarboxylase
CHRNA2	glutamic acid decarboxylase
HCN4	hyperpolarization activated cyclic nucleotide gated potassium channel 4
KCNN2	potassium calcium-activated channel subfamily N member 2
ADRA2A	adrenoceptor alpha 2A
KCNC1	potassium voltage-gated channel subfamily C member 1
KCND3	potassium voltage-gated channel subfamily D member 3

Genomes, gene expression, evolution, intellect

Florio et al.[34] noted that during evolution the hominin neocortex increased in size to become three times as large as the chimpanzee neocortex and to contain twice as many neurons. (The neocortex refers to the brain regions involved in sensory perception, cognition spatial reasoning, language, motor commands.)

The Neanderthal cranial volume exceeded that of modern humans. However, based on skull comparisons, the relative sizes of the parietal and temporal lobes of modern humans apparently exceed those of Neanderthals.

Florio et al., presented evidence that the increase in size of the human neocortex is primarily due to prolongations of neurogenesis during early development. They proposed that the increased neurogenesis in humans primarily arises due to the increased proliferation of neural progenitor cells that arise in the sub-ventricular region. These cells are sometimes referred to as the outer radial glial cells.

Questions arise related to which specific gene expression changes contribute to this expanded proliferation of neurogenesis in humans. A number of investigators have carried out studies to correlate genome alteration and gene expression changes that lead to prolonged early neurogenesis in humans.

Florio et al., presented evidence that the increase in size of the human neocortex is primarily due to prolongations of neurogenesis during early development. There is also evidence that in some cases, genomic changes may lead to alterations in the quantity of regulatory elements present in non-protein coding DNA. There are also examples of genomic changes that lead to increased production of inhibitors of certain functions.[32]

NOTCH2 gene expansion and impact on cortical neurogenesis

NOTCH signaling is an important determinant of neuronal precursor proliferation and in determination of the final neuron number. NOTCH binds to a specific receptor and this binding functions to promote proliferation and to delay maturations of neuronal progenitor derived cells.

In 2018 Fiddes et al.[35] and Suzuki et al.[36] studied and reported on the function of duplicates of NOTCH2 genes, paralogs that arose during human evolution as a result of duplication events and structural rearrangements of the genome.

Increased numbers of NOTCH2 genes that result from duplication events, result in longer periods of neuronal precursor proliferation and increased brain size. There is also evidence that deletion of NOTCH2 and reduced NOTCH signaling result in more rapid differentiation of neuronal precursors and less proliferation.

Studies of Fiddes et al. supported the conclusion the increased numbers of NOTCH2 encoding genes occurred through duplication events that occurred in humans and led to increased brain size and increased neocortical complexity. Fiddes et al., studies on cultured brain organoids revealed they increased copy numbers of the NOTCH2NL lead to increased expression of NOTCH2 protein that led to downregulation of genes known to be involved in differentiation of neuronal precursors. These products encoded by these genes included CNTN2 (Contactin 2), SOX10 (transcription factor), NEFL (neurofilament light) and GAP43 (growth associated protein).

Suzuki et al.[36] carried out RNA sequencing studies. They determined that the extra NOTCH protein provided from the NOTCH2NL genes inhibited the binding of a specific protein DELTA, to the NOTCH receptor.

NOTCH gene encodes a transmembrane protein with 3 domains, an extra-cellular domain, a short transmembrane domain and an intra-cellular domain.[37] The NOTCH receptor is activated when it interacts with protein encodes by DELTA1. In humans there are three Delta like proteins encoded by DLL1, DLL3 and DLL4 and in addition, proteins encoded by the JAGGED gene, JAG1 and JAG2 can activate NOTCH receptors.

Ligand interaction with NOTCH leads to proteolysis at the NOTCH protein segments that crosses the membrane. This proteolysis is carried out by proteases of the ADAM family and then by gamma secretase (presenilin). Following proteolysis, the intra-cellular cytoplasmic domain of NOTCH protein is released. The displaced intra-cellular form then interacts with specific transcription regulators, including RBPJ (CBF1) and releases them from inhibition so that transcription of specific genes commences.

In 2016 Tuand et al.[38] reported that the Neurobeachin protein also interacts with the NOTCH1 intracytoplasmic domain and the Neurobeachin protein likely acts as a transcription regulator. This finding is possibly relevant to autism and associated changes, since Neurobeachin has been reported to be deleted in some cases of autism.[39]

Brain development, genes and environment

Grigorenko et al.[40] noted evidence that 75% of all human genes are expressed in human brain at some time and that it is therefore likely that variations in genes can lead to changes in brain functions and behavior.

In addition, environmental factors clearly play roles in determining development, cognition, behavior and ultimately, human productivity and successful adaptation. In a comprehensive review Prado and Dewey[41] reviewed specific nutritional factors essential human development. It is also important that nutrition be adequate during pregnancy and throughout childhood. They also emphasized the importance of environment and of quality stimulation.

Nutritional deprivation can include protein and energy malnutrition. It can also include deficiency of necessary minerals (iron, calcium, zinc) and vitamin deficiency.

With respect to the importance of environmental stimulation, Prado and Dewey noted evidence that higher levels of education increase dendritic branching. There is also evidence from studies of children placed in minimal stimulation orphanages at early stages of life, that brain white matter development differs from that of normal developing stimulated children. White matter differences were measures by diffusion tensor brain imaging.

The advantages and disadvantages of extra-copies of genomic segments

In a 2004 publication Hurles[42] noted that Susumu Ohno in 1970[43] wrote of the potential for gene duplication to contribute to evolution. In 1970 Susumu Ohno[43] proposed that gene duplication was a key force in evolution. When a gene duplicated one copy was available to produce the product required for a specific function and a duplicate gene was available to undergo mutation and to possibly develop new function

By 2004 there was clear evidence in genome of humans, primates and other species for the existence of paralogous genes, i.e. the presence in the genome of a specific individual of several copies or in some cases of multiple copies of a particular gene. Thus, there was evidence for families of related genes.

Hurles noted that paralogous genes sometimes remained clustered within a specific genome segment. However, there was also evidence for dispersion of paralogous genes. In a specific genome, paralogous genes may be clustered together, they may be located at some distance from each other on a specific chromosome. However paralogous genes belonging to a specific family, are sometimes present on different chromosomes.

One advantage of the generation of duplicates of a gene is that the duplicate can serve to maintain product synthesis and function in situations where the parent gene is mutated and its function is impaired.

Hurles noted that paralogous genes sometimes undergo mutations and develop new functions. He noted that this apparently occurs more often in dispersed genes. In addition, dispersed genes may have different ranges of tissue expression than the parent gene and dispersed gene may be predominantly expressed at different stages of development than the parent gene and he presented examples.

Hurles noted that through duplication of the opsin encoding genes in humans and through mutations in the paralogous opsin genes, humans became able to distinguish light between light waves of three different wave lengths and thus developed color vision. Duplication and divergence of beta globin like genes, led to the generation of paralogous genes that are expressed at different stages of development.

Hurles documented different mechanisms which paralogous genes arise. The can arise during meiosis through unequal crossing over between sister chromatids of homologous chromosomes. Paralogous genes can also be derived through the reintegration of messenger RNA into the genome. In such cases the paralogous gene will be intron free.

Hurles emphasized that gene duplication events most likely played important roles in generation of evolutionary divergence of related species.

Structural variants and adaptation

De Grassi et al.[44] emphasized that duplication of genomic segments impacted both protein coding and non-protein coding segments. They noted that duplications arise primarily due to errors in meiosis. Genomic duplications over evolutionary time can lead to the generation of gene families that encode related proteins Degrassi et al., noted that individual products derived from the different members of a gene family sometimes interact to form multiprotein complexes. They considered the oxidative phosphorylation complexes in energy metabolism to represent and examples of evolutionary related proteins.

Radke and Lee[45] documented specific structural genomic variants in humans that contain genes that impact metabolic functions. Examples include duplication of the amylase genes AMY1 that impact starch digestion. These authors noted that with respect to immuno-logical responses increased copy numbers of the *CCL3L1* gene that encodes the chemokine receptors CCR5, have been reported to impact resistance to specific viral infections.

Radke and Lee also noted evidence that duplication of the promoter of the gene that encodes SLC45A2 were reported to predispose to olive skin color.

It is important to note that the structural genomic variants that are of evolutionary importance are structural variants transmitted through germ cells. However structural variants can also arise in somatic tissues. These variants can result from defects in mitosis. They may also result from DNA damage and DNA breaks and their subsequent repair.

Segmental duplications

In 2009 Marques-Bonet[46] reviewed segmental duplications and focused in part on highly identical blocks of duplicons that occur particularly in sub-telomeric and peri-centromeric regions of human chromosomes. They noted that specific duplication within genomes predispose to structural chromosome rearrangements including deletion and duplications that in some cases predispose to congenital defects.

Carvalho and Lupski[47] noted that at least 70 different disorders were due to structural genomic rearrangements that arose in regions of the genome where there are increased numbers of segmental duplications.

Whole genome sequencing and segmental duplications

Dennis et al.[48] utilized long-read single molecule sequencing to examine complex segmental duplications in humans. They analyzed genomes of humans and of non-human primates and included data of DNA derived from archaic humans, Neanderthal and Denisovan. Their analyses revealed that in 13 different human segmental duplication (HSD) regions there was evidence of polymorphisms so that different humans varied in the exact number of duplicons in particular regions. In six of the duplication regions the copy number of duplications was fixed in humans.

Dennis et al., also established that in 3 regions expansion of gene duplications were unique to humans and were not found in Neanderthal or Denisovan sequences. The youngest segmental duplication region unique to human occurred on chromosome 7q25 and included the *TCAF1* and *TCAF2* genes that encoded TRPM8 channel associated proteins (TRPMs form transient receptor cation channels).

There is evidence that variants in TRPM8 may be related to cold adaptation. Key et al.[48a] reported that a specific sequence variant in this gene rs 25,333,629 shows extreme population variation, ranging from frequencies of 5% in Nigerians to 88% in Finland.

Studies of Dennis et al., led to identification of specific segmental duplications that were present in primates and likely also existed in the most recent common ancestor of humans and primates. Dennis identified 218 segmental duplication in humans.

Particular human chromosome locations that are especially rich in segmental duplication tend to be prone to disease causing duplication and deletions. These include the following chromosome regions: 1q21.1, 7q11.2, 10q11.23, 15q13.3, 16p11.2. The chromosome 1q21.1 region includes several different types of duplicons.

With respect to chromosome 15 it is interesting to note that the GOLGA segmental duplication also occurs in primates, chimp, gorilla and orangutan. However, the ARHGAP11A repeat expansion is unique to archaic humans and to modern humans.

Comprehensive population studies of segmental duplications

It is important to emphasize that although unusual degree of segmental duplication and deletion in particular regions increase risk for certain complex common diseases, similar abnormalities also occur in individuals in the unaffected controls.

In a comprehensive study of 21,094 cases of schizophrenia and 20,227 controls Marshall et al.[49] reported an overall enrichment burden of copy number variants (CNVs) that minimally increased overall risk to OR 1.11 ($p = 5.7 \times 10^{75}$). It must be stressed that the low overall increase in risk reaches high significance because of the large numbers of cases and controls studied. CNVs in eight different chromosome regions were reported to increase schizophrenia risk: 1q21,1, 2p16.3; 3q29, 7q112, 15q13.3, 16p11.2, 22q11.

It is also important to note that duplications and deletions in 1q21.1, 7q11.2, 15q13.3, 16p11.2 and 22q11.2 have been reported to increase risk for neurodevelopmental defects, including intellectual disability, speech defects, seizures, developmental delay, autism, attention deficit hyperactivity disorders. However, these abnormalities also occur in controls.

Phenotypic plasticity and evolution

In 1896 Baldwin[50] published an article entitled, "A new factor in Evolution" in which he discussed phenotypic plasticity in response to environmental changes. Phenotypic plasticity has been defined as the ability of an organism to undergo phenotypic changes when environmental conditions changed. In 2018 in an article entitled "Buying Time" Pennisi[51] presented a number of interesting examples of phenotypic plasticity. It is important to note that phenotypic plasticity is also sometimes referred to as character displacement.[52]

Pfennig and coworkers have studied changes in the spadefoot toad in response to environmental changes. Pfennig first observed these toads in Texas. He noted that in the springtime in Texas toads emerged from burrows and they inhabited pools of rain water where they consumed algae and small crustaceans. He observed that as the pools shrank some tadpoles changed, they developed significant changes in the jaws and they ate larger crustaceans. These changes in a subset of toads then became fixed. Subsequent studies revealed that not only had the jaw muscles changed. These altered tadpoles also manifested changes in the expression of specific metabolic genes, Pfennig and Martin.[53] In 2018 Levis et al., reported that phenotypic plasticity (character displacement) proceeded through an initial phase of changed gene expression in altered environmental conditions, to a later phase when such differences became fixed through changes in gene expression.

Phenotypic plasticity has also been observed in a specific species of lizard that can survive either light colored sandy deserts or in dark lava rich regions. When lizards are moved from light colored sandy deserts to lava rich regions their skin color becomes darker. Cori et al.[54] reported the results of their studies revealed that change to darker color was initially due to plasticity that enable the lizards to respond to the environmental change to lava rich region through altered expression of genes associated with melanin production. Cori et al., reported evidence for later changes that involved development of genetic variants in the PREP and PRKAR1A regulators of melanin production. There was also evidence of natural selection of lizards with these changes in lava rich regions.

Dayan et al.[55] reported examples of phenotypic plasticity in response to temperature changes. They determined that altered temperature led to specific changes in muscle gene expression in the fish *Fundulus heteroclitus*.

Mitochondrial DNA variations and population divergence

Wallace[56] reviewed mitochondrial DNA variation. Mitochondrial DNA is maternally inherited, since mitochondria are present in the ovum but not in the sperm. Extensive studies determined that sequential mitochondrial DNA variation arose in maternal lineages.

As a result of the sequential variation in evolution a mitochondrial tree has evolved commencing in a root and stem and then branching to generate primary, secondary and tertiary branches with a defined set of nucleotide variants, specific haplotypes characteristic of specific branches.

Wallace[56] reported that the discernable mitochondrial root can be traced to Africa approximately 100,000 years before the present. The original mitochondrial DNA haplotype is designated L. Variants arose on L to generate L0 L1, L2 and L3 haplotypes in different regions of Africa. The L3 branch in North-east Africa branched to give rise to M and N haplotypes and individuals with these haplotypes then migrated to populate the world. N haplotypes dispersed primarily in Europe and M haplotypes were found in Asia.

From individuals with the M haplotype, haplogroups A, B, C and D emerged. Individuals with A, C and D haplogroups migrated across the Behring Strait from Asia to North America. Haplogroups B individuals colonized the Pacific Islands. N haplotype individuals migrated and gave rise to the S haplogroup in Australia.

James et al.[5] proposed that regional predominance of specific mitochondrial haplotypes related to selection under specific environmental conditions

Specific genes within the mitochondrial genome encode a number of important polypeptides in the mitochondria, including a number of components in the electron transfer system that are responsible for energy generation.

Wallace postulated that given the important roles of mitochondrial DNA encoded components in energy production, variants in mitochondrial DNA have significant effects on human biology.

Nutritional factors and genomic adaptations

In 2019 James et al.[5] reviewed the role of nutrition in human evolution. They emphasized the importance of adaptation to local environments through changes in nuclear and mitochondrial DNA. Amplification in the human genome of the gene that encodes fatty acid dehydratase leads to *FADS1* and *FADS2*. Presence of the two forms were reported by Mathias[57] to increase the ability of humans to transform oleic acid from plant digestion to molecules that could be more easily incorporated in triglycerides and long chain fatty acids. James et al. noted that in primates and humans, loss of the enzyme uricase proved advantageous in promoting fructose metabolism. Another population difference that arose included high frequency in certain Asian populations of a variant in the nuclear encoded enzyme acetaldehyde dehydrogenase that is much less efficient in metabolism of acetaldehyde generated from alcohol. Presence of this variant leads to low levels of alcohol consumption. James et al. also noted that increased numbers amylase genes facilitated adaptation to diets with high starch content.

James et al. also noted that specific variants in the methylene tetrahydrofolate reductase encoding *MTHFR* gene led to increased folate requirements in certain patients. They noted also that dark skinned individuals have higher need for vitamin D2 obtained either through sunlight exposure or diet, to avoid rickets.

Polygenic adaptations to changing environments

Analyses of polygenic variations have been facilitated by the availability of technologies for genome wide analyses of genetic variations. These technologies include microarrays to analyze single nucleotide polymorphisms and DNA sequencing methodologies. In addition, analysis of gene expression is now facilitated through availability of techniques for RNA sequencing. Transcriptome analysis along with DNA variant analysis enables definitive identification of quantitative trait loci.

Population genetic studies on plants and insects have revealed the impact of environmental changes and genotypes and phenotypes.[58] Potential mechanisms for adaptations in a specific population include selective sweeps based on existing genomic variations in that population or sometimes adaptations arise as a result of new mutations and selective sweeps based on the new mutations, Pritchard et al.[59] emphasized three separate modes of evolutionary adaptations: hard selective sweeps, soft selective sweeps and polygenic adaptations

Pritchard et al. also emphasized that polygenic variations could facilitate adaptation to environmental changes. They noted that Interesting examples of human adaptations to cnvironment include physiologic adaptations to low oxygen levels in individuals living at high altitudes. Some of these adaptations may have a polygenic basis. Pritchard et al. noted that genome wide single nucleotide polymorphism (SNP) analyses extend population variation studies beyond candidate gene loci.

Early studies in genetics led Falconer and Mackay[60] to propose that the most important traits were highly polymorphic and influenced by variables at many loci. Pritchard et al., noted that studies on human height supported this concept. In Europeans 50 loci were reported to have variants that influenced height. However, together these loci accounted for only 5% variation in height.[61] They noted that since height is highly heritable many loci involved in its determination had not yet been identified.

Pritchard et al.[59] expressed support for the concept that adaptation from polygenic existing variation best explained many but not all traits. They acknowledged that clearly there are examples of single genetic variants that play key roles to adaptation under certain circumstances e.g. DUFFY/DARC mutations and malaria resistance. They emphasized that to define genetic variants that influenced certain phenotypes would require analyses that combined data from large numbers of variants in large numbers of individuals.

Studies on expression quantitative trait loci (eQTLs)

These studies involve Genome wide variant association studies GWAS) and transcription analyses to identify specific nucleotide variants that impact expression of a specific gene or genes. In a 2014 report Westra and Franke[62] noted that cis eQTLs can have local effects and these QTLs impact genes within 1 megabase of the variant. QTLs can however impact expression of more distant genes on the same chromosome or impact expression of genes on a different chromosome, trans QTLs.

There are examples of eQTLs that impact cis-regulatory elements of genes. QTLs have also been identified in the introns of genes. One notable example is a QTL located in an intron of the FTO gene that does not impact expression of the FTO gene but alters transcription of genes several megabases away from FTO.

A SNP associated with systemic lupus erythrematosus was shown to act as a trans QTL that altered expression of several interferon response genes.[63]

Westra and Franke emphasized that since expression levels of a particular gene vary in different cell types, QTL mapping studies need to examine expression in different tissues and cell types. They noted that microarray technologies had primarily been used to examine gene expression. However increasingly RNA sequencing studies were being applied.

The question also arises as to what extent environmental factors influence gene expression.

References

1. Darwin C. *The descent of man and selection in relation to sex*. London, UK: John Murray; 1871.
2. de Menocal PB. Anthropology. Climate and human evolution. *Science* 2011;**331**(6017):540–2. https://doi.org/10.1126/science.1190683. 21292958.
3. Campisano CJ. Milankovitch cycles, paleoclimatic change and hominin evolution. *Nat Educ Knowl* 2012;**4**(3):5.
4. Forster P. Ice ages and the mitochondrial DNA chronology of human dispersals: a review. *Philos Trans R Soc Lond Ser B Biol Sci* 2004;**359**(1442):255–64. https://doi.org/10.1098/rstb.2003.1394. 15101581.

5. James WPT, Johnson RJ, Speakman JR, Wallace DC, Frubeck G, et al. Nutrition and its role in human evolution. *J Intern Med* 2019. https://doi.org/10.1111/joim.12878.

6. Pääbo S. Dorcas cummings lecture. *Cold Spring Harb Symp Quant Biol* 2015;**80**:291–4. https://doi.org/10.1101/sqb.2015.80.030171. 27325709.

7. Vernot B, Pääbo S. The Predecessors Within.... *Cell* 2018;**173**(1):6–7. https://doi.org/10.1016/j.cell.2018.03.023. 29570998.

8. Browning SR, Browning BL, Zhou Y, Tucci S, Akey JM. Analysis of human sequence data reveals two pulses of archaic Denisovan admixture. *Cell* 2018;**173**(1):53–61. e9. https://doi.org/10.1016/j.cell.2018.02.031. 29551270.

9. Wolf AB, Akey JM. Outstanding questions in the study of archaic hominin admixture. *PLoS Genet* 2018;**14**(5):e1007349. https://doi.org/10.1371/journal.pgen.1007349. eCollection 2018 May 29852022.

10. Dannemann M, Racimo F. Something old, something borrowed: admixture and adaptation in human evolution. *Curr Opin Genet Dev* 2018;**53**:1–8. https://doi.org/10.1016/j.gde.2018.05.009. Review, 29894925.

11. Morwood MJ, Soejono RP, Roberts RG, Sutikna T, Turney CS, et al. Archaeology and age of a new hominin from Flores in eastern Indonesia. *Nature* 2004;**431**(7012):1087–91. 15510146.

12. Détroit F, Mijares AS, Corny J, Daver G, Zanolli C, et al. A new species of Homo from the late Pleistocene of the Philippines. *Nature* 2019;**568**(7751):181–6. https://doi.org/10.1038/s41586-019-1067-9. 30971845.

13. Green RE, Krause J, Briggs AW, Maricic T, Stenzel U, et al. A draft sequence of the Neandertal genome. *Science* 2010;**328**(5979):710–22. https://doi.org/10.1126/science.1188021. 20448178.

14. Meyer M, Kircher M, Gansauge MT, Li H, Racimo F, et al. A high-coverage genome sequence from an archaic Denisovan individual. *Science* 2012;**338**(6104):222–6. https://doi.org/10.1126/science.1224344. 22936568.

15. Prüfer K, Racimo F, Patterson N, Jay F, Sankararaman S, et al. The complete genome sequence of a Neanderthal from the Altai Mountains. *Nature* 2014;**505**(7481):43–9. https://doi.org/10.1038/nature12886. 24352235.

16. Reher D, Key FM, Andrés AM, Kelso J. Immune gene diversity in archaic and present-day humans. *Genome Biol Evol* 2019;**11**(1):232–41. https://doi.org/10.1093/gbe/evy271. 30566634.

17. Sankararaman S, Mallick S, Patterson N, Reich D. The combined landscape of Denisovan and Neanderthal ancestry in present-day humans. *Curr Biol* 2016;**26**(9):1241–7. https://doi.org/10.1016/j.cub.2016.03.037. 2703249.

18. Dolgova O, Lao O. Evolutionary and medical consequences of archaic introgression into modern human genomes. *Genes (Basel)* 2018;**9**(7). pii: E358. https://doi.org/10.3390/genes9070358. Review, 30022013.

19. Quach H, Rotival M, Pothlichet J, Loh YE, Dannemann M, et al. Genetic adaptation and Neandertal admixture shaped the immune system of human populations. *Cell* 2016;**167**(3):643–656.e17. https://doi.org/10.1016/j.cell.2016.09.024. 27768888.

20. Deschamps M, Laval G, Fagny M, Itan Y, et al. Genomic signatures of selective pressures and introgression from archaic hominins at human innate immunity genes. *Am J Hum Genet* 2016;**98**(1):5–21. https://doi.org/10.1016/j.ajhg.2015.11.014. 26748513.

21. Quach H, Quintana-Murci L. Living in an adaptive world: genomic dissection of the genus Homo and its immune response. *J Exp Med* 2017;**214**(4):877–94. https://doi.org/10.1084/jem.20161942. Review. [28351985].

22. Andersen KG, Shylakhter I, Tabrizi S, Grossman SR. Genome-wide scans provide evidence for positive selection of genes implicated in Lassa fever. *Philos Trans R Soc Lond Ser B Biol Sci* 2012;**367**(1590):868–77. https://doi.org/10.1098/rstb.2011.0299. 22312054.

23. Fu Q, Posth C, Hajdinjak M, Petr M, Mallick S, et al. The genetic history of Ice Age Europe. *Nature* 2016;**534**(7606):200–5. https://doi.org/10.1038/nature17993. 27135931.

24. Nielsen R, Akey JM, Jakobsson M, Pritchard JK, et al. Tracing the peopling of the world through genomics. *Nature* 2017;**541**(7637):302–10. https://doi.org/10.1038/nature21347. 28102248.

25. Lazaridis I, Mittnik A, Patterson N, Mallick S, et al. Genetic origins of the Minoans and Mycenaeans. *Nature* 2017;**548**(7666):214–8. https://doi.org/10.1038/nature23310. 28783727.

26. Skoglund P, Thompson JC, Prendergast ME, Mittnik A, et al. Reconstructing prehistoric African population structure. *Cell* 2017;**171**(1). 59–71.e21. https://doi.org/10.1016/j.cell.2017.08.049. 28938123.

27. Dumas L, Sikela JM. DUF1220 domains, cognitive disease, and human brain evolution. *Cold Spring Harb Symp Quant Biol* 2009;**74**:375–82. https://doi.org/10.1101/sqb.2009.74.025. 19850849.

28. Dumas LJ, O'Bleness MS, Davis JM, Dickens CM, Anderson N, et al. DUF1220-domain copy number implicated in human brain-size pathology and evolution. *Am J Hum Genet* 2012;**91**(3):444–54. https://doi.org/10.1016/j.ajhg.2012.07.016.

29. Sikela JM, van Roy F. Changing the name of the NBPF/DUF1220 domain to the Olduvai domain. Version 2. *F1000Res* 2017;**6**:2185. https://doi.org/10.12688/f1000research.13586.2. [revised 2018 Jul 17]. eCollection 2017, 29399325.

30. Somel M, Liu X, Khaitovich P. Human brain evolution: transcripts, metabolites and their regulators. *Nat Rev Neurosci* 2013;**14**(2):112–27. https://doi.org/10.1038/nrn3372. Review, 23324662.

31. Deriziotis P, Fisher SE. Speech and language: translating the genome. *Trends Genet* 2017;**33**(9):642–56. https://doi.org/10.1016/j.tig.2017.07.002. Review, 28781152.

32. Heide M, Long KR, Huttner WB. Novel gene function and regulation in neocortex expansion. *Curr Opin Cell Biol* 2017;**49**:22–30. https://doi.org/10.1016/j.ceb.2017.11.008. 29227861.

33. Sousa AMM, Zhu Y, Raghanti MA, Kitchen RR, et al. Molecular and cellular reorganization of neural circuits in the human lineage. *Cell* 2017;**170**(2):226–47. https://doi.org/10.1016/j.cell.2017.06.036. Review, 28708995.

34. Florio M, Borrell V, Huttner WB. Human-specific genomic signatures of neocortical expansion. *Curr Opin Neurobiol* 2017;**42**:33–44. https://doi.org/10.1016/j.conb.2016.11.004. Review, 27912138.

35. Fiddes IT, Lodewijk GA, Mooring M, Bosworth CM, et al. Human-Specific NOTCH2NL Genes Affect Notch Signaling and Cortical Neurogenesis. *Cell* 2018;**173**(6). 1356–1369.e22. https://doi.org/10.1016/j.cell.2018.03.051. 29856954.

36. Suzuki IK, Gacquer D, Van Heurck R, Kumar D, et al. Human-specific NOTCH2NL genes expand cortical neurogenesis through Delta/NOTCH regulation. *Cell* 2018;**173**(6):1370–1384.e16. https://doi.org/10.1016/j.cell.2018.03.067. 29856955.

37. Lasky JL, Wu H. Notch signaling, brain development, and human disease. *Pediatr Res* 2005;**57**(5 Pt 2):104R–9R. 15817497, https://doi.org/10.1203/01.PDR.0000159632.70510.3D.

38. Tuand K, Stijnen P, Volders K, Declercq J, et al. Nuclear localization of the Autism candidate gene neurobeachin and functional interaction with the NOTCH1 intracellular domain indicate a role in regulating transcription. *PLoS ONE* 2016;**11**(3):e0151954. https://doi.org/10.1371/journal.pone.0151954. eCollection 2016, 26999814.

39. Nuytens K, Gantois I, Stijnen P, Iscru E, et al. Haploinsufficiency of the autism candidate gene Neurobeachin induces autism-like behaviors and affects cellular and molecular processes of synaptic plasticity in mice. *Neurobiol Dis* 2013;**51**:144–51. https://doi.org/10.1016/j.nbd.2012.11.004. 2315381.

40. Grigorenko EL, Urban AE, Mencl E. Behavior, brain, and genome in genomic disorders: finding the correspondences. *J Dev Behav Pediatr* 2010;**31**(7):602–9. https://doi.org/10.1097/DBP.0b013e3181f5a0a1. remediation approaches, 20814258.

41. Prado EL, Dewey KG. Nutrition and brain development in early life. *Nutr Rev* 2014;**72**(4):267–84. https://doi.org/10.1111/nure.12102. Review, 24684384.

42. Hurles M. Gene duplication: the genomic trade in spare parts. *PLoS Biol* 2004;**2**(7):E206. Review, 15252449.

43. Ohno S. Gene duplication and the uniqueness of vertebrate genomes circa 1970-1999. *Semin Cell Dev Biol* 1999;**10**(5):517–22. Review, 10597635.

44. De Grassi A, Lanave C, Saccone C. Genome duplication and gene-family evolution: the case of three OXPHOS gene families. *Gene* 2008;**421**(1-2):1–6. https://doi.org/10.1016/j.gene.2008.05.011. 18573316.

45. Radke DW, Lee C. Adaptive potential of genomic structural variation in human and mammalian evolution. *Brief Funct Genomics* 2015;**14**(5):358–68. https://doi.org/10.1093/bfgp/elv019. 26003631.

46. Marques-Bonet T, Eichler EE. The evolution of human segmental duplications and the core duplicon hypothesis. *Cold Spring Harb Symp Quant Biol* 2009;**74**:355–62. https://doi.org/10.1101/sqb.2009.74.011. Review, 19717539.

47. Carvalho CM, Lupski JR. Mechanisms underlying structural variant formation in genomic disorders. *Nat Rev Genet* 2016;**17**(4):224–38. https://doi.org/10.1038/nrg.2015.25. Review, 26924765.

48. Dennis MY, Harshman L, Nelson BJ, Penn O, et al. The evolution and population diversity of human-specific segmental duplications. *Nat Ecol Evol* 2017;**1**(3):69. https://doi.org/10.1038/s41559-016-0069. 28580430.

48a. Key FM, Abdul-Aziz MA, Manley R. Human adaptation of the TRPM cold receptor along a longitudinal cline. *PLoS Genet* 2018;**14**(5):e100728. 29723175.

49. Marshall CR, Howrigan DP, Merico D, Thiruvahindrapuram B, et al. Contribution of copy number variants to schizophrenia from a genome-wide study of 41,321 subjects. *Nat Genet* 2017;**49**(1):27–35. https://doi.org/10.1038/ng.3725. 27869829.

50. Baldwin JMA. New Factor of Evolution. *Am Nat* 1896.

51. Pennisi E. *Buying time Science Magazine December 6th*. 362; 2018. p. 988–91.

52. Levis NA, Isdaner AJ, Pfennig DW. Morphological novelty emerges from pre-existing phenotypic plasticity. *Nat Ecol Evol* 2018;**2**(8):1289–97. https://doi.org/10.1038/s41559-018-0601-8. 29988161.

53. Pfennig DW, Murphy PJ. How fluctuating competition and phenotypic plasticity mediate species divergence. *Evolution* 2002;**56**(6):1217–28. 12144021.

54. Corl A, Bi K, Luke C, Challa AS, Stern AJ, et al. The genetic basis of adaptation following plastic changes in coloration in a novel environment. *Curr Biol* 2018;**28**(18):2970-2977.e7. https://doi.org/10.1016/j.cub.2018.06.075. 30197088.

55. Dayan DI, Crawford DL, Oleksiak MF. Phenotypic plasticity in gene expression contributes to divergence of locally adapted populations of Fundulus heteroclitus. *Mol Ecol* 2015;**24**(13):3345–59. https://doi.org/10.1111/mec.13188. 25847331.

56. Wallace DC. Mitochondrial DNA variation in human radiation and disease. *Cell* 2015;**163**(1):33–8. https://doi.org/10.1016/j.cell.2015.08.067. 26406369.

57. Mathias RA, Fu W, Akey JM, Ainsworth HC, Torgerson DG, et al. Adaptive evolution of the FADS gene cluster within Africa. *PLoS ONE* 2012;**7**(9):e44926. https://doi.org/10.1371/journal.pone.0044926. 23028684.

58. Stetter MG, Thornton K, Ross-Ibarra J. Genetic architecture and selective sweeps after polygenic adaptation to distant trait optima. *PLoS Genet* 2018;**14**(11):e1007794. https://doi.org/10.1371/journal.pgen.1007794. eCollection 2018 Nov, 30452452.

59. Pritchard JK, Pickrell JK, Coop G. The genetics of human adaptation: hard sweeps, soft sweeps, and polygenic adaptation. *Curr Biol* 2010;**20**(4):R208–15. https://doi.org/10.1016/j.cub.2009.11.055. 20178769.

60. Falconer DS, Mackay TFC. *Introduction to quantitative genetics*. New York: Longmans; 1996.

61. Weedon MN, Frayling TM. Reaching new heights: insights into the genetics of human stature. *Trends Genet* 2008;**24**(12):595–603. https://doi.org/10.1016/j.tig.2008.09.006. Review, 18950892.

62. Westra HJ, Franke L. From genome to function by studying eQTLs. *Biochim Biophys Acta* 2014;**1842**(10):1896–902. https://doi.org/10.1016/j.bbadis.2014.04.024. Review, 24798236.

63. Voight BF, Kudaravalli S, Wen X, Pritchard JK. A map of recent positive selection in the human genome. *PLoS Biol* 2006;**4**(3):e72. Epub 2006 Mar 7. Biol. 2006;4(4):e154, 16494531.

GENE AND ENVIRONMENT INTERACTIONS AND PHENOTYPES

Introduction

The goal of this chapter is to present examples of phenotypic manifestations that are clearly impacted by gene environment interactions. Specific phenotypes to be considered include skin pigmentation, lactose tolerance, immune responses to pathogenic microorganisms, and adaptations to increasing altitude.

Skin pigmentation

Molecular genetic studies have increased our insights into the form of phenotypic variation that seems surprisingly to have played such a divisive role in human societies (Fig. 1).

Insights into factors involved in human skin pigmentation came initially from studies in the disorder albinism. This condition occurs in populations throughout the world but it is particularly striking in Africa, since dark skinned parents can give birth to a child with light skin and yellow hair.

Oculocutaneous albinism

This is the most frequently occurring form of albinism and as the name implies, it is characterized by abnormally low levels of dark types of melanin pigment in skin, hair and eyes and is frequently associated with impaired vision.

Insight into molecular defects has been provided by analysis of the function of gene products noted to be defective in albinism. Simeonov et al.[1] reported that 8 different genes had been found to have defects that lead to albinism.

Gene Environment Interactions. https://doi.org/10.1016/B978-0-12-819613-7.00004-9

Gene variants in humans confer resistance to pathogens	Malaria resistance	Hemoglobin variants HbS, HbC, G6PD deficiency
	P. vivax malaria resistance	Duffy blood group negative
	P. falciparum malaria resistance	Glycophorin gene variants
Specific variants in parasites lead them to overcome human resistance mechanisms	Trypano some resistance	Trypanolytic protein Apolipoprotein APOL1
	T. brucei gambiense	Resist APOL1 lysis,, variants in specific glycoproteins
	T. brucei rhodesiense	

Lactase persistence	lactase		Mutations upstream in lactase gene in enhancer	
	Lactose ———□———Galactose + Glucose		North European	LCT -13,910 T
evolutionary modification	Lactase persistence-the ability of adults to digest the lactose in milk		Nilo-Saharan Africa	LCT -14,010 G
	co-evolution with domestication of dairying animals		Saudi-Arabia	LCT -13,915 G

Melanin synthesis and sunlight

Tyrosine —○— Dopamine ——→ Dopaquinone+cysteine- ——CysteinylDOPA
|
Alanyl- hydroxy benzothiazine
|
Pheomelanin yellow red soluble

Dopaquinone —○- -Dopachrome ——DHICA +TRP1——DHICA melanin brown, poorly soluble

Fig. 1 *Gene variants and environmental adaptations.* Gene variants have arisen in certain human genes to promote resistance to Malaria and Trypanosome parasites. Some species of Trypanosomes evolved variants to overcome human resistance mechanisms. Specific mutations in the Lactase gene promote human ability to digest lactose beyond infancy. Humans have evolved mechanisms to synthesize brown poorly soluble melanin to resist damaging effects of sunlight.

Albinism type	Gene	Chromosome	Gene product function
OCA1A, 1B	*TYR*	11q14.3	Tyrosinase, conversion of tyrosine to melanin
OCA2	*OCA2*	15q12-q13	Melanosomal transmembrane protein
OCA2 mod	*MC1R*	16q24.3	Melanocortin receptor
OCA3	*TYRP1*	9q23	Tyrosinase related protein
OCA4	*SLC45A2*	5p13.2	Solute carrier, transporter of melanin
OCA5		4q24	Uncharacterized protein
OCA6	*SLC24A5*	15q21.1	Intracellular membrane protein Ca/Na exchange
OCA7	*LRMDA*	10q22.2	leucine rich melanocyte differentiation associated
OA1 (Eye only)	*GPR143*	Xp22.2	G-protein coupled receptor protein in melanosomes

Quillen et al.[2] reported that population differences in albinism occur. They reported the general population frequency of albinism to be 1 in 20,000. However, in certain populations in Mexico, Panama and Brazil the frequencies of albinism are much higher.

Syndromic albinism

Albinism also occurs in conditions in which there are additional systemic features. Montoliu et al.[3] reported that nine different forms of Hermansky Pudlak syndrome occur, each due to mutations in a different gene. In addition to oculocutaneous albinism, patients with this syndrome can manifest blood coagulation defects; they may also manifest respiratory difficulties due to pulmonary fibrosis. In this syndrome the defects occur in lysosome-endosome related organelles (LROs) that play roles not only in melanosomes, but also in platelets and in some other cell types. The 9 different genes that are defective in this syndrome are each involved in forming a multi-protein complex, the BLOC complex that plays a role in melanosome biogenesis.

Another form of syndromic oculo-cutaneous albinism is Chediak Higashi syndrome. Patients with this disorder may also manifest problems in immune system function and in blood clotting, Montoliu et al. The defective gene in this syndrome encodes lysosomal trafficking regulator, LYST that is involved in trafficking in lysosome endosome organelles.

Melanocytes and melanin

In humans, melanocytes occur in the epidermis, in hair follicles, in the eye, in the cochlear of the ear and also in mucosa in certain regions. Melanocytes are derived from neural crest cells. Yamaguchi and Hearing[4] reported four stages in melanocyte maturation. They also reviewed structural proteins within melanosomes, distinct enzymes required for melanin synthesis and proteins required for melanin and melanosome trafficking. A number of these proteins and enzymes have been found to be defective in pigmentation disorders, including hypopigmentation and hyperpigmentation disorders. Melanin pigments include dark brown and black pigments described as eumelanin and pheomelanins that are yellow, red and light brown.

Melanosomes the organelles that produce melanin, are defined as lysosome related organelles (LROs), since they also contain specific proteins that are found in lysosomes. The early stage melanosomes were reported to closely resemble lysosomes. Melanin production is regulated by the transcription factor MITF (melanogenesis associated transcription factor). Proteins involved in trafficking of melanin is important, since melanin is transferred from melanocytes to keratinocytes. Yamaguchi and Hearing listed key proteins in melanosomes and proteins involved in melanin production.

Melanosome structural proteins

PMEL (PMEL17)	premelanosomal protein a trans membrane glycoprotein
MRT (MLANA)	melan A
GPNMB	transmembrane glycoprotein

Proteins involved in melanin production

TYR	Tyrosinase
TYRP1	Tyrosinase related protein 1
DCT	Dopamine chrome tautomerase
BLOC1	Biogenesis of lysosomal organelle complex 1
GPR143	G-protein coupled receptor (OA1)
SLC45A2	Transporter protein

Proteins involved in intra-cellular transport and trafficking of melanin

Microtubules	F-actin, kinesins and dynein proteins
RAB27A	GTPase protein
MLPH	Melanophilin forms a complex with RAB27A
Ciliobrevin	Dynein related protein for organelle movement has several subunits
SLP2A (SYTL2)	RAB related vesicle transport protein
KIF13A	Kinesin related motor protein

Melanin and melanosome transcription factors

PAX9.
SOX9, SOX10.
LEF1.
CREB.
Dicer.
MITF1.

Hyper pigmentation disorders

Hyperpigmented skin lesions are referred to as nevi if raised and as lentigines if they are flat lesions. Yamaguchi and Hearing[4] noted that lentigines may occur without other clinical manifestations.

Two specific syndromes have been identified where patients have multiple lentigines in combination with other defects. Peutz Jeghers syndrome is characterized by hyperpigmented lesions around and in the mouth and with intestinal polyps. These patients are at increased risk for colon cancer. This syndrome has been reported to be due to mutations in the *STK11* gene that encodes a serine threonine kinase.

Noonan syndrome with lentigines, is associated with multiple lentigines and patients may have cardiac defects, short stature, unusual facial features an (formerly referred to as LEOPARD syndrome), and neurodevelopmental defects. This syndrome has been found to be due to defects in the *PTPN11* gene in approximately half of the case. Other genes that have been found to be defective in cases of this syndrome include *SOS*, *RAF*, and *RIT1*. All of these genes encode products involved in intracellular RAS signaling pathway.

Unusual pigmented lesions of the skin may also occur in the syndromic condition Neurofibromatosis. The *NF1* gene encodes the protein neurofibromin that participates in the RAS intracellular signaling pathway.

Acquired hyperpigmented lesions occur as an aging phenomenon. They may also occur in response to endocrine disorders, e.g. with excess adrenal hormone in Cushing syndrome.

Congenital hypopigmentation disorders

These include Albinism as described at the start of this chapter. Waardenburg syndrome is characterized by hypopigmentation of the frontal hair, eye pigment deafness and hearing impairments. Waardenburg syndrome was first described in 1951.[5] It is now known to be genetically heterogeneous and may result from defects in anyone of at least 9 different chromosomal loci. At least 7 genes with mutations that cause features of Waardenburg syndrome have been isolated.[6]

Pax3	Transcription factor
SNAI2	Transcription repressor
EDN3	Endothelin essential for development of neural crest-derived cell lineages, melanocytes
EDNR3	Endothelin receptor; reaction between EDN3 and EDNR3 are essential for melanocytes
MITF	melanogenesis associated transcription factor
SOX10	(SRY-box 10) transcription factor
KIT	Acts as a receptor tyrosine kinase
KITLG	Functions as a KIT ligand

Hypomelanosis of Ito is a congenital disorder associated with streaked mottled areas of hypopigmentation. It may also be associated with visual defects (strabismus) and seizures. This disorder may occur as a result of chromosome abnormality. It is thought to be manifestation of mosaicism for chromosome abnormality.

Tuberous sclerosis is characterized by hypomelanotic skin lesions that can include ash-leaf lesions and hypomelanotic freckling. Studies on cultured cells from these hypomelanotic lesions have revealed impaired melanogenesis due to over-activity of mTOR resulting from impaired inhibition of MTOR resulting from mutations in TSC1 or TSC2 genes.[7]

Acquired hypomelanosis referred to as Vitiligo is currently thought to represent an autoimmune disease. Variations in >30 different gene have been associated with this disorder. Particularly important is *NLRP1*, that encodes a product involved in protein-protein interactions.[8]

Quillen et al.[2] reported that geographic differences were reported in the frequency of variants in the gene that encodes MFSD12, the major facilitator domain protein that impacts melanogenesis. It is interesting to note that same variants in this gene are associated with vitiligo while other variants are associated with darker pigmentation.

Population studies on skin pigmentation

Crawford et al.[9] carried out extensive studies to identify and further investigate genetic variants associated with different degrees of skin pigmentation. They emphasized that variation in skin pigmentation is in part due to adaptation to different environments. In environments where humans are exposed to significant sun exposure, increased skin pigmentation is advantageous since it offers protection against ultra-violet (UV) radiation. In environments with low sunlight exposure light skin is advantageous since it promotes Vitamin D synthesis.

Crawford et al. assessed skin exposure using a DSMII colorimeter to evaluate degrees of skin pigmentation on the inner arm. They studied degree of pigmentation and DNA sequence variants in individuals from 3 different regions in Africa, Ethiopia, Tanzania and Botswana. They genotyped 1570 individuals. The San population in Southern Africa had the lightest skin pigmentation and Nilo-Saharan populations had the darkest pigmentation.

They noted that the strongest genetic association of skin color was with variants at or near the SLC24A5 encoding gene on chromosome 15. Studies have revealed that variants in this gene were introduced into East Africa >5000 years ago. A specific variant in this gene region, (rs1426654A) is associated with lighter pigmentation. It has a higher frequency in the San population in Botswana. The SLC24A5 is a solute carrier expressed in intracellular membranes.

The second strongest signal was on chromosome 19 in the vicinity of the *MFSD12* gene. The investigators also determined that decreased expression of the *MFSD12* gene was associated with darker pigmentation. *MFSD12* encodes a protein that is expressed in the lysosome related organelles and in melanosomes.

Crawford et al. reported that another important genome region associated with pigmentation occurs on human chromosome 11q12.1 in the vicinity of the gene that encodes damage specific DNA binding protein 1 (DDB1). The DDB1 protein complexes with two other proteins DDB2 (damage specific DNA binding protein 2) and XPC Xeroderma

Pigmentosum group C DNA repair gene homolog. This complex plays an important role in DNA repair following damage. Levels of expression of DDB1 were shown to be regulated by UV exposure. They noted the variants of this gene were found in African populations with high levels of European admixture. Analysis of RNA revealed that the associated variant alleles led to higher levels of gene expression. Other variants in the 11q12.1 region associated with pigmentation included variants in the transmembrane protein 138 (TMEM138).

A fourth genomic region significantly associated with pigmentation occurred in the 15q12-13 region in the vicinity of the *HERC2* and *OCA2* gene region. *OCA2* encodes a melanosomal transmembrane protein. The highest degree of association was with a variant in exon 10 of OCA2 and in intron 11 of HERC2 (HECT and RLD domain containing E3 ubiquitin protein ligase 2).

Crawford et al.[9] concluded that skin pigmentation is a complex trait impacted by variants in at least 8 different genes. They noted however, that together these variants contributed to 28.9% of the variation in skin pigmentation.

Complexities in determination of pigmentation

In a 2019 review Quillen et al.[2] noted that DNA sequencing studies on diverse populations and the applications of additional methods of assessment of skin pigmentation have revealed a growing number of genes that impact skin pigmentation. They noted that in addition to measurement of levels of pigmentation, assessment of skin sensitivity to sun exposure and tanning responses were included in studies.

They emphasized that the full range of factors that impact skin color include genetic diversity, epistatic gene interactions and environmental factors. They also emphasized that the same genetic variant in a specific gene may have somewhat different levels of impact in different populations.

Quillen et al.[2] considered the best characterized pigmentation related gene products in European populations to be the melanocortin receptor MC1R, SLC24A5 a transporter located in membranes and OCA2 a melanosomal transmembrane protein.

MC1R is a G-coupled receptor that occurs on melanocytes and it a receptor for melanocortin, Binding of melanocortin to this receptor stimulates production of melanin precursors. Defects in the MC1R and inadequate melanocortin binding results in decreased production of melanin and lead to reddish hair color and fair skin, as reported by Valverde et al.[10]

Quillen et al.[2] noted that studies in Eurasian populations revealed that a number of MC1R variants occurred and they had varying effect. MC1R variants were initially reported to be rare in African populations with dark skins. They noted that lighter pigmentation in some African populations have been more recently reported to be due to variants at a number of loci include loci that encode solute carriers SLC24A5, SLC45A2 and loci that encode KITLG that encodes a ligand for a tyrosine kinase receptor and SNX13 sorting nexin 13 that is involved in intra-cellular trafficking.

The SLC24A5 variants that impact skin color were reported to have impaired protein function that impacts melanosome function.

Quillen et al., summarized findings in 23 different reports related to skin color and genetic analyses done included melanin index determination in 19 studies tanning response in 3 studies and skin sensitivity to sun exposure in 1 study, It is important to note that findings in the different studies in, Africa, America, Asia, Europe and Melanesia demonstrated that variants in the same genes impacted skin color across countries. Six genes were most commonly impacted. The proteins encoded by these included OCA2 melanosomal transmembrane protein, solute carriers SLC24A5, SLC45A2, BC1R the melanosomal receptor HERC2 that has ubiquitin ligase activity, and TYR tyrosinase.

It is interesting to note that variants in the gene *ASIP* that encodes a signaling protein that stimulates melanin production were found to be associated with increased sun sensitivity in individuals in Iceland and in the Netherlands.

In addition, other genes were found to be associated with pigment differences in different populations. In the Koi San population in South Africa variants occurred in SNX13 sorting nexin 13 that is involved in intra-cellular trafficking were found to impact skin color.

Milk as nutrient and lactase persistence

Comprehensive studies have been carried out on different human populations to determine the nature of genetic variants that result in the persistence of intestinal lactase expression after infancy. In 2017 Ségurel and Bon[11] reviewed lactase persistence.

For many years, lactase persistence has been considered to be an example of gene-culture co-evolution that arose following domestication of milk producing animals. Olds and Sibley[12] reported that specific regulatory DNA variants upstream of the human lactase gene were associated with long-term persistence of lactase expression.

In 2017 Liebert et al.[13] published details of specific variants at 10 different sites in the regulatory region upstream of the lactase gene. They noted that the variants were likely located in an enhancer region. They also reported that there was a specific haplotype in which the alleles of the associated variants were located and indicated that there was little evidence of recombination (crossing-over) in this genomic region.

Ségurel and Bon[11] noted that there were several groups of herders in Central Asia where designated lactase persistence alleles occurred with low frequency. In addition, they noted that there were hunter gatherer populations who did not consume milk, yet the designates lactase persistence allelic variants occurred with relatively high frequency.

Ségurel and Bon noted that the advantage of lactase persistence alleles was based on the fact that in the presence of lactase lactose was digested high in the intestinal tract and little undigested lactose reached the colon. The problems that arise in cases with low lactase expression are due to the fact that lactose is not digested in the upper intestinal tract and it reaches the colon. There is can promote the influx of water. In addition, lactose in the colon is acted on by micro-organisms leading to the production of gas. This leads to discomfort and diarrhea. It is possible that the types of microorganisms present in the colon in different individuals likely influences the consequences of lactose there.

A main argument for positive selection of lactase persistence has been that milk ingestion is a good source of calories and protein. Ségurel and Bon noted that in many populations, methods evolved to generate lower lactose content products from milk, e.g. cheese These products would be less likely to cause symptoms in individuals with low or absent lactase.

It is interesting to note that content of lactose in yoghurt and in hard cheese in particular is much lower than in milk.

Recognizing non-self: The immune system

Innate immunity

Components of the innate immune system were first reported by Metchnikoff who received the Nobel prize for his discoveries in 1908.[14] Information has steadily grown since then on types of cells involved in innate immunity and more especially, on the mechanisms through which cell recognize invading organisms.[15]

Cells that play roles in innate immune responses include macrophages, neutrophils, mast cell eosinophils, dendritic cells and more recently, different types of killer cells. The cells involved in the innate immune are derived primarily from the hematopoietic systems and are present in the circulation but are also present in skin and in mucosa linings of gastrointestinal tract, respiratory tract and urogenital tracts.

In addition to the cellular components of the innate immune system, specific proteins in blood and other body fluids play roles in innate immunity related processes. These include various types of

Complement, C-reactive proteins and defensins. In addition, specific proteins produced by activated innate immune system cells facilitate responses designed to control infections. These include cytokines, histamines, pyrins include MEFV defined as an innate immunity regulator.

In the 2019 review Beutler[16] noted that early studies involved analysis of components within micro-organisms that stimulated an immune response. Many of these components were designated as endotoxins.[17] Lipopolysaccharides were also recognized as microorganism components that elicited an immune response. Subsequently additional components of invading micro-organisms were found to elicit an immune response. These included peptidoglycans, lipopeptides, double stranded RNA and micro-organism DNA.

Beutler emphasized that that in the early years of immunology, scientists were aware that in order to recognize invading microbes host cells required mechanisms to distinguish self from non-self. Efforts were initiated to identify specific receptors in the host that bound the inflammatory inducing microbial components. Other important research was designed to document the downstream signaling pathway activity that was induced following binding of microbial components to host cells.

Genetic analyses were designed to identify specific mutations in strains of mice that failed to have the usual immune response to lipopolysaccharides (LPS).[18] These studies led to identification of a specific gene locus, designated lps1 the was defective in LPS non-responsive mice. Studies were then initiated to identify the biologic functions of protein encoded at the Lps1 locus. These studies led to discovery that the locus encoded a specific receptor, referred to as the LPS receptor and later became known as the Toll receptor. In 1996 Hoffman et al.[19] reported that the Toll receptor was an essential component of the response of Drosophila to fungal infection. Immunologists including Beutler. then began working on isolating Toll-like receptors in eukaryotes.

Binding of micro-organism ligands to Toll receptors were found to activate a serine kinase that in turn activated intracellular transcription factors. Effective Toll signaling was found to required adaptor proteins. Subsequently a number of slightly divergent Toll receptors were identified. In human currently 9 different Toll receptors have been identified. Beutler noted that Toll receptors are located on different positions in the cells. Toll receptor proteins were shown to be component of multiprotein complexes.

In addition to Toll receptors specific innate immune systems cells such as natural Killer cells NK cells (a specific type of lymphocytes) were found to have specific receptors that could bind specific foreign proteins or foreign chemicals. Recognition of components of invading organism is one of the central functions of the immune system.

Kawai and Akira[20] noted that identification of Toll receptors that specifically recognize pathogenic associated molecular patterns (PAMPs) was a key advance in understanding immune responses. They described Toll receptors as transmembrane receptors in which the outer regions recognized PAMPS. The intra-cellular regions of the receptors were found to be involved in signal transduction. The specific pathogen associated molecular patterns that are recognized by Toll receptors included lipid, proteins and nucleic acids. Key organisms recognized by Toll receptors include viruses, parasites and fungi.

Toll receptors were subsequently found to be present not only on outer cell membranes but also on intra-cellular membranes of endosomes and lysosomes. The specific types of Toll receptors on cell surfaces differed from the types of Toll receptors expressed on intra-cellular vesicles.

In studying signaling processes connected with Toll receptors, investigators identified specific adaptors and cell type specific receptors. These included RIG-1 receptors (RLR) and Nod-like receptors (NLR). The RLR receptors specifically detect RNA, the NLR receptors when activated were found to trigger proinflammatory responses and interleukin secretion.

Kawai and Akira[20] noted that specific receptors later identified were found to recognize DNA and activation of those receptors led to production of interferon.

Jensen and Thomsen[21] reviewed the innate immune to RNA virus infection and activities of the Toll receptors and TLR receptors. They noted that toll receptor TLR7 sensed single stranded RNA derived from viruses. The RNA regions detected were frequently guanosine and uridine rich. They reported that TLR7 is usually located in the endoplasmic region (ER). Upon activation of the receptor by viral infection TLR7 moves to the endosome. Viral particles that invade cells are undergo autophagy and are then located in the endosomes and lysosome. TRL7 binds to a range of single stranded RNA viruses including influenza virus. Double stranded RNA viruses were recognized by TLR5 TLR8 recognizes single stranded RNA viruses.

Jensen and Thomsen reported that downstream signaling from Toll receptors involved activation of MAP kinase signaling pathway and NFkappaB and interferon regulatory factors and interleukin receptors.

They emphasized that the RLR receptors are present in the cytosol of the cells. The RIG1- RLR receptor was reported to have single stranded RNA and double stranded RNA binding domains at the C terminal end of the receptor. The viral sequences recognized included double stranded RNA and the 5′ triphosphate region of single stranded RNA, Binding of the RLR receptor to virus lead to activation of a caspase recruiting domain in the receptor.

O'Neill[22] reported that viral glycoproteins were detected by cell membrane bound receptors and that viral nucleic acids were detected primarily from Toll receptors on endosomes.

Additional pattern recognition receptors were identified in the cytoplasmic compartment. These included RLR, RIG-1 and MDA5 that recognized viral RNA genomes and viral transcripts.

Of particular interest were advances in structural studies of RNA that revealed how RLR pattern recognition receptors distinguished self RNA from viral RNA. The RLR pattern receptors when activated signaled via a molecule MAVS, mitochondrial antiviral signaling protein. The product of the MAVS gene is described as an intermediary protein necessary in the virus-triggered beta interferon signaling pathways. Activations of MAVS then activates transcription factors such as NFkappa B and interferon regulatory factors.

Dempsey and Bowie[23] reviewed DNA sensing by the innate immune system. They emphasized that key developments related to innate immunity included discovery of pattern recognition receptors (PRR) that respond to pathogen associated molecular patterns (PAMPs). Other key discoveries included those of Toll receptors and the induction of cytokines and interferons that follow receptor activation.

Dempsey and Bowie noted that DNA sensors were also identified in the cytoplasm. These include DAI also known as ZBP1. The function of protein is to bind to foreign DNA and to induce Type-I interferon production. A specific adaptor protein that functions as cytoplasmic DNA sensor STING is also known as TMEM173. It is a pattern recognition receptor that detects cytosolic nucleic acids and transmits signals that activate type I interferon responses. STING has also been shown to play a role in cell death responses.

Lassig and Hopfner[24] emphasized that viral RNAs present in the cell exist in the presence of host RNA, however, the immune response must be directed only by the viral RNA. They emphasized the importance of the RIG-1, (RLR1) protein and the structurally related proteins MDA5 and LGP2 (RLR3) in detecting RNA in the cells. MDA5 is also known as IGIF1 interferon induced with helicase C domain 1.

They reported that these proteins are pattern receptors that utilize energy from adenosine triphosphate ATP to carry out their function. RLR1 and MDA5 were found to have distinct profiles of RNA binding. RLR1 binds short segments of double stranded RNA the 5′ uncapped triphosphate RNA 7–10 bases in length. MDA5 binds longer regions of RNA >200 bp. RLR binds to RNA termini.

Lassig and Hopfner noted that RLR1 and MDA5 signal through the adaptor protein MAVs.

The mechanism that prevents self RNA from activating RLR1 are its requirements for uncapped 5′ triphosphate fragment. Editing of eukaryotic RNA by the enzyme ADAR1 changes adenosine to inosine and this modification apparently prevents MDA5 from binding to self RNA. The ATP hydrolysis activity of RLRs apparently also enhances their affinity for viral RNA.

Fine tuning of the innate immune response is also achieved through post-translational modifications in RLR receptors.

NOD-like receptors and Inflammasome

Jensen and Thomsen[21] noted that NOD-like receptors (NLR) located in the cytoplasm are also pattern associated receptors (PAMPs) and that they play key roles in regulating inflammatory and apoptotic responses. NLRs are composed of a leucine rich region (LRR) at their C-terminals and this region acts as a sensor. The central region NACAT serves to activate the N-terminal region that includes a Pyrin domain and a CARD domain (caspase associated domain.

There are different forms of NLRs. Upon activation of the NLR3 receptor was shown to initiate formation of the inflammasome complex that also plays a role in maturation of proinflammatory interleukins. The inflammasome complex is composed of caspase 1, of PYCARD an adapter protein that is a protein composed pf an N-terminal PYRIN related domain and a C-terminal CARD domain.

Ozen and Batu[25] reported the Familial Mediterranean fever, caused by mutations in the *MEFV* gene is a common autoinflammatory disease that primarily impacts individuals in populations with origins in the Mediterranean basin, The *MEFV* gene encodes Pyrins that were reported to recognize the specific molecular differences on bacterial RHOGTPase compared with host RHOGTPase and recognition of these differences led to inflammasome activation. Swanson et al.[26] reported that the protein encoded by the *MEFV* gene is a component of the inflammasome.

Adaptive immunity

In the early phases of studies on adaptive immunity, attention was focused primarily on B lymphocytes and their production of antibodies composed of immunoglobulins. Dryer and Bennett[27] described variable and constant regions of immunoglobulins. Edelman[28] and Porter[29] carried out studies that revealed the structure of antibodies and their content of heavy and light chains and of variable and constant regions. The structural studies revealed the variable region functions in antigen binding. Tonegawa[30] carried out studies that elucidated the structure of antibody encoding genes.

In 1975 Kohler and Milstein[31] published methods to derive cultures of B lymphocytes that produced antibodies of defined specificity (monoclonal antibodies).

In 1989 Janeway[32] presented his hypotheses on the likely existence of pattern receptors that recognized molecular patterns on pathogens and pattern receptors that elicited the immune response. He also proposed that the innate immune response was subsequently interpreted by the adaptive immune system.

Further studies revealed that the interactions of the innate and the adaptive immune systems were mediated by the dendritic cell discovered by Steinmann and Cohn.[33] The dendritic cells were first discovered in skin where they referred to as Langerhans cells based on earlier description. Dendritic cells were also found to be present at other sites where tissues interact with the environment, e.g. the nasopharynx, lungs and gastro-intestinal tract. Dendritic cells were found to have pattern receptors that bound to antigens. Following their activation by antigen binding, dendritic cells migrate to lymphoid tissues where they interact with T and B cells

Subsequently different subsets of dendritic cells were identified that are distinguished by the presence of different surface markers in the CD series CD103 dendritic cells exist in most tissues. Other dendritic cell subsets have CD11b, CD207 and CD24 subtypes. All of the subtypes can migrate to lymphoid tissues.

Components and functions of the adaptive immune system

Bonilla and Oettgen[34] reviewed adaptive immunity. The cells of the adaptive system, B cells and T-cells, originate in fetal life from the bone marrow and fetal liver. T cell precursors migrate to the thymus where they undergo maturation. T cell maturation in the thymus is dependent on the development of receptors that enable these cells to respond to interleukins. In the germline the T cell receptor gene contains different separated segments. Part of the T cell maturation process involves cleavage of DNA of this gene and recombination and repair. Portions of the gene that are excised remain in the nucleus and form structures known as TRECS (T cell receptor excision circles). TREC levels can be measured, e.g. in newborns and utilized to assess the efficacy of steps of T cell maturation, particularly migration from the thymus to lymphoid tissue.[35] Productive rearrangement of the germline T cell receptor gene gives rise to a genomic segment that encodes the mature T cell receptor.

T cell receptors respond to antigens or peptides that are bound to components encoded by the major histocompatibility gene complex (MHC). The cells that stimulate T cell receptors are known as antigen presenting cells. Steinman discovered that dendritic cells were antigen presenting cells.

Following antigen stimulation T cell can undergo further divergence based in part on the antigen encountered.

B cells and antibody production

B cells are derived from bone marrow where they undergo different developmental changes under influence of specific transcription factors. Bonilla and Oettgen[34] noted that B cells mature to develop specific cell surface markers including CD 19, 20, 21, 23 and CD40. Definitive maturation of B cells involves structural rearrangement of the germline immunoglobulin genes as described by Tonegawa.[30]

In response to antigen stimulation the variable region of the immunoglobulin gene undergoes hypermutation. This leads to production of immunoglobulins that specifically interact with particular antigens. B cells produce antibodies that can interact with antigens through recognition of the three-dimensional structure of the antigens.[36]

The specifically altered components of the adaptive immune system can endure in the organism for long periods and serve as immunologic memory.

Cytidine deaminases and their roles in the immune system

APOBEC proteins

Proteins in the APOBEC family are designated as apolipoprotein B MRNA editing catalytic polypeptides. Specific APOBEC forms have been found to inhibit viruses through several different mechanism, including nucleic acid editing.[37]

Salter et al.[38] reviewed the structure and divergent functions of proteins in the APOBEC family that in humans are the products of 11 genes. The initially described function of APOBECs was the editing of cytidine in MRNA to uridine. One specific protein APOBEC A3G was reported to edit nucleotides in the HIV virus by deaminating cytidine. This editing reduced the production of infectious virus; other forms of APOBEC A3 were later found to edit HIV and lead to inhibition of infection. Subsequently the A3subfamily of APOBEC was reported to edit retroviral elements in exogenous and endogenous retroviruses. Salter et al., noted that the A3G APOBEC protein binds to single stranded RNA and to single stranded DNA.

The different forms of APOBECS and the enzyme AID (activation induced cytidine deaminase) were found to contain a zinc dependent deaminase domain. However, the APOBEC proteins are longer than AID protein and contain several additional domains.

The AID enzymes play important roles in inducing somatic hypermutation during antibody generation.

ADAR enzymes adenosine deaminase RNA specific

In humans five genes encode ADAR enzymes that modify RNA through transformation of adenosine to inosine. The ADAR proteins contain a number of different domains including a double stranded RNA binding domain and a large deaminase domain.

ADAR1 proteins occur in the cytoplasm, nucleus and nucleolus[39] Specific damaging mutations in ADAR1 in homozygous or compound heterozygous forms, lead to a severe disease in human known as Aicardi-Goutieres syndrome. This syndrome may also be caused by mutations in at least four other genes. This syndrome can present as severe disease early in childhood. It is associated with increased levels of interferon. The manifestation of this syndrome mimic those of severe viral infections.

The normal functioning ADAR1 enzyme was found to prevent activation of the immune system that led to interferon production. ADAR1 activity acts to prevent the MDA5 (IFIH1) molecule from sensing self RNA. It therefore prevents MDA5 induction of interferon. A number of different conditions occur in humans that are associated with abnormal induction of interferon.

Genome diversity and protection against malaria infection

Haldane in 1949[40] proposed that the high frequency of thalassemia in the Mediterranean region arose since the alleles that caused thalassemia were protective against malaria. In 1954 Allison[41] proposed that the high frequency of the sickle cell allele in Africa was protective against malaria. In heterozygotes with these mutations red blood cells with these globin variants were resistant to infection by the malaria parasite.

In a 2014 review Mangano and Modiano[42] noted that Malaria parasites occur world-wide however the prevalence of specific sub-species of the malaria parasite differ in different regions. *Plasmodium falciparum* is the most prevalent form in Sub-Saharan Africa. *Plasmodium vivax* (*P. vivax*) is most prevalent in Asia and South America. *P. malaria* and *P. ovale* are reported to be responsible for between 3 and 8% of cases in sub-Saharan Africa. *P. knowlesi* infect macaques and is also reported to lead to malaria in certain regions of Asia.

Studies over the years have revealed that variant alleles in a number of different genes can lead to resistance to malaria. The resistance usually occurs in heterozygotes with the specific gene mutation.

Malaria resistant allelic variants have been reported in a number of different genes. The risk of malaria in HBS heterozygotes was reported to be reduced 20-fold from the population risk frequencies in regions of high HbS frequencies including sub-Saharan and parts of the Middle East.[43]

The Beta globin variants that decrease susceptibility to malaria include: HbS Glu6Val, HBC Glu6Lys and HBE Glu26Lys.

Kwiatkowski[44] in a review of genetic variants and decreased susceptibility to malaria reported that HBE is generally rare in Africa but is common in South-east Asia. One exception in Africa was in the Dogon population in Mali where HbE was common and HbS was rare. There is evidence from haplotype studies that both the HbS and the HbE mutations arose independently several different times. HbE homozygotes have only mild hemolytic anemia in contrast to the severe disease that occurs in HbS homozygotes. Hemoglobin C homozygotes and heterozygotes have increased resistance to malaria. HbC homozygotes have only mild hemolytic anemia. Travossos et al.[45] noted that the Hemoglobin C trait (carrier status) was known to protect against severe malaria in children. The study they undertook in Mali revealed that children with hemoglobin AC took a longer time than hemoglobin AA children to develop their first malaria episode, they had fewer episodes of clinical malaria and lower parasite burden.

Thalassemias

These conditions can result from specific mutations that lead to defects in production of functional globins, beta globin or alpha globin. There are two forms of the alpha globin encoding genes, Severe thalassemia occurs when there are deletions of 3or 4 of the alpha globin genes. Deletion of all four alpha globin genes results in death during fetal life. If some alpha globin is produced, e.g. in the case of absence of product from 3 of the 4 alpha globin genes survival is possible though severe anemia is usually present. There is evidence that individuals with alpha+ thalassemia associated with presence of reduced levels of alpha globin, have increased resistance to malaria.[46,47]

Flint et al.[48] reported that deletion of one of the four of alpha globin genes provided protection against malaria in Papua New Guinea and Melanesia.

Zimmerman et al.[49] reported that at the Duffy blood group locus there are two alleles A and B, (FyA and FyB). However, in certain populations Duffy antigens are absent and this condition is referred to as Fy-. The Fy- allele is reported to provide resistance to *P. vivax* malaria. This may not be the case in all populations however.[50]

Malaria parasite erythrocyte interactions

Kwiatkowski[44] reviewed information on the malaria parasite erythrocyte interactions. Specific erythrocyte membrane proteins facilitate parasite adherence. These include glycophorins. The anion exchanger SLC4A1 also impacts parasite adherence. A specific 9 amino-acid deletion in this protein induces ovalocytosis a condition that changes the shape of red blood cells and results in them being more resistant to invasion. Heterozygotes for this abnormality have increased resistance to malaria. Rosanas-Urgell et al.[51] reported that this mutation occurs in certain South-East Asian populations and protects against cerebral malaria caused by *P. falciparum*. This mutation also protects against *P. vivax* malaria.

There is evidence that following its entrance into the human erythrocyte, the malaria parasite begins to synthesize proteins that bind to the red cell surface and increase the adherence of the infected red cells to the endothelium of small blood vessels. Kwiatkowski reported that variants in 4 different proteins impact cyto-adherence of infected red cells, these include.

CD36	thrombospondin receptor
CRI	complement receptor
ICAM1	intracellular adherence molecule
PECAM1	platelet endothelial cell adherence molecule

Glucose-6-phosphate dehydrogenase (G6PD) variants and malaria resistance

One of the earlier reports of G6PD deficiency and malaria resistance was from Bienzle et al.[52] Kwiatkowski[44] noted that the malaria parasite must degrade hemoglobin in red cells in order for the parasite to survive and proliferate. The breakdown of hemoglobin leads to iron release and oxidative stress within the red blood cells. This is counteracted by the production of nicotinamide adenine dinucleotide phosphate (NADPH).

The enzyme G6PD encoded on the X chromosome, converts glucose-6-phosphate to 6-phosphogluconate using NADP as coenzyme and in the reaction NADP is converted to NADPH. Adequate function of G6PD is necessary to provide sufficient NADPH to counteract oxidative stress in the red blood cell.

The G6PD on the X chromosome has many different allelic variants. Some of these predispose to hemolytic anemia, particularly on exposure to certain oxidative compounds. Specific G6PD variants offer selection against malaria in female heterozygotes and in male hemizygotes.[53]

Hedrick[54] reported that the G6PD variants that provide selection against malaria frequently include variants that decrease levels of G6PD in the host cells. Important G6PD reducing variants include G6PD Mediterranean c.653C>T and G6PD Mahidol c.577G>A (ClinVar) https://www.ncbi.nlm.nih.gov/clinvar/?term=G6PD%5Bgene%5D.

Awareness of G6PD mutations leading to low levels of G6PD is important since Primaquine, given for Malaria prevention can be very damaging to individuals with such variants. G6PD deficiency also results in increased sensitivity to a number of other medications than induce oxidative stress, G6PD patients are also intolerant of Fava beans.

Mangano and Modiano[42] presented evidence that ATP2B4, a protein that functions as a calcium pump has specific variants that provide protection against malaria.

Malaria infestation and immune response

Kwiatkowski[44] reported that variants in several immune response related genes impact the outcome of malaria infestation. These genes include HLAB, HLADR, interferon and interleukin genes. Other genes with variants reported of impact outcome in malarial infections include CD36 that encodes a glycoprotein; it impacts cytoadherence of *P. falciparum* parasitized erythrocytes.

Trypanosome infection and adaptations

Capewell et al.[55] reviewed aspects of Trypanosomiasis that causes Sleeping Sickness in humans and Nagana in quadrupeds in Africa and aspects of co-evolutionary adaptations in hosts and parasites.

They noted that different sub-species of Trypanosome exist in Africa. Human infections are caused by Trypanosoma brucei gambiensi that occurs in west and central sub-Saharan Africa and Trypanosoma rhodesiense that occurs in Eastern Sub-Saharan Africa. They noted that there is evidence that Tr. Brucei rhodesiense is spreading in Africa. A defense mechanism against Trypanosomiasis developed in some humans and animals. This defense mechanism was reported in 1989 by Hajduk et al.[56] and it involves two protein complexes that lead to lysis of the trypanosome parasite. These complexes are referred to as Trypanosome lysis fractions TLF1 and TLF2. These complexes both contain apolipoproteins APOA1, APOL1 and haptoglobin related protein, HPR. Capewell et al., reported that these complexes are involved in lysis of the trypanosome.

Evidence is however now emerging that resistance to this lysis mechanism is developing in Trypanosome organism. Investigators have established that in Tr. Brucei rhodesiense a specific mutation in a specific surface glycoprotein encoded by a parasite gene SRA confers resistance of the parasite to lysis. The specific SRA mutation involved is a deletion in the SRA gene that results in synthesis of a proteins with an altered conformation.

There is also evidence that resistance to apolipoprotein lysis has developed in some Tr. brucei gambiensi organisms. This resistance was shown to be dependent on changes in a different glycoprotein TgsGP.

Selective forces of pathogens on human populations

Karlsson et al.[57] emphasized the important selective forces that pathogens have exerted on human populations. They noted that pathogen exposures varied as humans migrated to different regions and also as domestication of animals and growth of agriculture increased. Clearly survival of humans exposed to pathogen is dependent on immune response and to some extent on genetic variants that influence response to infection. Karlsson noted estimated timing of exposures of humans to specific pathogens, timing is given for each as approximate years before the present: Malaria 150,000; Tuberculosis 100,000–50,000; Small pox 25,000; Leprosy 12,000; Cholera 5000; HIV Aids present.

Karlsson et al. noted that studies of population in the Ganges delta where epidemics of Cholera had previously occurred that there was evidence for selection on specific potassium channel encoding genes. The channels encoded by these gene specific impacted handling of chloride. In addition, in this population there was evidence for selection of specific NFkappaB1 variants, expression of this gene is triggered by specific extra-cellular stimuli and the NFKB1 protein regulates gene transcription.

Immune response, pathogens and adaptations in different environments

There are also other genetic variants present in modern humans that encode proteins that apparently impact susceptibility or response to specific pathogens, Deschamps et al.[58] documented several of these. Variation in MERTK (MER tyrosine kinase) involved in inflammatory response was reported to lead to fibrosis following hepatitis C infection, Specific variants in LARGE (that encodes an *N*-acetylglucosaminyltransferase and IL21(interleukin 21) were reported to increase resistance to Lassa fever in West African population.[59]

Iron metabolism in macrophages impacts the growth of *Mycobacterium tuberculosis* organisms.[60] ZFPM2 protein impacts intracellular iron metabolism and there is evidence that variants in this protein influence susceptibility to Tuberculosis.

Quintana-Murci and Clark[61] reviewed innate immunity and natural selection based on pathogen exposures. They reported that the Toll receptor cluster TLR6-TLR1-TLR10 was a hot spot for selection based on analysis of single nucleotide polymorphisms. They also reported that adapter molecules for the receptors showed evidence for variants that were positively selected. They include MYD88 innate immune signal transduction adaptor, TIRAP Toll-interleukin 1 receptor (TIR) domain, and TRIF adapter molecules. Selection was also evident for specific variants in interferon alpha encoding gene IFNA6, A8 A13 and A14, and in interferon gamma encoding genes.

Quintana Murci and Clark proposed that genetic variants in the TLR and adapter pathways impacted response to pyogenic bacteria and the herpes simplex virus HSV1.

Karlsson et al., emphasized that variants in some genes impacted sensitivity to pathogens in some populations but not in others. Examples include variants in genes that influence susceptibility to malaria infection, e.g. DARC that encodes the Duffy blood cell antigen, Hemoglobin beta variant and variants in SLC4A1 an ion channel protein. Heterozygotes for a deletion in SLC4A1 have a condition referred to as red cell ovalocytosis associated with mild anemia but ovalocytosis apparently decreases susceptibility to *Plasmodium falciparum* malaria.

Selective variants reported to impact pathogen sensitivity include the following:

Gene/Protein	Function of gene product	Organism impacted by variants
CCR5	Chemokine receptor	HIV AIDS
TLR	Toll receptor	Flagellated bacteria
APOL1	Apolipoprotein L1	Trypanosomes
CYLD	CYLD lysine 63 deubiquitinase	Leprosy
TLR1	toll like receptor 1	Leprosy
Potassium channels	cAMP-activated/ chloride balance	Cholera
O blood group	glycosyl transferase	Increased sensitivity Cholera
NFKB	nuclear factor-kappa-B transcription factor	Cholera
FUT2	Fucosyl transferase 2	Noro virus
IFITMs	Interferon induced trans-membrane proteins	Influenza

Natural selection and infectious diseases

In 2014 Fumagalli and Sironi[62] reported that studies on immune function related genes had revealed evidence that pathogens had served as selective pressures during evolution. They also noted evidence that pathogen exerted selection often occurred in gene regulatory regions. They noted strong evidence in human species for selection of specific variants in the Toll like receptor genes. The cluster TLR10-TLR1-TLR6 showed particular evidence for selection. The Toll like receptors play fundamental role in pathogen recognition and activation of innate immunity. They also noted evidence for positive selection of variants in genes that encode specific molecular adapters of Toll signaling. These include RIG-1 like receptors; NOD like receptors that act as pattern recognition receptors and AIM2-like receptors with expression induced by interferon. Other loci with evidence of variant selection included:

MYD88	Innate immune signal transduction adaptor
SARM1	TIR motif containing (Toll interleukin receptor)
IFIHI	Interferon induced with helicase C domain 1
NLRP14	NLR family pyrin domain containing 1 apoptosis inducing
IFNG	Interferon gamma

Another interesting example of gene variant selection related to variants that were shared by two populations with different ancestries who inhabited the same specific geographic region and who were exposed to the same pathogen. Laayouni et al.[63] reported that Rroma and European population who inhabited the same geographic region where they were exposed to Bubonic plague (Yersinia pestis) epidemics shared specific variants in Toll receptor like genes.

Polygenic adaptations to pathogens

Daub et al.[64] reported that although much attention has been focused on individual genes with variants that have significant effects on human adaptation to pathogens, it is likely that cumulative effects of variants each of small effect, on many genes have influenced adaptation to pathogens. They emphasized that variants on gene that together operate within a specific pathway, may collectively act to influence adaptation.

Daub et al., used single nucleotide polymorphism data (SNPs) from the Human Diversity Panel that contains information on 660,918 SNPs in 53 populations and they classified data into gene set nodes. Each node corresponded to a specific functional pathway. Nodes were classified as significant if they harbored specific variants that showed evidence of positive selection in a specific population.

Daub et al., tested 1043 gene sets and evidence for positive selection was evident in 70 sets. Results of their studies revealed enrichment for positive selection in genes that were involved in the immune response. Immune related pathways of particular significance that emerged from their analyses included cytokine and cytokine receptor pathways known to be key factors in adaptive immune responses. This pathway and interferon play key roles in anti-mycobacterial immunity and in anti-viral signaling.

Another pathway containing positively selected variants included the genes known to carry variants that influence resistance to malaria, these included:

DARC (ACKR1)	atypical chemokine receptor 1 (Duffy blood group)
CRI	Complement C3b/C4b receptor 1
IFNG	Interferon gamma
CD40LG	CD40 ligand, regulates B cell function by engaging CD40 on the B cell surface
CD36	Glycoprotein, impacts adherence of *Plasmodium falciparum* parasitized erythrocytes
ICAM1	Intercellular adhesion molecule 1
HBB	Hemoglobin subunit beta
HBA1	Hemoglobin subunit alpha 1
TNF	Proinflammatory cytokine that belongs to the tumor necrosis factor (TNF) superfamily

Adaptation of organisms to different environments

Pritchard and di Renzo[65] noted that different mechanism have been proposed to account for the adaptation of organisms to different environments and to environmental changes. Many studies have provided evidence for "selective sweeps". This describes the situation where a specific mutation arises and because of the advantage it provides under specific environmental conditions, over time the mutation achieves high frequency in a specific population. They noted that examples include the Duffy blood group locus where the null allele Fy- has achieved high frequency in populations exposed to the *Plasmodium vivax* malaria parasite.

Polygenic adaptation

Pritchard and di Renzo[65] proposed that modest changes in allele frequencies at many loci may play roles in adaptation. They referred to this as polygenic adaptation and emphasized that this form of adaptation would include small differences in allele frequencies at multiple loci.

They emphasized the importance of identifying phenotypic variations in specific population groups exposed to different environments.

Evolutionary adaptations of microorganisms and their hosts

Hoal et al.[66] noted that there are reports of evolutionary changes in humans that influenced resistance to pathogens and that there is evidence of sequence changes in the tuberculosis bacillus Mtb.

With respect to the incidence of tuberculosis disease there is abundant evidence that specific environmental factors, particularly poverty and over-crowding, play key roles in causation.

The organism Mtb has existed and caused human disease over extended periods of human history. Hoal et al., noted that skeletal changes considered to be suggestive of Mtb infections have been found in skeletons in Egypt dated to between 3500 and 2650 BCE. And in skeleton found in Sweden and dated to between 3200 and 2300 BCE.

Hoal et al., proposed that as the human population increased in number from the Neolithic period on, and migrated to different sites expansion of MTB disease also occurred.

There is also evidence for evolution of Mtb organism Comas et al.[67] reported that 285 different strains of Mtb occurred in Ethiopia. Other clear evidence for evolution of Mtb comes from studies on the emergence of new strains of Mtb that are no longer sensitive to antibiotics that were previously found to be efficacious in controlling Mtb infections. One multidrug resistant strain that recently emerged in Mtb Beijing.

In a comprehensive review of gene variants associated with altered resistance to Tuberculosis van Tong et al.[68] reported that many of the identified genes impacted the immune function. Implicated genes encoded Toll receptors, Interferon gamma, Interleukin (IL17A, immunity related GTPase, killer cell immunoglobulin like receptor. Also implicated were variants in a NADPH oxidase gene and the AGMO gene that encodes alkylglycerol mono-oxygenase.

There is clear evidence that compromised immune function is major factor that led to a great increase in Tuberculosis infections in recent years. Higher rates of tuberculosis infections occur in individuals with HIV AIDS.

Human adaptations to high altitude

High altitude areas are technically defined as areas 2500 m above sea level. Population genetic studies have been carried out in key high-altitude areas, including the Qinhai Tibetan Plateau. The Andean Altiplano, and the Semian Plateau in Ethiopia.

Bigham[69] reviewed key physiologic adaptations that have been observed in these populations. Physiologic adaptations included altered pulmonary function, altered arterial oxidation (O_2) saturation levels and altered hemoglobin concentrations. Bigham compared physiologic changes across populations from the 3 high altitude areas and reported that only some of these were shared among the populations. Resting ventilation changes were found to be increased only in the Tibetan high- altitude population. Arterial O_2 saturation levels were increased in the Andean and Ethiopian high-altitude population levels. In the Tibetan high-altitude population, the arterial O_2 saturation levels were lower than average.

In addition, there are specific genes that show selected variants that are common to more than one of the 3 high altitude population. Variants in the Hypoxia inducible (HIF) pathway were of particular importance. HIF1 is a transcription factor that controls expression of a large number of genes.

Bigham reviewed and illustrated the genes that showed variants that were apparently selected by high-altitude conditions in the three different populations.

Shared genes and products with selected variants in Andes Altiplano and Tibetan plateau populations

ADRA1B	Adrenoreceptor B G protein coupled receptor
COPS4	COP Signalosome subunit, signaling pathway related
IGBP1, IGBP2	Immunoglobin binding B lymphocyte receptor signaling
ILI1A ILIB	Interleukins involved in hematopoiesis
IL6	Interleukin 6, involved in maturation of B cells
MDM2	Encodes a ubiquitin ligase involved in protein degradation
NOS2	Nitric oxide synthase 2
POLR2A	RNA Polymerase 2 (RNA synthesis)

Shared genes and products with selected variants in Andes Altiplano and Semian populations

ARNT2	Aryl-hydrocarbon receptor complexes with HIF
CXR4	Chemokine receptor
EDNRB	Endothelin receptor signaling protein
EGLN2	EGL9 family modifies HIF
KCNM1	Potassium activated calcium channel
MMP2	Matrix metallo- proteinase
SPRY2	Receptor tyrosine kinase inhibitor

References

1. Simeonov DR, Wang X, Wang C, Sergeev Y, et al. DNA variations in oculocutaneous albinism: an updated mutation list and current outstanding issues in molecular diagnostics. *Hum Mutat* 2013;**34**(6):827–35. https://doi.org/10.1002/humu.22315. Review. 23504663.

2. Quillen EE, Norton HL, Parra EJ, Lona-Durazo F, Ang KC, et al. Shades of complexity: new perspectives on the evolution and genetic architecture of human skin. *Am J Phys Anthropol* 2019;**168**(Suppl 67):4–26. https://doi.org/10.1002/ajpa.23737. 30408154.

3. Montoliu L, Grønskov K, Wei AH, Martínez-García M, et al. Increasing the complexity: new genes and new types of albinism. *Pigment Cell Melanoma Res* 2014;**27**(1):11–8. https://doi.org/10.1111/pcmr.12167. 24066960.

4. Yamaguchi Y, Hearing VJ. Melanocytes and their diseases. *Cold Spring Harb Perspect Med* 2014;**4**(5). pii: a017046. https://doi.org/10.1101/cshperspect. a017046. 24789876.

5. Waardenburg, P. A new syndrome combining developmental anomalies of the eyelids, eyebrows and nose root with pigmentary defects of the iris and head hair and with congenital deafness. Am J Hum Genet, 1951 3: 195–253. PubMed: 14902764.

6. Saleem MD. Biology of human melanocyte development, Piebaldism, and Waardenburg syndrome. *Pediatr Dermatol* 2019;**36**(1):72–84. https://doi.org/10.1111/pde.13713. 30561083.

7. Møller LB, Schönewolf-Greulich B, Rosengren T, Larsen LJ, et al. Development of hypomelanotic macules is associated with constitutive activated mTORC1 in tuberous sclerosis complex. *Mol Genet Metab* 2017;**120**(4):384–91. https://doi.org/10.1016/j.ymgme.2017.02.008. 28336152.

8. Jin Y, Mailloux CM, Gowan K, Riccardi SL, LaBerge G. NALP1 in vitiligo-associated multiple autoimmune disease. *N Engl J Med* 2007;**356**(12):1216–25. 17377159.

9. Crawford NG, Kelly DE, Hansen MEB, Beltrame MH, et al. Loci associated with skin pigmentation identified in African populations. *Science* 2017;**358**(6365). pii: eaan8433. https://doi.org/10.1126/science.aan8433. 29025994.

10. Valverde, P., Healy, E., Sikkink, S., Haldane, F., Thody, A. J., et al. The Asp84Glu variant of the melanocortin 1 receptor (MC1R) is associated with melanoma. Hum Mol Genet, 1996 5: 1663–1666. PubMed: 8894704.

11. Ségurel L, Bon C. On the evolution of lactase persistence in humans. *Annu Rev Genomics Hum Genet* 2017;**18**:297–319. https://doi.org/10.1146/annurev-genom-091416-035340. Review, 28426286.

12. Olds Lynne C, Eric S. Lactase persistence DNA variant enhances lactase promoter activity in vitro: functional role as a cis regulatory element. *Hum Mol Genet* 2003;**12**(18):2333–40. https://doi.org/10.1093/hmg/ddg244. 12915462.

13. Liebert A, López S, Jones BL, Montalva N, et al. World-wide distributions of lactase persistence alleles and the complex effects of recombination and selection. *Hum Genet* 2017;**136**(11−12):1445–53. https://doi.org/10.1007/s00439-017-1847-y. 29063188.

14. Metchnikoff I. *1908 from Nobel lectures, physiology or medicine 1901–1921.* Amsterdam: Elsevier Publishing Company; 1967.

15. Turvey SE, Broide DH. Innate immunity. *J Allergy Clin Immunol* 2010;**125**(2 Suppl 2):S24–32. https://doi.org/10.1016/j.jaci.2009.07.016. 19932920.

16. Beutler BA. TLRs and innate immunity. *Blood* 2009;**113**(7):1399–407. https://doi.org/10.1182/blood-2008-07-019307. Review, 18757776.

17. Condie ERM, Zak SJ, Good RA. Effect of meningococcal endotoxin on the immune response. *Proc Soc Exp Biol Med* 1955;**90**(2):355–60. 13273447.

18. Sultzer BM. Genetic control of leucocyte responses to endotoxin. *Nature* 1968;**219**(5160):1253–4. 4877918.

19. Hoffmann JA. *Nobel lecture: The host defense of insects: A paradigm for innate immunity.* Nobelprize.org. 28 November, 2013.

20. Kawai T, Akira S. The role of pattern-recognition receptors in innate immunity: update on Toll-like receptors. *Nat Immunol* 2010;**11**(5):373–84. https://doi.org/10.1038/ni.1863. Review, 20404851.

21. Jensen S, Thomsen AR. Sensing of RNA viruses: a review of innate immune receptors involved in recognizing RNA virus invasion. *J Virol* 2012;**86**(6):2900–10. https://doi.org/10.1128/JVI.05738-11. 22258243.

22. O'Neill LA, Golenbock D, Bowie AG. The history of toll-like receptors—redefining innate immunity. *Nat Rev Immunol* 2013;**13**(6):453–60. https://doi.org/10.1038/nri3446. Review, 23681101.

23. Dempsey A, Bowie AG. Innate immune recognition of DNA: a recent history. *Virology* 2015;**479-480**:146–52. https://doi.org/10.1016/j.virol.2015.03.013. 25816762.

24. Lässig C, Hopfner KP. Discrimination of cytosolic self and non-self RNA by RIG-I-like receptors. *Biol Chem* 2017;**292**(22):9000–9. https://doi.org/10.1074/jbc.R117.788398. 28411239.

25. Ozen S, Batu ED. The myths we believed in familial Mediterranean fever: what have we learned in the past years? *Semin Immunopathol* 2015;**37**(4):363–9. https://doi.org/10.1007/s00281-015-0484-6.

26. Swanson KV, Deng M, Ting JP. The NLRP3 inflammasome: molecular activation and regulation to therapeutics. *Nat Rev Immunol* 2019. https://doi.org/10.1038/s41577-019-0165-0. Review. 31036962.

27. Dryer WJ, Bennett JC. The molecular basis of antibody formation: a paradox. *Proc Natl Acad Sci USA* 1965;**54**:864.

28. Edelman GM. Antibody structure and molecular immunology. *Science* 1973;**180**(4088):830–40. 4540988.

29. Porter RR. Structural studies of immunoglobulins. *Science* 1973;**180**(4087):713–6. 4122075.

30. Tonegawa S, Brack C, Hozumi N, Matthyssens G, Schuller R. Dynamics of immunoglobulin genes. *Immunol Rev* 1977;**36**:73–94. 408266.

31. Köhler G, Milstein C. Continuous cultures of fused cells secreting antibody of predefined specificity. *Nature* 1975;**256**(5517):495–7. 1172191.

32. Janeway Jr CA. Pillars article: approaching the asymptote? Evolution and revolution in immunology. *Cold Spring Harb Symp Quant Biol* 1989;**54**:1–13. J Immunol. 2013;191(9):4475–87, 24141854.

33. Steinman RM, Cohn ZA. Identification of a novel cell type in peripheral lymphoid organs of mice. I. Morphology, quantitation, tissue distribution. *J Exp Med* 1973;**137**(5):1142–62. 4573839.

34. Bonilla FA, Oettgen HC. Adaptive immunity. *J Allergy Clin Immunol* 2010;**125**(2 Suppl 2):S33–40. https://doi.org/10.1016/j.jaci.2009.09.017.12. 20061006.

35. Chan K, Puck JM. Development of population-based newborn screening for severe combined immunodeficiency. *J Allergy Clin Immunol* 2005;**115**(2):391–8. https://doi.org/10.1016/j.jaci.2004.10.012. 15696101.

36. Paul WE. Bridging innate and adaptive immunity. *Cell* 2011;**147**(6):1212–5. https://doi.org/10.1016/j.cell.2011.11.036. 22153065.

37. Vieira VC, Soares MA. The role of cytidine deaminases on innate immune responses against human viral infections. *Biomed Res Int* 2013;**2013**:683095. https://doi.org/10.1155/2013/683095. 23865062.

38. Salter JD, Bennett RP, Smith HC. The APOBEC Protein Family: United by Structure, Divergent in Function. *Trends Biochem Sci* 2016;**41**(7):578–94. https://doi.org/10.1016/j.tibs.2016.05.001. Review, 27283515.

39. Gallo A, Vukic D, Michalík D, O'Connell MA, Keegan LP. ADAR RNA editing in human disease; more to it than meets the I. *Hum Genet* 2017;**136**(9):1265–78. https://doi.org/10.1007/s00439-017-1837-0. 28913566.

40. Haldane JB. The rate of mutation of human genes. *Hereditas* 1949;**1949**(35):267–73.

41. Allison AC. The distribution of the sickle-cell trait in East Africa and elsewhere, and its apparent relationship to the incidence of subtertian malaria. *Trans R Soc Trop Med Hyg* 1954;**48**(4):312–8 PMID:13187561.

42. Mangano VD, Modiano D. An evolutionary perspective of how infection drives human genome diversity: The case of malaria. *Curr Opin Immunol* 2014;**30**:39–47. https://doi.org/10.1016/j.coi.2014.06.004. Review, 24996199.

43. Ackerman H, Usen S, Jallow M, Sisay-Joof F, et al. A comparison of case-control and family-based association methods: the example of sickle-cell and malaria. *Ann Hum Genet* 2005;**69**(Pt 5):559–65. https://doi.org/10.1111/j.1529-8817.2005.00180.x. 16138914.

44. Kwiatkowski DP. How malaria has affected the human genome and what human genetics can teach us about malaria. *Am J Hum Genet* 2005;**77**(2):171–92. Review, https://doi.org/10.1086/43251916001361.

45. Travassos MA, Coulibaly D, Laurens MB, Dembélé A, et al. Hemoglobin C trait provides protection from clinical falciparum malaria in Malian children. *J Infect Dis* 2015;**212**(11):1778–86. https://doi.org/10.1093/infdis/jiv308. 26019283.

46. Modiano G, Morpurgo G, Terrenato L, Novelletto A, et al. Protection against malaria morbidity: near-fixation of the alpha-thalassemia gene in a Nepalese population. *Am J Hum Genet* 1991;**48**(2):390–7. 1990845.

47. Williams TN, Mwangi TW, Wambua S, Peto TE, et al. Negative epistasis between the malaria-protective effects of alpha+−thalassemia and the sickle cell trait. *Nat Genet* 2005;**37**(11):1253–7. https://doi.org/10.1038/ng1660. 16227994.

48. Flint J, Hill AV, Weatherall DJ, Clegg JB, Higgs DR. Alpha globin genotypes in two north European populations. *Br J Haematol* 1986;**63**(4):796–7. 3730299.

49. Zimmerman PA, Woolley I, Masinde GL, Miller SM, et al. Emergence of FY*a(null) in a plasmodium vivax-endemic region of Papua New Guinea. *Proc Natl Acad Sci U S A* 1999;**96**(24):13973–7.

50. Zimmerman PA, Ferreira MU, Howes RE, Mercereau-Puijalon O, et al. Red blood cell polymorphism and susceptibility to plasmodium vivax. *Adv Parasitol* 2013;**81**:27–76. https://doi.org/10.1016/B978-0-12-407826-0.00002-3. 23384621.

51. Rosanas-Urgell A, Lin E, Manning L, Rarau P, et al. Reduced risk of plasmodium vivax malaria in Papua new Guinean children with southeast Asian ovalocytosis in two cohorts and a case-control study. *PLoS Med* 2012;**9**(9):e1001305. https://doi.org/10.1371/journal.pmed.1001305. 22973182.

52. Bienzle U, Ayeni O, Lucas AO, Luzzatto L. Glucose-6-phosphate dehydrogenase and malaria. Greater resistance of females heterozygous for enzyme deficiency and of males with non-deficient variant. *Lancet* 1972;**1**(7742):107–10. 4108978.

53. Ruwende C, Khoo SC, Snow RW, Yates SN, et al. Natural selection of hemi- and heterozygotes for G6PD deficiency in Africa by resistance to severe malaria. *Nature* 1995;**376**(6537):246–9. https://doi.org/10.1038/376246a0. 7617034.

54. Hedrick PW. Resistance to malaria in humans: the impact of strong, recent selection. *Malaria J* 2012;**11**:349. https://doi.org/10.1186/1475-2875-11-349. 23088866.

55. Capewell P, Cooper A, Clucas C, Weir W, Macleod A. A co-evolutionary arms race: Trypanosomes shaping the human genome, humans shaping the trypanosome genome. *Parasitology* 2015;**142**(Suppl 1):S108–19. https://doi.org/10.1017/S0031182014000602. 25656360.

56. Hajduk SL, Moore DR, Vasudevacharya J, Siqueira H, Torri AF. Lysis of Trypanosoma brucei by a toxic subspecies of human high density lipoprotein. *J Biol Chem* 1989;**264**(9):5210–7. 2494183.

57. Karlsson EK, Kwiatkowski DP, Sabeti PC. Natural selection and infectious disease in human populations. *Nat Rev Genet* 2014;**15**(6):379–93. https://doi.org/10.1038/nrg3734. 24776769.

58. Deschamps M, Laval G, Fagny M, Itan Y, et al. Genomic signatures of selective pressures and introgression from archaic hominins at human innate immunity genes. *Am J Hum Genet* 2016;**98**(1):5–21. https://doi.org/10.1016/j.ajhg.2015.11.014. 26748513.

59. Andersen KG, Shylakhter I, Tabrizi S, Grossman SR, Happi CT, Sabeti PC. Genome-wide scans provide evidence for positive selection of genes implicated in Lassa fever. *Philos Trans R Soc Lond B Biol Sci* 2012;**367**(1590):868–77. https://doi.org/10.1098/rstb.2011.0299. 22312054.

60. Boelaert JR, Vandecasteele SJ, Appelberg R, Gordeuk VR. The effect of the host's iron status on tuberculosis. *J Infect Dis* 2007;**195**(12):1745–53. https://doi.org/10.1086/518040. 17492589.

61. Quintana-Murci L, Clark AG. Population genetic tools for dissecting innate immunity in humans. *Nat Rev Immunol* 2013;**13**(4):280–93. https://doi.org/10.1038/nri3421. 23470320.

62. Fumagalli M, Sironi M. Human genome variability, natural selection and infectious diseases. *Curr Opin Immunol* 2014;**30**:9–16. https://doi.org/10.1016/j.coi.2014.05.001. Review, 24880709.

63. Laayouni H, Oosting M, Luisi P, Ioana M, et al. Convergent evolution in European and Rroma populations reveals pressure exerted by plague on toll-like receptors. *Proc Natl Acad Sci USA* 2014;**111**(7):2668–73. https://doi.org/10.1073/pnas.1317723111. 24550294.

64. Daub JT, Hofer T, Cutivet E, Dupanloup I, Quintana-Murci L, et al. Evidence for polygenic adaptation to pathogens in the human genome. *Mol Biol Evol* 2013;**30**(7):1544–58. https://doi.org/10.1093/molbev/mst080. 23625889.

65. Pritchard JK, Di Rienzo A. Adaptation—not by sweeps alone. *Nat Rev Genet* 2010;**11**(10):665–7. https://doi.org/10.1038/nrg2880. 20838407.

66. Hoal EG, Dippenaar A, Kinnear C, van Helden PD, Möller M. The arms race between man and mycobacterium tuberculosis: time to regroup. *Infect Genet Evol* 2018;**66**:361–75. https://doi.org/10.1016/j.meegid.2017.08.021. 28843547.

67. Comas I, Hailu E, Kiros T, Bekele S, Mekonnen W, et al. Population genomics of *Mycobacterium tuberculosis* in Ethiopia contradicts the virgin soil hypothesis for human tuberculosis in Sub-Saharan Africa. *Curr Biol* 2015;**25**(24):3260–6. https://doi.org/10.1016/j.cub.2015.10.061. 26687624.

68. van Tong H, Velavan TP, Thye T, Meyer CG. Human genetic factors in tuberculosis: an update. *Trop Med Int Health* 2017;**22**(9):1063–71. https://doi.org/10.1111/tmi.12923. 28685916.

69. Bigham AW. Genetics of human origin and evolution: high-altitude adaptations. *Curr Opin Genet Dev* 2016;**41**:8–13. https://doi.org/10.1016/j.gde.2016.06.018. 27501156.

5

SIGNALS, EPIGENETICS, REGULATION OF GENE EXPRESSION

Conveying signals into cells to modify gene expression

Signaling pathways

The first step involves the binding of an activator molecule to a specific receptor on the cell surface. The step that follows requires the activation of transducers. The transducers are proteins coupled to guanosine nucleotide molecules designated as G proteins The G protein molecule GTP Guanosine tri-phosphate can then phosphorylate components of intra-cellular signaling pathways. There is evidence that early steps in the signaling pathway activations involve conversion of proteins in the RAS kinase family, RASGDP to RASGTP through the activity of Ras guanine nucleotide exchange factor. RAS GTP can then phosphorylate and activate proteins in downstream signaling pathways. Gilman and Rodbell[1] were awarded a Nobel Prize in 1994 for their work in describing the steps in transmission of signal from the cell surface to the interior (Fig. 1).

The MAPK pathway (mitogen activated signaling pathway) is a key downstream pathway. It is comprised of several families of related proteins Yang et al.[2] described major MAPK signaling pathways in mammals. These include ERK, extra-cellular signal regulated kinase, JNK proteins that respond to extra-cellular stress stimuli and MAPK kinases (MAP2K1, MAP2K2).

Anchoring proteins in the cytoplasm may serve to sequester the MAPK proteins under conditions when receptors are not activated, Yang et al. noted that the activated MAP kinases can phosphorylate transcription factors that can then move to the nucleus. There is also evidence that MAPkinases can directly enter the nucleus. MAPKs can also bind to coregulatory complexes prior to entering the nucleus. MAPKs can also enter organelles in the cytoplasm, e.g. Golgi,

Gene Environment Interactions. https://doi.org/10.1016/B978-0-12-819613-7.00005-0

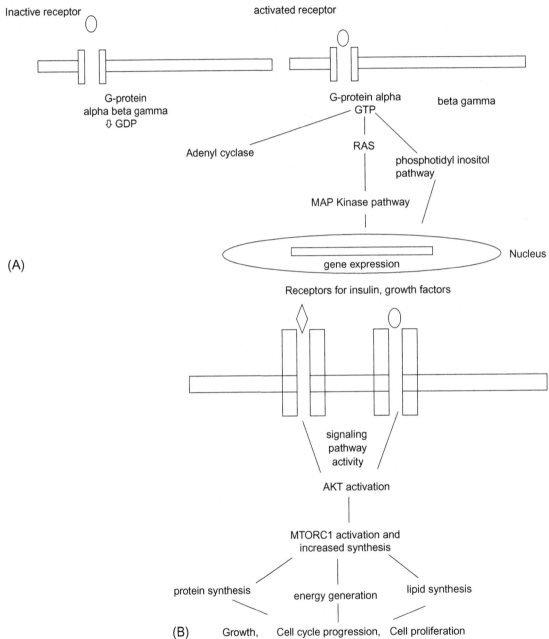

Fig. 1 (A) *Signaling pathway to gene expression*. G protein coupled receptors are activated by binding of ligands. Activation can trigger signaling in different signaling pathways to activate gene expression. (B) *MTORC1 activation and growth*. Binding of insulin and growth factors to specific receptors triggers signaling and downstream activation of MTORC1 complex that in turn activates energy generation, protein and lipid synthesis and promotes cell growth and proliferation.

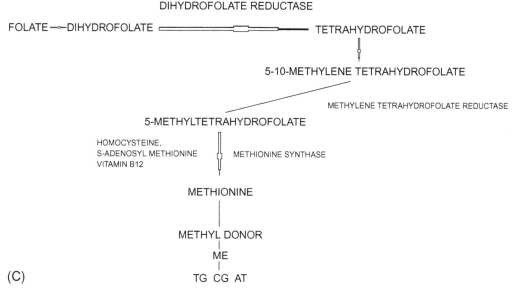

Fig. 1, Cont'd (C) Epigenetics CpG methylation. Folate, Vitamin B12 and S-adenosyl methionine play key roles in the generation of methyl donors that can bind cytosine guanine dinucleotides in DNA and alter gene expression.

endoplasmic reticulum and mitochondria and can influence activity of specific proteins in those organelles.

Yang et al. also noted that within the nucleus MAPkinases can facilitate reactions involved in chromatin modification.

The phosphatidyl inositol kinase signaling pathway first described by Cantley in 2002[3] is activated when receptor activation and G-protein activity lead to phosphorylation of the lipid phosphatidylinositol. Several different classes of phosphatidyl inositol kinases exist. These molecules go through successive stages of phosphorylation eventually leading to the generation of phosphatidylinositol-3-4-5 phosphate PI3K. Phosphatidyl inositol kinases can activate the enzyme AKT serine threonine kinases. The PI3K AKT pathway is responsible for activation of key components in metabolism and cell functions including MTOR.

Ion channels and transmission of signals to alter gene expression

Yeh and Parekh[4] reported on the role of calcium ions in conveying information from the cell surface to the nucleus to modify gene expression. They reviewed these processes in relation to neurons and immune cells.

In these two cell types increases in intra-cellular calcium ion concentrations occur predominantly through two different mechanisms and in the two cell types the downstream effects of increases in calcium levels differ.

Voltage gated ion channels in the outer membrane of neuronal cells facilitate entry of calcium ions. Yeh and Parekh reported that this calcium then leads to activation of calcium calmodulin kinases. The activated kinases phosphorylate the transcription factor CREB (cyclic AMP response element binding protein).

They report that the rise in intra-cellular calcium in cells of the immune system takes place primarily through release of calcium from intra-cellular stores, e.g. in the endoplasmic reticulum. Store operated calcium release takes place through calcium release channels, known as CRAC channels. The calcium thus released activates calcineurin, also known as protein phosphatase 3 catalytic subunit alpha, that leads to dephosphorylation of NFAT transcription factors (nuclear factors of activated T cells). Dephosphorylated NFAT can then enter the nucleus.

Yeh and Parekh noted that stimulation of immune cells through increased calcium levels also took place from external sources and that these cells have a number of different calcium entry cells. However, in immune cells the influx of calcium is not mediated by voltage changes.

Detailed studies of the endoplasmic reticulum have revealed that specific proteins act as sensors of calcium levels in the endoplasmic reticulum. *STIM1* and *STIM2* genes encode transmembrane proteins that mediates Ca^{2+} influx after depletion of intracellular Ca^{2+} stores by gating of store-operated Ca^{2+} influx channels (SOCs), the CRAC channels. The STIM1 and 2 proteins function to regulate calcium concentrations in the cytosol and endoplasmic reticulum, and are involved in the activation of plasma membrane Orai Ca (2+) entry channels (calcium release-activated calcium modulators). The ORAI1 channel is the primary way for calcium influx into T-cells.

The rise in intra-cellular calcium activates NFAT transcription factors and other cytoplasmic proteins can enter the nucleus where they can also activate the AP1 transcription factor subunit.

It is important to note that calcium can also enter cells through TRPC1 and TRPC3 channels transient receptor potential cation channel subfamily C members.

In a review in 2016 Liu et al.[5] emphasized the key role of signal transduction pathways in converting environmental stimuli to alterations in cell function. They also concentrated on altered signaling that occurs in cancer. And emphasized that insights into altered signaling in cancer have given rise to a number of useful therapies. Earlier studied revealed that signaling pathway activation in cancer was primarily associated with altered kinase activity and altered phosphorylation of transcription regulators. The altered phosphorylation was initially thought to be primarily due to kinases that occurred in the cytoplasm that then impacted transcription factors. Liu et al. noted however the certain kinases, including JAK2 and AKT were subsequently shown to also be active in the nucleus and to be active in histone modification.

MTORC1 and MRNA translation

MTOR a protein kinase, plays particularly important roles as a downstream sensor of signaling pathway activities. MTOR exists in two complexes MTORC1 and MTORC2. MTORC1 activity phosphorylates the binding protein, EIF4EBP1 and releases it from EIF4E in this way translation of MRNAs to protein is facilitated.

MTORC1 and MRNA translation

MTOR a protein kinase, plays particularly important roles as a downstream sensor of signaling pathway activities. MTOR exists in two complexes MTORC1 and MTORC2. MTORC1 phosphorylates and activates S6 kinases. S6K1, S6K2. Between 1997 and 1999 the connection between MTOR and S6 kinase S6K1 was established.[6]

Magnuson et al.[7] reviewed regulation and function of S6 kinases. He reported that S6 kinases were first identified in the 1970's as 40S and 90S ribosomal bound proteins that were produced in response to mitogenic stimulation of cells. In 1995 the compound Rapamycin was found to inhibit phosphorylation and activity of S6 kinases. In 1996 Rapamycin was shown to inhibit phosphorylation of EIF4EBP1 and in this way the connection of Rapamycin to mRNA translation and protein synthesis was established.

The MTORC1 facilitated phosphorylation of EIF4EBP1 promotes its release from eukaryotic translation initiation factor 4E (EIF4E) enables interaction of other EIF4E with other components of the translation machinery and initiation of translation and protein synthesis. In their 2012 review Magnusson et al. noted that there is clear evidence that protein synthesis is co-ordinated through activity of MTORC1 S6K1 and MTORC1 and EIF4EBP1.

Transcription factors

Transcription factors within the cytoplasm were reported to have a signaling domain and can be activated by binding of a ligand or they may be activated by phosphorylation.[8] When they pass into the nucleus transcription factors recognize specific sequences in DNA and they serve to modify chromatin and facilitate interaction of DNA with RNA polymerase to transcribe DNA into RNA.

Vaquerizas et al. in 2009[9] documented data on 1291 human transcription factors and classified these into 23 families. Transcription factors can be classified according to specific domains they contain. Transcription factors with zinc finger domain are most common, other common domains present include homeodomains and helix loop helix domains. Transcription factors are also classified according to the DNA sequences to which they bind.

In 2018 Lambert et al.[10] reviewed transcription factors. They reported that approximately 8% of human genes encode transcription factors. They noted the precise binding sites of all transcription factors have not yet been clearly defined. In addition, the precise mechanisms through which transcription factors function have not yet been clearly defined.

Lambert et al. noted that approximately three-quarters of the human transcription factors have defined DNA binding domains. However, some transcription factors may recognize a specific DNA sequence and closely related sequences.

They noted that some transcription factors can directly bind to promoter sequences and recruit RNA polymerase while others require accessory factors to recruit RNA polymerase. The transcription factor co-factor interactions are reported to be transient.

It is important to note that in addition to MRNA derived from protein coding genes, long non-coding RNAs are transcribed from the genome. Transcription factors involved in their production are under investigation.

An et al.[11] reported establishment of a database LIVE that documents interactions of transcription factors, generation of long non-coding RNAs (Lnc RNAs). The goal of this database is also to facilitate exploration of the regulatory roles on LncRNAs.

NFkappaB transcription factor complex

In 2017 Zhang et al.[12] reviewed information generated during the previous 30 years on the function of NFkappaB (NFKB). NFKB was first reported in 1986 as a protein that bound a specific sequence in DNA. It was found to bind to DNA with the specific sequence near the kappa light chain immunoglobulin gene in B lymphocytes. Sen and Baltimore reported that the binding of NFKB to the DNA at specific sites impacted B cell development. Furthermore, NFKB binding was shown to initiate rapid response activation of immunoglobulin production in response to infection.

Subsequently NFKB was shown not to be a single protein but to be a complex composed of different subunits encoded on different chromosomes. Specific subunits were derived from the NFKB1 gene on human chromosome 4q24. A gene on chromosome 10q24.32, NFKB2 can also contribute subunits to the complex. Additional subunits are derived from REL transcription family genes, including REL on 2p16.1 RELA on 11p13.1 and RELB on 19q13.32. Subsequently the NFKB protein complex was found to bind to hundreds of target genes. The DNA sequence to which NFKB multiprotein complex bound was defined as 5′ GGGRNWYYCC3′ where R is a purine base, N is any base, W is adenine or thymine and Y is a pyrimidine.

Zhang et al.[12] reported that NFkappaB B is primarily located in the cytoplasm in unstimulated cells. The cytoplasmic location was found to be due to the binding of specific inhibitors. IKBKB (inhibitor of Nuclear factor kappa B NFKB) and IKBKA (CHU) ubiquitous kinase, and IKBKG (also known as NEMO). Phosphorylation of these proteins, particularly of NEMO could promote their destruction through a ubiquitin process.

Following destruction of the inhibitory complex NFKB can become active can move to the nucleus and can activate transcription of thousands of genes, many of these genes encode proteins involved in fighting infection.

Yang et al.[13] established a database of N FKB target genes. Categories included were genes that encode cell adhesion proteins, cell surface receptor, growth factors immunoreceptors, proteins involved in antigen presentation, stress response genes, transcription factor encoding genes and microRNA genes.

In 2006 Courtois and Smahi[14] reviewed NFKB related genetic diseases. They emphasized that there are multiple members of the complex defined as the RELNFKB family. They also documented the multiple components involved following activation of the Tumor Necrosis Factor (TNFR) receptor) the Interleukin receptor ILR) and the Toll receptors (TLRs). The molecules that bind to and activate these receptors are derived from micro-organisms and from factors released during inflammation. Downstream of these receptors are receptor related kinases.

Key factors in the upper reaches of the NFKB pathway include TRADD (TNFR associated), TRAF2 inhibitor of apoptosis, IRAK1 (interleukin receptor associated kinase) TAB1 and TAB2 (TGF associated kinases and TAK1 (MAP3K7) kinase.

Specific factors that can induce release of NFKB from the inhibitors include proinflammatory cytokines. These factors lead to modifications, primarily phosphorylation of the inhibitory proteins. NF Kappa B activity can also be induced by factor released from B and T lymphocytes CARD11-BCL10—MALT1. Zhang et al. reported that NFKB plays a particularly important role in immune function.

The NFKB transcription factor can induce expression of hundreds of genes because of the versality of DNA elements to which it can bind.

Pathologies associated with defective functions of these proteins

Incontinentia pigmenti is a skin disorder associated with skin blistering, development of wart-like hyperkeratotic lesions and inflammatory responses. Altered pigmentation may appear at the site of healed lesions. Neurological problems may be present. The gene locus for Incontinentia pigmenti was mapped to chromosome Xq28 and the disease-causing defect was discovered to most commonly be due to structural defects in NEMO; in some cases, nucleotide mutations were present in the gene. Mutations in NEMO may also lead to Incontinentia pigmenti. Other mutations in Nemo lead to a disease associated with impaired ability to sweat, anhidrotic ectodermal dysplasia with immunodeficiency.

Defects in TRAF2 TNF receptor associated factor 2 which activates the NFKB pathway may lead to cylindromatosis and tricoepithelioma, disorders associated with benign tumors and lesions of hair follicles and sweat glands.

Defects in the IRAK kinases have been found in some cases of immunodeficiency.

Nuclear receptors

Nuclear receptors are ligand activated transcription factors. The ligands include lipid soluble molecules, such as retinoic acid and steroid hormones, including estrogen and progesterone. These ligands can cross the plasma membrane of the cell. In the cytoplasm specific ligands bind to specific transcription factors known nuclear receptors within the cytoplasm. Mangelsdorf[15] reported that 48 nuclear receptors are encoded in the human genome and that different nuclear receptors share a common structure.

Sever and Glass[16] noted that nuclear receptors function as transcription factors but also have additional functions within the cytoplasm, Within the cytoplasm the nuclear receptors are bound to HSP90 chaperone proteins. Upon binding of ligand to the nuclear receptor, these receptors are freed from HSP90 binding.

Nuclear receptors with bound ligand then cross from the cytoplasm to the nucleus. Within the nucleus the nuclear receptor then interacts with cofactors and then binds to DNA to activate expression of specific genes. An important feature of nuclear receptors is that a specific nuclear receptor can activate different genes in different cell types.

The nuclear pore complex

In recent decades details have emerged regarding the complex structure of the nuclear pore complex that serves as a passage of exchange between the cytoplasm and the nucleus. Knockenhauer and Schwartz[17] reviewed these structures that exist at junctions where inner and outer nuclear membranes fuse and form circular opening. The structure of the nuclear pore complexes has been revealed by electron microscopy and proteomic analyses have also been carried out on isolated pore complexes. These studies have revealed that each nuclear pore complex is composed of 456 different proteins that include 34 distinct nucleoporins.

The structural components of the nuclear pore complex include an outer cytoplasmic region and cytoplasmic filaments, a cytoplasmic ring, a central region, a nuclear ring and a nuclear basket. In addition, specific filaments extend from the cytoplasmic ring, through the central region and the nuclear ring to the nuclear basket.

Oka and Yoneda[18] reviewed aspects of transport between cytoplasm and nucleus. They noted that in eukaryotes the transport of macromolecules through the nuclear pore requires the GTPase RAN system, and nuclear transport receptor protein, importin.

Proteins with a nuclear targeting signal are recognized by an adaptor molecule importin a and transport of the molecule through the nuclear pore is facilitates by binding of importin a to importin b and interaction with RANGTP. After translocation through the nuclear pore is complete the RANGTP is converted to RAN GDP through the activity of RAN-GTPase and the protein is released into the nucleus.

Transport of mRNA from the nucleus to the cytoplasm

Transport of messenger RNA from the nucleus to the cytoplasm also takes place through the nuclear pores. The mRNA is packaged into mRNA-protein complexes for transport. Stewart[19] reported that linking of mRNA to transporters requires accessory proteins and that export is an energy requiring process that is provided by RANGTP.

Stewart noted that important checkpoint processes are present in the nucleus to ensure that only mature messenger RNA is transported from nucleus to cytoplasm. Splicing needs to be completed and polyadenylation of mRNA need to be completed.

Epigenetics

Epigenetic mechanisms

Holiday and Pugh[20] first proposed that DNA methylation could alter gene expression. Initial studies on DNA methylation focused on specific changes in DNA methylation patterns during early development. Subsequent studies revealed that DNA methylation includes primarily methylation of cytosine at the 5th position and in the CpG context Dense islands of CpG residues are referred to as CpG islands. CpG islands frequently occur close to transcription start sites of genes. CpG islands close to transcription sites were shown to be unmethylated in expressed genes.[21]

DNA methylation represents one of the mechanisms through which one X chromosome is inactivated in humans. DNA methylation also plays a role in maintenance of imprinting in specific genomic regions; in imprinted regions genes are expressed from one member of a homologous chromosome pair, i.e. only from the maternally derived chromosome or only from the paternally derived chromosome.[22]

The term epigenetics refers to molecules, biochemical processes and mechanisms that modify DNA and histone proteins that together constitute chromatin in the nuclei of cells. Modifications of DNA and chromatin are key factors in control of gene expression. In addition to molecules that modify histones and DNA there are protein complexes that act to modify the structure and architecture of chromatin and to thereby impact gene expression. Chromatin is composed of DNA strands that are wrapped around nucleosomes that are composed of histones. In addition, there are histones that surround DNA between the nucleosomes, these are referred to as linker histones. Nucleosomes are composed of core histones H2A, H2B, H3 and H4. H1 histones form the links between the nucleosomes. Multiple copies of histone genes are present in the human genome. Exchange of histones within chromatin also occurs to replace degraded histones.[23]

Key to chromatin modifications are enzymes that add modifiers, e.g. methyl, phosphate and acetyl groups to histone tails and enzymes that remove these modifiers when appropriate. Specific enzymes modify DNA primarily through addition of methyl groups to cytosine residues in DNA. The secondary modifications of DNA and histones impact gene expression, DNA and histone modification are influenced in part by environmental factors including nutrient availability, including availability of vitamins such as folic acid and vitamin B12.

Key elements that impact chromatin function are referred to as writers, that induce chromatin modifications, readers of the modifications include specific DNA and histone binding proteins and erasers, enzymes that remove the DNA and chromatin modifications.[24]

Chromatin remodeling is carried out by multi protein complexes that derive energy from ATP to move nucleosomes. These complexes include the SWI/SNF complexes known in mammals as BAF complexes that act to move nucleosomes and to generate regions of open chromatin that facilitate gene transcription. Polycomb proteins are proteins that modify chromatin at specific sites and bring about epigenetic changes so that those gene are silenced. Simon and Kingston[25] reported that hundreds of genes are silenced by binding of polycomb proteins, sometimes referred to as polycomb repressive complexes. They reported that genes that normally encode products important in development are later silenced through activities of polycomb proteins. Teschendorff et al.[26] reported that polycomb gene targets were more likely to undergo hypermethylation with advancing age. Rakyan et al.[27] carried out a genome wide study of DNA methylation and they reported that aging was particularly associated with hypermethylation at genes that played a key role in development.

These genes are often referred to as polycomb gene targets. The PRC2 complex promotes methylation of lysine 27 in histone H3 and this serves to reduce gene transcription.[28] A particularly important component of PRC2 is a catalytic subunit EZH2.

Epigenetics and aging

Horvath[29] analyzed dataset of samples of different tissues that had been examined using the Illumina DNA methylation array. He developed an ANOVA based analysis of variances in methylation in different tissues at different ages. These studies led to development of a DNA methylation- based predictor of age.

Horvath and Raj[30] described a parameter referred to as the age correlation, the Pearson correlation coefficient between methylation age and age in years. They reported that analyses of the degree of methylation of DNA provided an accurate estimate of the biological aging across the life span. They noted that the existence of epigenetic clock potentially provides insights into the underlying biological processes in aging.

Questions arise as to the specific genomic location that undergo age related methylation. In the Horvath initial study,[29] 353 CpG sites were analyzed, 193 of these showed correlation with chronological age.

Following further studies on many individuals and more extensive data analysis. Horvath and Raj[30] determined that outlier individuals occurred, in these individuals, DNA methylation epigenetic age and chronological age differed. Some individuals manifested age acceleration. Other manifested age deceleration. Horvath and Raj proposed that age acceleration indicated that the increased CpG methylation indicated that the underlying tissue aged more rapidly than expected.

Questions arise as to how epigenetic change and function are correlated. Horvath suggested that the increased methylation with age was possibly beneficial at sites where DNA damage has occurred, and the silencing of these sites prevents generation of abnormal transcripts from mutated DNA.

Accelerated aging in down syndrome role of methylation

Horvath et al.[31] utilized quantitative methylation analysis to study individuals with trisomy 21 (Down syndrome). Their studies revealed that trisomy 21 lead to accelerated aging of approximately 6.6.

Two genomic markers significantly associated with age acceleration occurred on chromosome 21. Horvath et al. did not however have transcription data to investigated whether the altered methylation of these markers influenced transcription of nearby genes.

Horvath and Raj noted that the methylation DNA age of somatic cells can be reduced by treatment of these cells with the Yamanaka factors, i.e. factors that can reverse differentiated cells to pluripotent cells. They proposed that DNA methylation increases consistent with chronological age may reflect the decline of stem cells in a particular tissue.

Edwards et al.[32] reviewed the different genomic compartments in which CpG methylation occurs. They noted that 3×10^7 CpG dinucleotides exist in the human genome. Dense methylation was found to occur in regions of the genome that harbored remnants of ancient retroviral integrations, Sines, Lines and long terminal repeat (LTRs). They noted that variable methylation occurs throughout the genome. Promoter CpG islands and first exons were found to most frequently be unmethylated.

The DNA methyltransferase enzymes play key roles in methylation of DNA and pathogenic mutations in the genes that encode these enzymes have been shown to give rise to specific developmental disorders.

Removal of methylation from DNA also occurs through activity of TET enzymes. These enzymes recognize 5-methylcytosine and bring about interactive oxidation and demethylation reactions.

Epigenetics, nutritional resources and metabolism

Methyl groups for DNA methylation, histone methylation and methyl requiring metabolic reactions are generated in the one carbon metabolism. Key nutrients required for the one carbon metabolism include folate (vitamin B9), cobalamin (Vitamin B12), pyridoxal phosphate (Vitamin B6), betaine and choline. Ducker and Rabinowitz[33] noted that folates include a series of molecules that share a pteridine ring, para-aminobenzoic acid and polyglutamate. Tetrahydrofolate that has a reduced pteridine is the most biologically active form of folate in humans. The form of folate that is present in dietary supplement is folic acid that requires modification in the body to generate dihydrofolate or tetrahydrofolate.

In a review of the one carbon metabolism, Friso et al.[34] emphasized that since epigenetic processes are influenced by environmental
factors including nutrition, these factors can therefore influence gene
expression.

Folate synthesis occurs in microorganisms and plants; however,
humans require nutrients that supply folate. Agents that inhibit folate
metabolism are used in cancer chemotherapy since folate inhibition
impairs cell proliferation and growth. Folate is essential for the one
carbon metabolism and the one carbon metabolism plays essential toles for purine and thymidine synthesis and for homocysteine
metabolism.

The first reaction in the one carbon cycle involved the transfer
of a one carbon unit CH3 from serine to tetrahydrofolate through
activity of the enzyme serine hydroxymethyltransferase with pyridoxal phosphate (vitamin B6) a cofactor. This reaction generates
5–10-methylenetetrahydrofolate.

The next reaction converts 5–10-methylenetetrahydrofolate to
5-methyltetrahydrofolate through activity of the enzyme methylene
tetrahydrofolate reductase (MTHFR) that requires as cofactor flavin
adenine dinucleotide (Vitamin B2).

Methyltetrahydrofolate serves as methyl donor to convert homocysteine to methionine. This reaction requires Vitamin B12 as cofactor. Homocysteine can also me methylated in a reaction that utilizes
methyl groups from choline and betaine.

In summary, folate (vitamin B9) accepts methyl groups from serine
and through a series of reactions generates the essential amino acid
methionine.

There is evidence that folate depletion decreases DNA methylation. Maternal folate depletion in pregnancy is known to lead to neural
tube defects in infants.[35] Friso et al.[34] noted that the levels of folate in
pregnant women were reported to influence the degree of methylation
of the imprinted gene IGF2. There are also reports that Vitamin B12
deficiency can lead to DNA hypomethylation. Friso et al. noted that
there is evidence that folate and methionine deficiency in human can
also reduce levels of histone methylation.

Choline and betaine influence one carbon metabolism since
they can donate methyl groups that convert homocysteine to methionine. Studies in mice revealed that dietary deficiency of choline altered methylation in the mouse fetal brain.[36] Pauwels et al.[37] reported
that choline deficiency can impact hydroxymethylation in humans.
5-hydroxymethylcytosine was reported to be abundant in the nervous
system particularly in the cortex and cerebellum and the post-natally
levels of hydroxymethyl cytosine increase in brain.[38]

Importantly there is evidence, based on studies in rat that supplementation fo diet with choline during pregnancy reduced the damaging impact of alcohol on neurons.[39]

DNA methylation and environmental exposures

In 2017 van der Plaat et al.[40] reported results of a DNA methylation study on individuals in the Dutch population: they noted that during the course of their work agricultural workers can be exposed to pesticides insecticides and herbicides. The study they designed included 1392 individuals with no history of pesticide exposure, 108 individuals with low pesticide exposure and 61 individuals with high pesticide exposure. DNA was isolated from each individual and analyzed using the Illumina 450 bead chip microarray to analyze CpG methylation at 420,938 sites in the genome.

In their analyses van der Plaat et al. stratified analyses by gender smoking status and pesticide exposure. The also obtained data on differential white blood cell counts. Results revealed that high pesticide exposure was associated differential methylation that impacted CpG sites related to 29 genes. Women who had pesticide exposure were found to have increased methylation at the NKAIN3 gene locus, sodium/potassium transporting ATPase interacting 3.

Pesticide exposure and smoking were found to interacts to impact CpG methylation. A CpG methylation site that impacted expression at 3 loci was found to show high methylation following pesticide exposure in smokers. Mostafalou et al.[41] and deJong et al.[42] had previously reported that pesticide exposure was associated with progressive impairment of lung function.

Several of the genes found in the van der Plaat study to have altered methylation related to pesticide exposure had previously been shown to have altered expression in individuals exposed to pesticides or in individuals with altered lung function. These genes included:

RYR1	Ryanodine receptor 1 encoded protein functions as a calcium release channel
ALLC	Allantoicase functions in the uric acid metabolic pathway
PTPRN2	Protein tyrosine phosphatase receptor type N2
LRRC3B	Leucine rich repeat containing 3B, encodes a tumor suppressor
PAX2	Paired Box 2 encodes a transcription factor
VTRNA2-1	Vault RNA2-1 This gene produces an RNA polymerase III transcript
VTRNA2-1	alterations were particularly found in association with smoking

Epigenetic status and behavior following early childhood deprivation

Naumova et al.[43] reported results of studies carried out to investigate the impact of social and parental deprivation in young children. They analyzed physical and mental health, cognition and behaviors and in addition they analyzed DNA methylation patterns on blood cells. Their study involved 29 children raised in orphanages and 29 children raised in biological families. The maximum age of the children was 4 years.

Naumova et al. noted that in institutional care there is often profound psychosocial deprivation. They also noted that there is evidence that relationships between a child and a stable adult caregiver are critical for socio-emotional development. There is also evidence that children raised in institutions may have impairment in growth, in motor abilities and delays in cognitive and language development. Several investigators, including Van Tieghem et al.[44] have implicated stress as one of the main pathways implicated in impaired neurobehavioral development in institutionalized young children. Naumova et al. noted that there is evidence from animal models that social stress leads to altered expression of specific genetic networks. In addition, the altered gene activity in these networks may be related to altered epigenetic modifications.

In their study Naumova et al. assessed adaptive behavior skill using the Vineland Adaptive Behavior Scales. Methylation analyses were carried out using the Illumina Infinium Methylation microarray that analyzed 850,000 CpG sites across the genome.

Results revealed that the institutionalized children showed significantly lower adaptive behavior skills than the biological family raised groups. The largest differences in scores were in the communications and motor skills domains.

Methylation analyses on blood cells revealed that 164 specific CpG sites in the genome showed differences between the two groups, In the institution raised children 82 CpG sites showed increased methylation and 82 CpG sites had decreased methylation compared with levels in the groups raised in biological families. Use of genomic data bases revealed that most of the sites with differential methylation were found at sites that regulates expression of specific genes. The specific genes proposed to have altered expression as a result of differential methylation included 5 gives involved in immune functions and cytokine production and cell proliferation. These genes that showed decreased methylation included:

ENO	Enolase 1, immunoregulatory role in dendritic cells
GNB1	G protein subunit beta 1, integrate signals between receptors and effector proteins
SOCS3	suppressor of cytokine signaling 3
RB1	RB transcriptional corepressor 1, regulator of the cell cycle
HSPA8	Heat shock protein family A (Hsp70) member 8, facilitates correct protein folding

In addition, The MAPK4 showed increased expression. MAPK4—Mitogen activated protein kinase regulates several processes including transcription.

Naumova et al. reported that there was evidence that the duration of institutionalization correlated with the degrees of epigenetic alteration at particular sites. Correlations were also found with institutionalization and specific immune functions. In institutionalized children the granulocyte count was higher and counts of B and T cells were lower. It is important to noted that there is growing evidence for neuro-immune interaction and evidence that malfunctioning of immunity impacts neurodevelopment.

Epigenetics and cell type specificity

Gu et al.[45] reported results of analyses of CpG methylomes in 54 different human cell types. Their study involved the analyses of 26,000,000 autosomal CpG methylation sites and they identified 660,000 sites that showed variable methylation across cell types. They also examined genomic sequence data that correlated with the methylation sites and they determined that the regions that showed variable regulation were likely related in regulatory regions particularly in enhancer sequences. Variable methylation was also found at transcription factor binding sites.

Gu et al. noted that methylation status was also dependent on the stage of differentiation of the specific cell types. In any specific differentiated cell types approximately 25.4% of the CPG sites were unmethylated and these sites were likely involved in regulating expression of specific genes important for the functions of those cells.

Epigenetic differences in monozygotic twins

Castellani et al.[46] carried out methylation studies on two sets of monozygotic twins who were discordant for schizophrenia. They noted that there were several published reports regarding phenotypic differences between the two members of specific monozygotic twin pairs. These differences can arise as a result of post-zygotic genomic events and as a result of epigenetic processes

In the Castellani et al. study the two members of each pair were female, only one in each pair was diagnosed with schizophrenia. Members of the first twin pair were 43 years old at the time of study and one member had been diagnosed with schizophrenia at 27 years of age. The twins in the second pair were 31 years old at the time of study and one member had been diagnosed with schizophrenia at the age of 22 years.

Methylation analyses were carried out on blood cell derived DNA using methylation microarrays. In the first pair, methylation differences between the two members occurred on 138 genes. In second pair methylation differences between the pairs occurred at 330 genes.

The methylation differences covered all the chromosomes however there were specific genomic regions that showed methylation differences in both twin pairs.

The shared sites at which the affected members of each pair differed included 27 genes. Of particular interest was the finding that the affected individuals both showed methylation changes in 5 genes in the HIST2H histone gene cluster on chromosome 1 and in the SNORD gene cluster on chromosome 15q, this cluster is located between the SNRPN and the UBE3A genes.

Environmental agents and epigenetic changes

Specific data designed to evaluate the relevance of epigenetic changes in cancer include studies published by Jones et al.[47] who reported that in more than 50% of cancers there are mutations in genes that impact chromatin organization. Jones et al. proposed that mutations in drivers of epigenetic changes may play key roles in cancer.

Herceg et al.[48] presented evidence that specific environmental factors impact DNA methylation. There was evidence that tobacco smoke impacts DNA methylation in newborn, children and adults. Specific environmental contaminants, including arsenic and nitrous oxide have been shown to associated with methylated sequences in specific genes. Herceg et al. also noted evidence that specific infectious agents and chronic inflammatory processes impact epigenetic modifications and predispose to cancer. Evidence for these influences come from studies of Helicobacter pylori infections and predisposition to gastric cancer, and from studies on Hepatitis B viral infections and development of liver cancer. There is also evidence of increased susceptibility to colorectal cancer in individuals with a history of inflammatory bowel diseases.

Metabolism derived substrates and cofactors involved in epigenetic modifications

These factors were reviewed by Reid et al.[49] and are summarized in the table below.

Enzymes	Substrates and cofactors that can be used in modifications
Histone acetyl transferases	Acetyl coenzyme A
Methyl transferases	S-adenosyl methionine, S-adenosylhomocysteine
Histone deacetylases	Nicotinamide adenine dinucleotide (NAD)
Histone demethylases	alpha-ketoglutarate, succinate, fumarate, Vitamin C
	FAD flavine adenine dinucleotide

Epigenetic states and gene expression

In a report describing a new analysis system Chrom HMM, Ernst and Kellis[50] emphasized the importance of mapping epigenomic markers and regions of open chromatin to identify regulatory element in the genome. They noted that identification of chromatin state could have important implications for understanding cell functions of all types and alteration in disease states.

Transcription factor binding to DNA and impact of epigenetic processes

Hughes and Lambert[51] noted that more than 1600 proteins have been identified that function as transcription factors, nevertheless the exact DNA binding specificities of many of these have not been identified. Furthermore, the impact of DNA modifications on transcription factor binding had not been clearly defined.

Binding of transcription factors to specific DNA sites can be impacted by methylation of cytosine residues at that site. Yin et al.[52] carried out analyses of cytosine methylation and transcription factor binding. They determined that many transcription factors do not bind readily to DNA with methylated cytosine; however, they identified some transcription factors that do bind to DNA with methylated cytosine The OCT4 pluripotency transcription factor was shown to bind to methylated CpG cytosine motifs in DNA. They also determined that other specific transcription factors involved in developmental processes bound to DNA with methylated cytosine.

Yin et al. noted that hypomethylation of DNA is characteristic of DNA that flanks active regulatory elements that bind transcription factors. Earlier studies focused on the binding of transcription factors to specific nucleotide sequences in DNA. These binding sites were frequently associated with enhancers sequences and promoter regions of genes. Through efforts of the ENCODE project more than 10 thousand enhancer elements have been identified on the base of their epigenomic modifications.

Inukai et al.[53] reported that expanded insights into transcription factor binding have been obtained through application of newer techniques. Chromatin immunoprecipitation (e.g. using antibodies to specific transcription factors) followed by high throughput sequencing have yielded information on transcription factor binding sites on the DNA and chromatin modifications present at those sites.

Transcription regulation and the mediator complex

The multi-subunit Mediator complex is required for transcription of DNA by RNA polymerase II. Soutourina[54] reviewed the mediator complex and its assembly at promoters. Following interaction of enhancer elements with gene promoters through chromatin looping and assembly of the pre-initiation complex that includes transcription factors and the Mediator complex, RNA polymerase can bind and transcription can be initiated.

In humans the Mediator complex is composed of 30 different subunits. Soutourina noted that mutations in the subunit MED12 encoded by a gene on the X chromosome have been found in some male patients with X-linked mental retardation, mutations in MED17, encoded on chromosome 11 were reported in a case of infantile cerebral atrophy. Patients with intellectual disability were reported to have a deleterious mutation in the MED23 subunit, encoded by a gene on chromosome 6.[55]

Activity dependent transcription

In a review of transcription and neural activity in 2018 Yap and Greenberg[56] emphasized the key differences between non-dividing neurons and other cells in the body. They noted that neurons coordinate uses of their genomes and synapses to store requisite information and to perform computations on demand, and to also adapt as new experiences are encountered. They attributed these capacities in part to the properties of neuronal synapses. They noted further that the interactions of neuronal synapses and neuronal nuclei play key roles in memory consolidation and storage.

Yap and Greenberg[56] emphasized the importance of designing studies to establish how activity dependent transcription impacts neuronal assemblies and generates specific behaviors. Evidence for the importance of such studies derives in part from observation that specific defects in transcription pathways and variants in non-protein coding genomic elements have been shown to be involved in specific neuro-developmental disorders. They noted further that there is evidence that evolutionary acquisition of cognitive abilities is due to adaptations in the genome that control transcription.[57]

Calcium dependent gene transcription

1. Neurotransmitter signaling leads to opening of L type voltage sensitive calcium channels. Calcium can also enter cells through activated NMDA glutamate receptors
2. Influx of calcium ions into the cell leads to activation of RAS and MAPK signaling pathways and also to activation of calmodulin dependent protein kinases and to calcineurin activity.
3. Activated signaling pathways that activate transcription factors are already parent in the cells, e.g. CREB and SRF and MYEF2.
4. These transcription factors stimulate expression of immediate early genes (FOS and JUN). FOS and JUN and other component to form the activating complex 1, AP-1, that also activated heat response elements.
5. AP-1 elements then activate late response genes.

Yap and Greenberg noted that the late response genes encode products that promote cellular processes including growth and maturation of dendritic spines and products that impact the excitation inhibition balance. They emphasized that the effects of the products of late response genes are likely cell type specific.

They noted further that recent evidence indicates that the AP-1 activating complex binds to enhancer elements. They proposed further that enhancers that have been activated by the AP-1 complex remain primed so that they can be more readily activated by further stimuli. In addition, they noted that transcription factor activation might lead to alteration in the genome of the neuron, for examples changes in the methylation status of CG or other cytosine dinucleotides such as described by Lister et al.[58]

Lister et al.[58] examined DNA methylation in human and mouse brains. They specifically studied the frontal cortex at different ages from fetal to young adult life. Their studies revealed that non-cytosine-guanine methylation (CA, CT, CC) methylation, accumulated in post-natal life during the period when synaptogenesis increased. The non -CG methylation was referred to them as mCH. It is sometimes referred to as mCA by other investigators.

Lister et al. also determined that levels of 5-OH methyl cytosine were present during the same period of post-natal life and changes in parallel with synaptic development. These changes including increases in non-CG methylation and increases in levels of hydroxy methylation increased specifically in neurons as the brain matured. In a brief review of this work by Lister, Gabel and Greenberg[59] noted that the post-natal frontal cortex development occurs in parallel with input from the environment and that it seemed possible then that neuronal specific methylation patterns contributed to gene expression and synaptic development.

Studies on epigenetic modifications, gene expression and memory

In the review published in 2015 Lister and Mukhamel[60] traced developments of concepts related to the establishment of memory. They noted that since the original reports of Hebb in 1949 the synaptic centered framework of memory has predominated. Long-term potentiation and long-term repression of synaptic functions have been considered to be important factors in memory. Gene expression and related signaling pathways have also received attention as important factors in memory.

Lister and Mukamel presented evidence that epigenetic mechanisms likely play roles in memory. They noted that methylation of DNA and cytosine-guanine (CG) dinucleotides have been most intensely studied. Earlier studies by Lister et al. revealed the occurrence of methylation at non-CG dinucleotides and the occurrence of 5-hydroxymethyl cytosine in the brain during early development in the brain during early post-natal life.

Questions that arise relate to the role of these unusual forms of DNA methylation in the brain and the development of cognition. Lister and Mukamel noted that one step toward establishing such connection would be to determine if perturbations in these methylation processes lead to impaired cognitive development. There is evidence from studies in mice that disruption of enzymes responsible for methylation, e.g. DNA methyltransferases either Dnmt1a or Dnmt3a in forebrain can disrupt cognition, memory and behavior.

It is important to emphasize that there is abundant evidence from human studies of the importance of adequate functioning of epigenetic machinery for appropriate neurocognitive development and functioning. The epigenetic machinery includes not only proteins that impact DNA methylation, it includes the writers, readers and erasers of both DNA and histone modifications and factors involved in establishing appropriate chromatin architecture.

Regulation of gene expression

Much progress has been made in the elucidation of mechanisms of gene expression through advances in techniques of nucleic acid analyses and through large scale endeavors such as the ENCODE project https://www.encodeproject.org/. In addition, evidence has accumulated that indicates that many important risk variants for common human diseases are located in regulatory regions of the genome.

The key regulatory elements in the genome include gene promoter region, enhancers elements, insulators and repressors of gene expression.[61, 62] Enhancers of gene expression are located in cis to the gene or genes they regulate. Enhancers have been found in non-proteins coding regions, sometimes at significant distances from the loci they regulate. They sometimes occur in introns. There is some evidence for species conservation of the nucleotide sequences of enhancers that impact expression of a specific gene. Active enhancers are present in open chromatin and may also be recognized by the presence of specific histone modification H3K27 acetylation and H3K4 monomethylation.

An important discovery in recent decades had been the demonstration that in region of active transcription chromatin looping occurs that brings specific enhancers into contact with the promoter region of the genes that they activate. Methods to identify open chromatin regions that facilitate transcription have advanced.

One important new method is ATAC sequencing. The activity of putative enhancer sequence elements can be investigated by transferring these into specific cell or model organism and analyzing their effects on transcription.[63]

In some regions of the genome regulatory elements, such as enhancers are clustered together in non-protein coding regions to form a locus control region for a specific gene.

Other key elements in regulation of gene expression include DNA sequences that act as transcription factor binding sites. There is some evidence that combinations of transcription factors bind to enhancers. Transcription factor binding to promoter sequences are essential for activation of gene expression.

Insulator sequences in DNA can bind specific proteins that then act to block transcription of a specific gene. Ghirlando et al.[64] described two key functions of insulators. They act as blockers to prevent the interaction of enhancers with promoters and they block the propagation of heterochromatin into euchromatin. In heterochromatin DNA is tightly packed. There is also evidence that heterochromatin differs from euchromatin with respect to methylation modifications. Heterochromatin is rich in H3K9 methylation. Together these alterations reduce the accessibility of DNA to binding proteins, e.g. transcription activators.

Alternative splicing of mRNA

Alternative mRNA splicing leads to the generation of different protein isoforms from a specific gene. Temporal or tissue specific functions may require specific protein isoforms to be generated from a specific gene.[65]

Specific mechanisms that regulate alternative splicing include the binding of specific proteins to mRNA. Examples of such proteins include the muscle blind like protein MBNL.[66] Lee et al.[67] identified additional proteins that bind to mRNA and alter splicing these include HNRNPA1 and RNA helicase.

The epithelial splicing regulator protein ESRP1 was reported to have mutation that led to sensorineural hearing loss in humans. Rohacek et al.[68] then carried out studies and discovered that the loss of Esrp in mice led to altered cochlea development and altered auditory function in mice.

Weyn-Vanhentenryck et al.[69] reported that the RBFOX 1 and 2 proteins are RNA binding proteins that are key regulators of alternative exon splicing particularly in the nervous system. These proteins bind to conserved elements in mRNA UGCAUG. Weyn-Vanhentenryck et al. identified 1059 RBFOX fox alternate splice sites encoded in the human genome.

Chromatin architecture folding and modeling

Chromatin folding and looping have been shown to play important roles in gene regulation in recent decades.[70] There is evidence that separate compartments of chromatin exist within the nucleus at a specific time point compartment A contains active chromatin regions with expressed genes while compartment B contains repressed chromatin and silenced genes.

In addition, specific domains within chromatin referred to a transcriptionally active domains (TADS), are separated from inactive domain by specific proteins referred to as boundary elements. Boundary element proteins include CCCTC-binding factor (CTCF) and cohesin, Specific structural rearrangements in chromosomes that impact chromatin structural domains can lead to impairments in gene expression even if the protein coding sequence of genes are not impaired.[71]

Chromatin remodeling involves energy utilization and specific protein complexes to move nucleosomes to promote regions of open chromatin that promote gene expression or to move nucleosomes closer together to form closed chromatin in which gene expression is silenced.

Ronan et al.[72] reviewed chromatin regulators and particularly noted their importance in neural development. Key chromatin regulator complexes include the BAF family of proteins, (also referred to as SWI/SNF complexes. At least 11 genes encode subunits of BAF complexes, these genes include ARID1A and ARID1B and 5 different SMARC genes. Mutations in these genes have been found to lead to intellectual disability.

The CHD8 protein, chromodomain helicase, acts as an ATP dependent chromatin remodeler and has been found to harbor pathogenic mutations in some cases of autism.

Other important proteins involved in chromatin modification include the polycomb repressor complexes that often act in the presence of co-repressor complexes to decrease gene expression. These complexes act to trimethylate histone H3 on lysine 27.

Writers of chromatin modification include histone methyltransferases and histone acetyltransferase.

Removal of methylation marks on histone include histone demethylase include KDM4A and KDM4C.

Genetic variations in regulators of gene expression

Studies carried out by numerous investigators including the Genotype Tissue Expression Consortium https://gtexportal.org/home/, have identified sequence variants located in promoters, enhancers and repressors that impact levels of gene expression. These variants are referred to as expression quantitative trait loci QTLs. These loci are frequently located in cis to the genes they impact (cisQTLs), sometimes they are located on different chrom0somes (transQTLs).[73]

Some of the genomic loci revealed by genome wide association studies (GWAS) to be impact disease risk, act as QTLs that influence gene expression.[74]

RNA modifications

Specific important modifications were discovered more than 5 decades ago. These included polyadenylation at the 3'end of mRNA,[75] and additions that constituted the 5' cap of mRNA. The 5' cap is comprised of guanine linked to mRNA in a triphosphate linkage and this guanosine is then methylated, leading to the formation of m7G. Specific enzyme complexes that participated in this process were later identified.[76] Modifications also occur in the internal sequences of mRNA transcripts.

Another key modification involves conversion of adenosine to inosine (A to I) that is then read as guanosine, through activity of ADAR enzymes. Editing of cytosine to uridine (C to U) takes place through activity of AID/APOBEC enzymes.[77] Another modification includes methylation of cytosine in RNA and modifications of cytosine to 5-hydroxymethylcytosine.

Roundtree et al. reported that in addition to modification of the bases in RNA methylation of ribose in mRNA has also been reported. Key enzymes involved in methylation of RNA include METTL3, methyltransferase like 3. This enzyme is involved in the posttranscriptional methylation of internal adenosine residues in eukaryotic mRNAs, forming N6-methyladenosine, METTL14 methyltransferase like 14 is another important methyltransferase. A specific demethylase alkB homolog 5, (ALKBH5) RNA demethylase, was shown to be involved in N6 methyladenosine demethylation.

Roundtree et al. noted that chemical modification of RNA altered its charge, base-pairing potential and its capacity to bind proteins. These alterations impact RNA localization and translation.

In 2017 Tan et al.[78] presented data on RNA editing and noted that it differed in different tissues.

The epitranscriptome

Modified RNA is sometimes referred to as the epitranscriptome. Thomas et al.[79] presented evidence that alteration in cellular metabolism and altered concentration of specific metabolites can disrupt RNA modification and the epitranscriptome and thereby alter gene expression. They noted that earlier studies by Cai et al.[80] had demonstrated that altered levels of acetylCoA influenced histone acetylation. These findings demonstrated a link between environmental mechanism, metabolism and regulation of gene expression.

Thomas et al. emphasized that RNA modifications play key roles in development and may influence disease.

Lerner et al.[81] presented data on families of enzymes that play key roles in RNA modification. Enzymes in the adenosine deaminase RNA (ADAR) group catalyze conversion of adenosine to inosine conversion and inosine gets read as guanosine. Tomaselli et al.[82] reported that this modification can alter splicing and translation and can impact binding of proteins to RNA

Enzymes in the AID APOBEC family catalyze conversion of cytosine to uridine. Conversion of cytosine to inosine was first reported to modify the pre-mRNA derived from the apoliprotein B gene (*APOB*). This conversion of cytosine to uridine led to generation of a stop codon and this resulted in a truncated form of the APOB protein. This finding led to the generation of a name for this family of enzymes, apolipoprotein B mRNA editing catalytic polypeptide (APOBEC). These enzymes are also referred to as AID (AICDA) activation induced cytidine deaminase. These enzymes function as RNA-editing cytidine deaminases.

A key enzyme in the activation dependent cytidine deaminase family plays a key role in antibody diversification.

The APOBEC1 enzymes was found to requires specific co-factors for optimal function. APOBEC1 has also been found to edit cytosines in the 3' RNA region.

References

1. Gilman AG. *G proteins and regulation of adenyl cyclase in Nobel lectures physiology or medicine 1991–1995*. Singapore: World Scientific Publishing; 1997. Rodbell MA, (1994) Signal transduction. Evolution of an idea. Nobel lecture https://www.nobelprize.org/prizes/medicine/1994/rodbell/lecture/.
2. Yang SH, Sharrocks AD, Whitmarsh AJ. MAP kinase signalling cascades and transcriptional regulation. *Gene* 2013;**513**(1):1–13. https://doi.org/10.1016/j.gene.2012.10.033. Review, 23123731.
3. Cantley LC. The phosphoinositide 3-kinase pathway. *Science* 2002;**296**(5573): 1655–7. Review, 12040186.
4. Yeh YC, Parekh AB. CRAC channels and Ca^{2+}-dependent gene expression. In: Kozak JA, Putney Jr JW, editors. *Calcium entry channels in non-excitable cells*. Boca Raton (FL): CRC Press/Taylor & Francis; 2018 [chapter 5]. 30299657.
5. Liu X, Li Z, Song Y, Wang R, Han L, et al. AURKA induces EMT by regulating histone modification through Wnt/β-catenin and PI3K/Akt signaling pathway in gastric cancer. *Oncotarget* 2016;**7**(22):33152–64. https://doi.org/10.18632/oncotarget.8888. 27121204.
6. Isotani S, Hara K, Tokunaga C, Inoue H, Avruch J, Yonezawa K. Immunopurified mammalian target of rapamycin phosphorylates and activates p70 S6 kinase alpha in vitro. *J Biol Chem* 1999;**274**(48):34493–8. 10567431.

7. Magnuson B, Ekim B, Fingar DC. Regulation and function of ribosomal protein S6 kinase (S6K) within mTOR signalling networks. *Biochem J* 2012;**441**(1):1–21. https://doi.org/10.1042/BJ20110892. 22168436.

8. Bohmann D. Transcription factor phosphorylation: a link between signal transduction and the regulation of gene expression. *Cancer Cells* 1990;**2**(11):337–44. Review, 2149275.

9. Vaquerizas JM, Kummerfeld SK, Teichmann SA, Luscombe NM. A census of human transcription factors: function, expression and evolution. *Nat Rev Genet* 2009;**10**(4):252–63. https://doi.org/10.1038/nrg2538. 19274049.

10. Lambert SA, Jolma A, Campitelli LF, Das PK, Yin Y, et al. The human transcription factors. *Cell* 2018;**172**(4):650–65. https://doi.org/10.1016/j.cell.2018.01.029. Review. Erratum in: Cell. 2018 Oct 4;175(2):598–599, 29425488.

11. An G, Sun J, Ren C, Ouyang Z, Zhu L, et al. LIVE: a manually curated encyclopedia of experimentally validated interactions of lncRNAs. *Database (Oxford)* 2019;2019. https://doi.org/10.1093/database/baz011. 30759219.

12. Zhang Q, Lenardo MJ, Baltimore D. 30 Years of NF-κB: a blossoming of relevance to human pathobiology. *Cell* 2017;**168**(1–2):37–57. https://doi.org/10.1016/j.cell.2016.12.012. Review, 28086098.

13. Yang Y, Wu J, Wang J. A database and functional annotation of NF-κB target genes. *Int J Clin Exp Med* 2016;**9**(5):7986–95. www.ijcem.com/ISSN:1940-5901/IJCEM0019172.

14. Courtois G, Smahi A. NF-kappa B-related genetic diseases. *Cell Death Differ* 2006;**13**(5):843–51. https://doi.org/10.1038/sj.cdd.4401841. 16397577.

15. Mangelsdorf DJ, Thummel C, Beato M, Herrlich P, Schütz G, et al. The nuclear receptor superfamily: the second decade. *Cell* 1995;**83**(6):835–9. Review, 8521507.

16. Sever R, Glass CK. Signaling by nuclear receptors. *Cold Spring Harb Perspect Biol* 2013;**5**(3):a016709. https://doi.org/10.1101/cshperspect.a016709. 23457262.

17. Knockenhauer KE, Schwartz TU. The nuclear pore complex as a flexible and dynamic gate. *Cell* 2016;**164**(6):1162–71. https://doi.org/10.1016/j.cell.2016.01.034. Review, 26967283.

18. Oka M, Yoneda Y. Importin α: functions as a nuclear transport factor and beyond. *Proc Jpn Acad Ser B Phys Biol Sci* 2018;**94**(7):259–74. https://doi.org/10.2183/pjab.94.018. Review, 30078827.

19. Stewart M. Polyadenylation and nuclear export of mRNAs. *J Biol Chem* 2019;**294**(9):2977–87. https://doi.org/10.1074/jbc.REV118.005594. Review, 30683695.

20. Holliday R, Pugh JE. DNA modification mechanisms and gene activity during development. *Science* 1975;**187**(4173):226–32. 1111098.

21. Bird AP. CpG-rich islands and the function of DNA methylation. *Nature* 1986;**321**(6067):209–13. https://doi.org/10.1038/321209a0. 2423876.

22. Bartolomei MS, Tilghman SM. Genomic imprinting in mammals. *Annu Rev Genet* 1997;**31**:493–525. Review, 9442905.

23. Kouzarides T. Chromatin modifications and their function. *Cell* 2007;**128**(4):693–705. Review, 17320507.

24. Borrelli E, Nestler EJ, Allis CD, Sassone-Corsi P. Decoding the epigenetic language of neuronal plasticity. *Neuron* 2008;**60**(6):961–74. https://doi.org/10.1016/j.neuron.2008.10.012. Review, 19109904.

25. Simon JA, Kingston RE. Mechanisms of polycomb gene silencing: knowns and unknowns. *Nat Rev Mol Cell Biol* 2009;**10**(10):697–708. https://doi.org/10.1038/nrm2763. Epub 2009 Sep 9. Review, 19738629.

26. Teschendorff AE, Menon U, Gentry-Maharaj A, Ramus SJ, Weisenberger DJ, et al. Age-dependent DNA methylation of genes that are suppressed in stem cells is a hallmark of cancer. *Genome Res* 2010;**20**(4):440–6. https://doi.org/10.1101/gr.103606.109. 20219944.

27. Rakyan VK, Down TA, Maslau S, Andrew T, Yang TP, et al. Human aging-associated DNA hypermethylation occurs preferentially at bivalent chromatin domains. *Genome Res* 2010;**20**(4):434–9. https://doi.org/10.1101/gr.103101.109. 20219945.

28. Kadoch C, Copeland RA, Keilhack H. PRC2 and SWI/SNF chromatin remodeling complexes in health and disease. *Biochemistry* 2016;**55**(11):1600–14. https://doi.org/10.1021/acs.biochem.5b01191. 26836503.

29. Horvath S. DNA methylation age of human tissues and cell types. *Genome Biol* 2013;**14**(10):R115. Erratum in: Genome Biol. 2015;16:96, 24138928.

30. Horvath S, Raj K. DNA methylation-based biomarkers and the epigenetic clock theory of ageing. *Nat Rev Genet* 2018;**19**(6):371–84. https://doi.org/10.1038/s41576-018-0004-3. Review, 29643443.

31. Horvath S, Garagnani P, Bacalini MG, Pirazzini C, Salvioli S, et al. Accelerated epigenetic aging in Down syndrome. *Aging Cell* 2015;**14**(3):491–5. https://doi.org/10.1111/acel.12325. 25678027.

32. Edwards JR, Yarychkivska O, Boulard M, Bestor TH. DNA methylation and DNA methyltransferases. *Epigenetics Chromatin* 2017;**10**:23. https://doi.org/10.1186/s13072-017-0130-8. eCollection 2017, 28503201.

33. Ducker GS, Rabinowitz JD. One-carbon metabolism in health and disease. *Cell Metab* 2017;**25**(1):27–42. https://doi.org/10.1016/j.cmet.2016.08.009. Review, 27641100.

34. Friso S, Udali S, De Santis D, Choi SW. One-carbon metabolism and epigenetics. *Mol Aspects Med* 2017;**54**:28–36. https://doi.org/10.1016/j.mam.2016.11.007. Review, 27876555.

35. Smithells RW, Sheppard S, Schorah CJ. Vitamin deficiencies and neural tube defects. *Arch Dis Child* 1976;**51**(12):944–50. 1015847.

36. Niculescu MD, Craciunescu CN, Zeisel SH. Dietary choline deficiency alters global and gene-specific DNA methylation in the developing hippocampus of mouse fetal brains. *FASEB J* 2006;**20**(1):43–9. 16394266.

37. Pauwels S, Duca RC, Devlieger R, Freson K, Straetmans D, et al. Maternal methyl-group donor intake and global DNA (hydroxy)methylation before and during pregnancy. *Nutrients* 2016;**8**(8). pii: E474. https://doi.org/10.3390/nu8080474. 27509522.

38. Kinde B, Gabel HW, Gilbert CS, Griffith EC, Greenberg ME. Reading the unique DNA methylation landscape of the brain: Non-CpG methylation, hydroxymethylation, and MeCP2. *Proc Natl Acad Sci U S A* 2015;**112**(22):6800–6. https://doi.org/10.1073/pnas.1411269112. Review, 25739960.

39. Bekdash RA, Zhang C. Sarkar DK Gestational choline supplementation normalized fetal alcohol-induced alterations in histone modifications, DNA methylation, and proopiomelanocortin (POMC) gene expression in β-endorphin-producing POMC neurons of the hypothalamus. *Alcohol Clin Exp Res* 2013;**37**(7):1133–42. https://doi.org/10.1111/acer.12082. 23413810.

40. van der Plaat DA, de Jong K, de Vries M, van Diemen CC, Nedeljković I, et al. Occupational exposure to pesticides is associated with differential DNA methylation. *Occup Environ Med* 2018;**75**(6):427–35.

41. Mostafalou S, Abdollahi M. Pesticides and human chronic diseases: evidences, mechanisms, and perspectives. *Toxicol Appl Pharmacol* 2013;**268**(2):157–77. https://doi.org/10.1016/j.taap.2013.01.025. Review, 23402800.

42. de Jong K, Boezen HM, Kromhout H, Vermeulen R, Postma DS, et al. Pesticides and other occupational exposures are associated with airway obstruction: the LifeLines cohort study. *Occup Environ Med* 2014;**71**(2):88–96. https://doi.org/10.1136/oemed-2013-101639. 24142985.

43. Naumova OY, Rychkov SY, Kornilov SA, Odintsova VV, Anikina VO, et al. Effects of early social deprivation on epigenetic statuses and adaptive behavior of young children: a study based on a cohort of institutionalized infants and toddlers. *PLoS ONE* 2019;**14**(3):e0214285. https://doi.org/10.1371/journal.pone.0214285. eCollection 2019, 30913238.

44. VanTieghem MR, Tottenham N. Neurobiological programming of early life stress: functional development of amygdala-prefrontal circuitry and vulnerability for stress-related psychopathology. *Curr Top Behav Neurosci* 2018;**38**:117–36. https://doi.org/10.1007/7854_2016_42. 28439771.

45. Gu J, Stevens M, Xing X, Li D, Zhang B, et al. Mapping of variable DNA methylation across multiple cell types defines a dynamic regulatory landscape of the human genome. *G3 (Bethesda)* 2016;**6**(4):973–86. https://doi.org/10.1534/g3.115.025437. 26888867.

46. Castellani CA, Laufer BI, Melka MG, Diehl EJ, O'Reilly RL, Singh SM. DNA methylation differences in monozygotic twin pairs discordant for schizophrenia identifies psychosis related genes and networks. *BMC Med Genomics* 2015;**8**:17. https://doi.org/10.1186/s12920-015-0093-1. 25943100.

47. Jones PA, Issa JP, Baylin S. Targeting the cancer epigenome for therapy. *Nat Rev Genet* 2016;**17**(10):630–41. https://doi.org/10.1038/nrg.2016.93. Review, 27629931.

48. Herceg Z, Ghantous A, Wild CP, Sklias A, et al. Roadmap for investigating epigenome deregulation and environmental origins of cancer. *Int J Cancer* 2018;**142**(5):874–82. https://doi.org/10.1002/ijc.31014. Epub 2017 Sep 13. Review, 28836271.

49. Reid MA, Dai Z, Locasale JW. The impact of cellular metabolism on chromatin dynamics and epigenetics. *Nat Cell Biol* 2017;**19**(11):1298–306. https://doi.org/10.1038/ncb3629. Review, 29058720.

50. Ernst J, Kellis M. Chromatin-state discovery and genome annotation with ChromHMM. *Nat Protoc* 2017;**12**(12):2478–92. https://doi.org/10.1038/nprot.2017.124. 29120462.

51. Hughes TR, Lambert SA. Transcription factors read epigenetics. *Science* 2017;**356**(6337):489–90. https://doi.org/10.1126/science.aan2927. 28473550.

52. Yin Y, Morgunova E, Jolma A, Kaasinen E, et al. Impact of cytosine methylation on DNA binding specificities of human transcription factors. *Science* 2017;**356**:6337. pii: eaaj2239, https://doi.org/10.1126/science.aaj2239.PMID:28473536.

53. Inukai S, Kock KH, Bulyk ML. Transcription factor-DNA binding: beyond binding site motifs. *Curr Opin Genet Dev* 2017;**43**:110–9. https://doi.org/10.1016/j.gde.2017.02.007. Epub 2017 Mar 27. Review, 28359978.

54. Soutourina J. Transcription regulation by the Mediator complex. *Nat Rev Mol Cell Biol* 2018;**19**(4):262–74. https://doi.org/10.1038/nrm.2017.115.

55. Trehan A, Brady JM, Maduro V, Bone WP, Huang Y, et al. MED23-associated intellectual disability in a non-consanguineous family. *Am J Med Genet A* 2015;**167**(6):1374–80. https://doi.org/10.1002/ajmg.a.37047. 25845469.

56. Yap EL, Greenberg ME. Activity-regulated transcription: bridging the gap between neural activity and behavior. *Neuron* 2018;**100**(2):330–48. https://doi.org/10.1016/j.neuron.2018.10.013. Review, 30359600.

57. Hardingham GE, Pruunsild P, Greenberg ME, Bading H. Lineage divergence of activity-driven transcription and evolution of cognitive ability. *Nat Rev Neurosci* 2018;**19**(1):9–15. https://doi.org/10.1038/nrn.2017.138. 29167525.

58. Lister R, Mukamel EA, Nery JR, Urich M, et al. Global epigenomic reconfiguration during mammalian brain development. *Science* 2013;**341**(6146):1237905. https://doi.org/10.1126/science.1237905. 23828890.

59. Gabel HW, Greenberg ME. Genetics. The maturing brain methylome. *Science* 2013;**341**(6146):626–7. https://doi.org/10.1126/science.1242671. 23929975.

60. Lister R, Mukamel EA. Turning over DNA methylation in the mind. *Front Neurosci* 2015;**9**:252. https://doi.org/10.3389/fnins.2015.00252. eCollection 2015. Review, 26283895.

61. Pennacchio LA, Bickmore W, Dean A, Nobrega MA, Bejerano G. Enhancers: five essential questions. *Nat Rev Genet* 2013;**14**(4):288–95. https://doi.org/10.1038/nrg3458. Review, 23503198.

62. Chatterjee S, Ahituv N. Gene regulatory elements, major drivers of human disease. *Annu Rev Genomics Hum Genet* 2017;**18**:45–63. https://doi.org/10.1146/annurev-genom-091416-035537. 28399667.

63. Long HK, Prescott SL, Wysocka J. Ever-changing landscapes: transcriptional enhancers in development and evolution. *Cell* 2016;**167**(5):1170–87. https://doi.org/10.1016/j.cell.2016.09.018. 27863239.

64. Ghirlando R, Giles K, Gowher H, Xiao T, Xu Z, et al. Chromatin domains, insulators, and the regulation of gene expression. *Biochim Biophys Acta* 2012;**1819**(7):644–51. https://doi.org/10.1016/j.bbagrm.2012.01.016. 22326678.

65. Raj B, Blencowe BJ. Alternative splicing in the mammalian nervous system: recent insights into mechanisms and functional roles. *Neuron* 2015;**87**(1):14–27. https://doi.org/10.1016/j.neuron.2015.05.004. Review, 26139367.

66. Xiong HY, Alipanahi B, Lee LJ, Bretschneider H, Merico D, et al. RNA splicing. The human splicing code reveals new insights into the genetic determinants of disease. *Science* 2015;**347**(6218):1254806. https://doi.org/10.1126/science.1254806. 25525159.

67. Lee YJ, Wang Q. Rio DC Coordinate regulation of alternative pre-mRNA splicing events by the human RNA chaperone proteins hnRNPA1 and DDX5. *Genes Dev* 2018;**32**(15–16):1060–74. https://doi.org/10.1101/gad.316034.118. 30042133.

68. Rohacek AM, Bebee TW, Tilton RK, Radens CM, McDermott-Roe C, et al. ESRP1 mutations cause hearing loss due to defects in alternative splicing that disrupt cochlear development. *Dev Cell* 2017;**43**(3). 318–331.e5. https://doi.org/10.1016/j.devcel.2017.09.026.

69. Weyn-Vanhentenryck SM, Feng H, Ustianenko D, Duffié R, Yan Q, et al. Precise temporal regulation of alternative splicing during neural development. *Nat Commun* 2018;**9**(1):2189. https://doi.org/10.1038/s41467-018-04559-0. 29875359.

70. Benabdallah NS, Bickmore WA. Regulatory domains and their mechanisms. *Cold Spring Harb Symp Quant Biol* 2015;**80**:45–51. https://doi.org/10.1101/sqb.2015.80.027268. 26590168.

71. Spielmann M, Lupiáñez DG, Mundlos S. Structural variation in the 3D genome. *Nat Rev Genet* 2018;**19**(7):453–67. https://doi.org/10.1038/s41576-018-0007-0. Review, 29692413.

72. Ronan JL, Wu W, Crabtree GR. From neural development to cognition: unexpected roles for chromatin. *Nat Rev Genet* 2013;**14**(5):347–59. https://doi.org/10.1038/nrg3413. 23568486.

73. Ward MC, Gilad Y. Human genomics: Cracking the regulatory code. *Nature* 2017;**550**(7675):190–1. https://doi.org/10.1038/550190a. 29022577.

74. Hauberg ME, Fullard JF, Zhu L, Cohain AT, Giambartolomei C, et al. Differential activity of transcribed enhancers in the prefrontal cortex of 537 cases with schizophrenia and controls. *Mol Psychiatry* 2018. https://doi.org/10.1038/s41380-018-0059-8. 29740122.

75. Edmonds M, Abrams R. Polynucleotide biosynthesis: formation of a sequence of adenylate units from adenosine triphosphate by an enzyme from thymus nuclei. *J Biol Chem* 1960;**235**:1142–9. 13819354.

76. Shatkin A. Capping of eucaryotic mRNAs. *Cell* 1976;**9**(4):645–53. https://doi.org/10.1016/0092-8674(76)90128-8.

77. Roundtree IA, Evans ME, Pan T, He C. Dynamic RNA Modifications in Gene Expression Regulation. *Cell* 2017;**169**(7):1187–200. https://doi.org/10.1016/j.cell.2017.05.045. Review, 28622506.

78. Tan MH, Li Q, Shanmugam R, Piskol R, Kohler J, et al. Dynamic landscape and regulation of RNA editing in mammals. *Nature* 2017;**550**(7675):249–54. https://doi.org/10.1038/nature24041. 29022589.

79. Thomas JM, Batista PJ, Meier JL. Metabolic regulation of the epitranscriptome. *ACS Chem Biol* 2019;**14**(3):316–24. https://doi.org/10.1021/acschembio.8b00951. 30653309.

80. Cai L, Sutter BM, Li B, Tu BP. Acetyl-CoA induces cell growth and proliferation by promoting the acetylation of histones at growth genes. *Mol Cell* 2011;**42**(4):426–37. https://doi.org/10.1016/j.molcel.2011.05.004.PMID:21596309.

81. Lerner T, Papavasiliou FN, Pecori R. RNA Editors, cofactors, and mRNA Targets: an overview of the C-to-U RNA editing machinery and its implication in human disease. *Genes (Basel)* 2018;**10**(1). pii: E13. https://doi.org/10.3390/genes10010013.

82. Tomaselli S, Locatelli F, Gallo A. The RNA editing enzymes ADARs: mechanism of action and human disease. *Cell Tissue Res* 2014;**356**(3):527–32. https://doi.org/10.1007/s00441-014-1863-3. Review, 24770896.

6

MAINTAINING HOMEOSTASIS AND MITIGATING EFFECTS OF HARMFUL FACTORS IN THE INTRINSIC OR EXTRINSIC ENVIRONMENT

Concepts of homeostasis

Claude Bernard[1] first wrote of the physiological balance and the concept of "milieu intérieur". Walter Cannon[2] introduced the term homeostasis in his book entitled, "The wisdom of the body."

The metabolic state in the body can be considered as a key component of the internal environment.

Metabolism and homeostasis

Key aspects of metabolism include sensing levels of specific molecules to limit ATP consumption when nutrients are limited and sensing when nutrients are abundant to promote tissue maintenance cell proliferation and tissue growth (Fig. 1).

During periods of nutrient adequacy cells and tissue respond to a variety of different growth factors and stimulants and MTOR complexes serve to promote processes of protein synthesis. The MTORC1 complex has turned out to be a major controller of metabolism. The key component of MTORC1 is a protein defined initially on the basis its ability to bind to a specific soil derived macrolide rapamycin, that was being investigated for its antibiotic properties. In 1991 Hall and associates (see Heitman, Movva and Hall)[3] discovered that rapamycin binds to a specific cellular protein FKBP12 that in turn binds to a specific domain within a kinase enzyme that was designated as Target of Rapamycin (TOR). The TOR molecule present in mammals is designated MTOR. The binding of rapamycin to MTOR was found to inhibit its kinase activity.

Gene Environment Interactions. https://doi.org/10.1016/B978-0-12-819613-7.00006-2

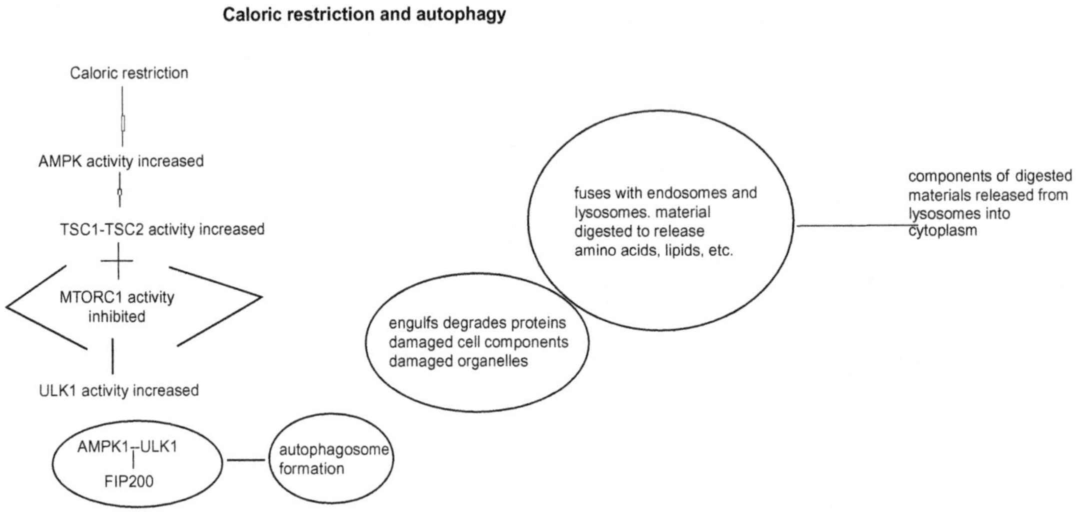

Caloric restriction and autophagy

Fig. 1 *Caloric restriction and autophagy.* Caloric restriction can trigger activity of the TSc1 TSC2 complex that leads to inhibition of the MTORC1 complex and to activation of the ULK1 complex that in turn promoted autophagy and degradation of cellular material including proteins and lipids.

Subsequently MTOR protein was found to be present in two different complexes MTORC1 and MTORC2 and the various components of these two complexes were identified. The MTORC1 complex was shown to be active when nutrient levels were adequately high. MTORC1 activity was also stimulated by binding of specific growth factors to their receptors, these included insulins, insulin like growth factors and specific cytokines.[4] Insulin and growth factor stimulated signaling through the phosphoinositide kinase pathway.

In 2017 Saxton and Sabatini[5] reviewed mTOR signaling in growth and metabolism. The key functions of MTORC1 are to stimulate synthesis of proteins, lipids and nucleotides to promote cell growth and cell division. In addition, MTORC suppresses autophagy. They noted that early studies revealed that MTORC1 activity facilitated mRNA translation. This occurred in part through MTORC1 phosphorylation of S6K also known as ribosomal protein S6 kinase B1 and through impact on eukaryotic translation initiation factor EIF4B. MTORC1 also phosphorylates and inactivates specific inhibitors of translation.

Recent studies have revealed that the MTORC1 complex interacts with the lysosomal membrane and promotes efflux of amino acids from the lysosome. Under conditions where MTORC1 activity is reduced, the efflux of amino acids from the lysosome is reduced. Under these conditions amino acids remain stored in the lysosome. There is evidence that a specific transporter SLC38A9 interacts with the MTORC1 complex to facilitate efflux of amino-acids from the lysosome.

Saxton and Sabatini reported that MTORC1 activity also plays important roles in nucleotide synthesis. MTORC1 stimulates activity of ATF4 (activating transcription factor 4), that increases expression of MTHFD2 (methylenetetrahydrofolate dehydrogenase (NADP+ dependent) 2, methenyltetrahydrofolate cyclohydrolase) that functions in the mitochondrial tetrahydrofolate cycle that plays a key role in generating one carbon units for purine synthesis. In addition, S6K1 phosphorylates and activates the enzyme CAD (carbamoyl-phosphate synthetase 2, aspartate transcarbamylase, and dihydroorotase) that is involved in de novo synthesis of pyrimidines.

MTORC1 activity was found to promote glycolysis rather than oxidative phosphorylation. Glycolysis is enhanced through activity of MTORC1 mediated enhanced translation of HIF1A (hypoxia inducible factor 1A) that stimulates transcription of specific glycolytic encoding genes. TORC1 activity also impacts lipid activity as it leads to increased levels of SREBP (Sterol regulatory element binding protein. In addition, it increases activity in the pentose phosphate pathway that increases generation of NADPH nicotinamide adenine dinucleotide that acts as a coenzyme in many enzyme reactions.

In summary, the MTORC1 complex has been found to play key roles in anabolic, metabolism, the synthesis of new proteins and promotion of cell growth.

Saxton and Sabatini noted that MTORC1 phosphorylates and inhibits activity of ULK1 (unc-51 like autophagy activating kinase 1) that complexes with autophagy related proteins ATG13 and ATG101 to inhibit autophagy.

Under conditions of cellular stress, induced by ATP depletion, glucose depletion, hypoxia and/or DNA damage, MTORC1 activity was shown to be inhibited. This inhibition takes place through activity of TSC2 tuberous sclerosis protein and through inhibitory phosphorylation of Raptor that occurs in the MTORC1 complex.

Saxton and Sabatini reported that DNA damage inhibited MTORC1 activity, through induction of enzymes including AMPK (also known as protein kinase AMP-activated catalytic subunit alpha 1). AMPK increases TSC2 activity leading to decreased MTORC1 activity. This is important, since cell proliferation should be inhibited in the presence of DNA damage.

The MTORC2 complex was reported to function primarily in stimulating the PI3K (phosphatidylinositol-4,5-bisphosphate 3-kinase) insulin signaling pathway. The activity and cellular role of MTORC2 is less well defined. There is however evidence that it is involved in cell metabolism and actin cell organization. There is some evidence that MTORC2 associates with membranes at the interface of mitochondria and endoplasmic reticulum and that its activity there can positively impact membrane integrity and mitochondrial metabolism.

Saxton and Sabatini noted that MTOR and pathways are highly conserved in eukaryotic species.

The MTOR system as sensor and as molecular integrator of cellular environmental conditions

In a 2013 review Efeyan et al.[6] emphasized the important roles of MTORC1 as a sensor of cellular amino acid levels and as a sensor of glucose levels.

Giguère[7] emphasized that the key roles of MTOR containing complexes are to link cellular and energy and nutrition with growth control. There is evidence that this is achieved primarily through impact on processes involved in the translation of mRNA transcripts to protein.

Early information on the function of MTORC1 revealed that it acts to phosphorylate the ribosomal protein S6 kinase and the eukaryotic translation initiation factor binding protein, both of which are keys to translating mRNA. There is now also evidence that MTORC1 is involved in sensing the degree to which specific aminoacids are available for amino-acyl TRNA synthetase to charge TRNAs.

Efeyan referred to earlier studies by several groups who established that MTORC1 activity is dependent on the presence of Rheb GTP. When nutrient resources and energy are insufficient Rheb GDP predominates since the activity of the Tuberous sclerosis complex, TSC1 TSC2 and TBC1D7, inhibits conversion of Rheb GDP to its active form Rheb GTP. Activation of the TSC complex is achieved through activity of specific kinases including AMPK (adenosine monophosphate kinase).

When nutrients are adequate the enzyme adenylate kinase AKT binds to specific amino acids in TSC proteins. This phosphorylation inhibits activity of TSC complex. Sufficient RHEB GTP is then available to activate MTOR, and synthesis of new proteins and cell proliferation can take place.

During the past decade intense research has been carried out on the MTOR complexes. MTORC1 is composed of 5 components MTOR, Raptor, Deptor, Pras 40 and mlST8. The MTORC2 complex is composed of MTOR, Rictor, Deptor Protor, mSin1 and mlST8.

A number of different cellular locations of MTOR have been described. These include the cytoplasm, different organelles including lysosomes and the nucleus. There is evidence for localization of Rheb on lysosomes and evidence that specific proteins RAGGTPases are required to shuttle MTORC1 to the lysosome surface where it can interact with RHEBGTP. The lysosome is emerging as a key site for amino acid storage and contains aminoacids derived from protein breakdown and amino acids obtained as nutrients.[8,9] Demetriades et al.[10]

studied the cellular location of the TSC complex under different metabolic conditions. They reported that under conditions of amino acid sufficiency and growth factor sufficiency, the TSC complex was cytoplasmic. They demonstrated that under different conditions when there was evidence of energetic stress, the TSC complex translocated to lysosomal membranes.

The Ragulator complex serves to anchor RAG GTPases to the lysosomal membrane. Huang and Fingar reported that a specific enzyme vATPase (vacuolar H+ adenosine triphosphate) senses aminoacids within lysosomes and activates RAG GTPase components.

Giguère[7] noted evidence that MTORC1 in the nucleus impacts chromatin and gene transcription.

Energy stress

This can be induced by glucose deprivation, glycolysis inhibition or hypoxia and factors that act to impair mitochondrial function and reduce cellular levels of ATP.[8] Under these conditions the levels of adenosine monophosphate (AMP) are higher. The AMP then activates AMP kinase, AMPK can phosphorylate TSC2 on amino acid S1245 and this stimulates its ability to block conversion of GDP to Rheb GTP activity and this leads to decrease in MTORC1 activity. The active AMPK also serves to phosphorylate the enzyme ULK1 that increases catabolism and the breakdown of proteins.[11]

MTOR as a major link between nutrient availability, and anabolic and catabolic processes

Sabatini[12] presented evidence that has accumulated on the growing repertoire of functions of MTOR containing complexes. He noted that lysosome membranes form the scaffold for MTOR. Menon et al.[13] reported than when levels of insulin fall the TSC complex moves to the lysosomal surface and there it suppresses RHEB GTP production thereby inhibiting MTORC1 activity.

Sabatini et al., demonstrated that RAG GTPase activities are impacted not only by the lysosomal amino-acid concentrations but also by cytoplasmic amino acid concentrations.

Sabatini[12] reported that at least 26 different proteins constitute the nutrient sensing arm of TORC1, and these proteins impact RAG GTPase activity. RAG GTPase regulators include GATOR1 GATOR2 FLCN-FNIP, KICSTOR. Of particular importance is the fact specific proteins have been identified that constitute sensors for specific nutrients. SLC38 is a sensor of lysosomal arginine. Sestrin senses cytosolic leucine. Castor 1 senses cytosolic and lysosomal arginine, Samtor constitutes a specific sensor of S-adenosyl methionine.

More recently evidence has also been obtained on the importance of MTORC1 and associated proteins in control of lipid metabolism.

Studies have revealed that starving cells for glucose or amino acids led MTORC1 to leave the lysosomal surface.

Importance of TSC complex in inhibiting MTORC1 activity during endoplasmic reticulum stress

A specific enzyme that can enhance activity of the TSC complex is REDD1, REDD1 activity is apparently induced by a transcription factor that is produced in greater quantities under conditions of endoplasmic reticulum stress. Through activation of TSC complex activity and consequent suppression of MTORC1 activity under conditions of endoplasmic reticulum stress, the production of additional proteins is suppressed in order to reduce the endoplasmic reticulum stress.[14]

MTOR signaling and mitochondrial function

Insights into the role of TORC1 related function and mitochondrial processes have been obtained in part through studies of impaired mitochondrial function in cases of tuberous sclerosis due to deletions or mutations in either the TSC1 or TSC2 genes.

Ebrahimi-Fakhari et al.[15] analyzed mitochondria in TSC deficient neurons. They determined that in TSC deficient neurons mitochondrial respiratory chain function and mitochondrial membrane potential were decreased. Autophagic processes were shown to be defective. Microscopy revealed that mitochondria in TSC deficient neurons were small and had altered internal configuration. Importantly rapamycin treatment was shown to restore mitochondrial homeostasis.

Sensors of specific metabolites

In a review of metabolic regulation in 2013 Metallo and Vander Heiden[16] noted that adenosine monophosphate (AMP) activated protein kinase (AMPK) acts as a sensor that responds to changes in levels of ATP relative to ADP and AMP.

AMPK is a complex with different subunits encoded by different genes AMPK subunits including 2 catalytic subunits and 5 noncatalytic subunits. AMPK phosphorylates a number of different enzymes and glucose transporters thereby increasing their activity and increasing glucose uptake and thereby ultimately ATP production is increased.

Metallo and Vander Heiden also noted that MTORC1 (target of rapamycin complex) is down regulated by AMPK. Biosynthetic processes are decreased if MTORC1 is suppressed and this suppression therefore makes ATP available for bioenergetic processes. Stimulation of activity of ULK1 through phosphorylation can stimulate autophagy and also increase mitochondrial ATP production.

AMPK was reported to inactivate the transcription factor SREBP1, thereby reducing lipid metabolism.

AMPK stimulates activity of PGC1alpha (PPARG coactivator 1 alpha) and also activates PPARG that increases mitochondrial biogenesis and regulates transcription of various genes.

Metabolism and co-factors

Metallo and Vander Heiden emphasized the important role of the oxidative pentose phosphate pathway in maintaining adequate cellular concentration of NADPH (nicotinamide adenine dinucleotide phosphate reduced form). Reduction in the ratio of NADPH to NADP activity stimulate this pathway. Sufficient quantities of NADPH are necessary to counteract reactive oxygen species.

Nicotinamide adenine dinucleotide (NAD)

Nicotinic acid niacin was discovered in the latter part of the nineteenth century. In 1906 Harden and Young[17] demonstrated the importance of NAD in enzyme reactions. Goldberger in 1906[18] demonstrated that niacin was a vitamin substance that cured pellagra and was designated as Vitamin B3. Warburg and Christian[19] isolated nicotinamide adenine dinucleotide phosphate (NADP) and demonstrated its importance in enzymatic reactions.

Canto et al.[20] reviewed the roles of NAD and its importance in control of energy homeostasis. NAD plays key roles in metabolic processes in the cytoplasm, primarily at specific points in glucose metabolism. Within the mitochondria NADH, the reduced form of NAD is oxidized by respiratory complex 1 activity and electrons released are transmitted via ubiquinone Coenzyme Q10 to successive mitochondrial respiratory complexes.

NAD is also a cofactor in reactions in which acetyl groups are removed from proteins through the activity of sirtuins that function as protein de-acylase enzymes. Sirtuins modify NAD to yield nicotinic acid and O-acetyl-ADP-ribose.

NAD is also an essential co-factor for activity of poly-ADP-ribose polymerase (PARP). PARP and related proteins play roles in numerous processes, including repair of DNA damage. PolyADP ribose polymerases are defined as NAD consuming enzymes.

Canto et al., noted that NAD therapeutics have gained attention. Nicotinic acid and niacin have been used in the treatments of hyperlipidemia. Several investigations have provided evidence that lower NAD levels promote aging.

Tissue NAD levels were reported to rise with adequate exercise and with intake of low-calorie diets. Some investigators have indicated that maintenance of adequate NAD levels may be of value in reducing the impact of neurodegenerative diseases.

Flavin adenine dinucleotide, a condensation of riboflavin (vitamin B2) and adenosine dinucleotide phosphate functions as an electron acceptor. The reduced form, FADH functions as an electron donor.

Molecular machinery and cholesterol homeostasis

Studies over the past 21 years, primarily by Brown and Goldstein and associates, have revealed details of the molecular machinery that maintains cholesterol homeostasis. In 2018 Brown et al.[21] reviewed cholesterol homeostasis and the specific molecular components involved. Initial progress in elucidation of mechanisms of cholesterol homeostasis focused on uptake of cholesterol from low density lipoproteins (LDLs) through the LDL receptor and this uptake was found to be influenced by the concentration of cholesterol in the cell. The concentration of cholesterol in the cell was also shown to influence the transcription of the gene encoding the cholesterol biosynthetic enzyme.

Detailed analyses of the LDL receptor gene by Sudhof et al.[22] led to identification of a specific sequence element in the LDL receptor gene that activated expression of the gene when cholesterol levels were low in the cell. This element was referred to as a sterol response element SRE.

Subsequently specific proteins in the Golgi were found to bind to the sterol response element and were designated as sterol element binding proteins SREBP1 and SREBP2 and these proteins were found to contain domains that function as transcription factors.

It turned out that specific domains need to be released by proteolytic cleavage of SREBP. The specific proteins required for cleavage was designated SCAP and it was found to also contain a specific sterol sensing domain.

Sterol element binding proteins (SREBPs)

SREBPs were reported to activate transcription of genes required for cholesterol synthesis and also transcription of genes that encode products required for cholesterol receptor synthesis and activity and proteins associated with cholesterol uptake from low-density lipoprotein.

Brown et al.[21] described the more recently identified molecule SCAP (SREBF chaperone) this protein binds to sterol regulatory element binding proteins (SREBPs) and mediates their transport from the ER to the Golgi apparatus of the cell. In the Golgi SREBPs are proteolytically cleaved to release the specific domains that can pass to the nucleus and act as transcription factors.

Brown et al., noted that in addition to SCAP, six other membrane proteins have been found to contain sterol sensing domains. These proteins include:

HMG Co Reductase (HMGCR) HMG-CoA reductase is the rate-limiting enzyme for cholesterol synthesis.

7-dehydrocholesterol reductase (DHCR7).

NPC1 intracellular cholesterol transporter.

NPC1L1 NPC1 like intracellular cholesterol transporter 1.

PTCH1 protein is the receptor for the secreted hedgehog signaling ligands.

DISPATCHED (DISP1) dispatched RND transporter family member 1.

TRC8 (RNF39) encodes a multi-membrane spanning protein containing a RING-H2 finger.

A specific protein was subsequently discovered that anchors SCAP and the SREBP complex to endoplasmic reticulum. The protein encoded by the INSIG1 gene (insulin induced gene) was found to disassociate from the endoplasmic reticulum SCAP-SREBP complex when cellular cholesterol levels were low. The INSIG1 encoded protein was found to regulate cholesterol metabolism, lipogenesis, and glucose homeostasis.

Free fatty acid receptors and sensing

Free fatty acids are classified as short chain when they have carbon length between 1 and 6, medium chain when lengths are between 7 and 12 and as long chain when carbon chain lengths are greater than 12. Free fatty acids are examples of nutrients that require direct binding of nutrient to sensor to trigger subsequent reactions. Free fatty acids bind to receptors and this binding leads to activation of G-protein coupled receptor signaling. Hara et al.[23] reported that medium and long chain fatty acid bind to and activate GPR40 and GPR120 G protein coupled receptors. Short chain fatty acids activate different receptors, namely GPR43 and GPR41. Impairment of free fatty acid binding to receptors has been reported to been reported to play roles in type 2 diabetes and in obesity.

Free fatty acid receptors are present in particular tissues and have been reported to play important roles in those tissues. Fatty acid receptors GPR40 and GPR120 are present in pancreas and there they were important to play roles in the control of glucagon secretion. Glucagon is defined as a preproprotein that is cleaved into four distinct mature peptides. One of these, glucagon, is a pancreatic hormone that counteracts the glucose-lowering action of insulin by stimulating glycogenolysis and gluconeogenesis. Binding of free fatty acids to their receptors in the intestine impacts secretion of incretins that are glucagon related proteins. These findings indicate the relationship between fatty acid and glucose metabolic pathways. Ichimura et al.[24] reported on the basis of studies in mice that GPR120 activation by fatty acids influenced insulin sensitivity and played a key role in sensing fat in white adipose tissue and in regulating energy metabolism. Adipose tissue is rich in cells of the immune system including neutrophils and macrophages. Free fatty acid receptors occur on these cells and impact the function of these cells in response to the presence of free fatty acids.

It is interesting to note that specific mutations in the fatty acid receptors were reported to lead to increased risk for diabetes and obesity. These observations have stimulated efforts to develop agonists that increase the expression of GPR120 and can potentially be used in the treatment of diabetes and obesity.[23] The mutation p. R270H led to loss of function of the GPR120 receptor and to high fasting glucose levels in plasma.

GPR120 is now designated FFAR4 (free fatty acid receptor, since it binds unsaturated fatty acids including omega-3 fatty acid. Ichimura et al.[24] reported that FFAR4 is expressed in many different tissues, including taste buds, intestine and pancreas. FFAR4 is also expressed in lung, thymus, pituitary and in macrophages.

Activation of FFAR4 in the intestine leads to expression of incretins including glucagon like peptide 1 (GLP1/GCG), cholestokinin (CCK). Cholestokinin stimulates gall bladder contraction and release of bile into the intestine. The presence of FFAR4 in taste buds likely impacts the oral perceptions of fat.

Im et al.[25] reported that following the binding of ligand to the receptor FFAR4 (GPR 120), there is activation of downstream signaling. This signaling leads to decreased release of ghrelin and decreased release of somatostatin. Im noted that ghrelin is the hunger signal. Somatostatin is a regulator of endocrine gene expression. FFAR4 activation also leads to downstream signaling that increases expression of incretin GLP1 and CCK. GLP1 promotes release of insulin by the pancreas and it also stimulates insulin sensitivity and uptake by cells. FFAR4 stimulation of cholecystokinin release acts to stimulate bile acid release and also acts as an appetite suppressor.

Sheng et al.[26] reported that selective agonists of GPR120 (FFAR4) have been shown to have antidiabetic activity.

Mitochondria and metabolism, new insights

New insights into altered mitochondrial metabolism in insulin deficiency and insulin resistance were reported by Ruegsegger et al.

in 2018.[27] They reported that insulin levels play key roles in maintaining mitochondrial protein abundance and in stimulating mitochondrial biogenesis.

They noted that in diabetes there is inefficient coupling in mitochondria and levels of reactive oxygen species are increased. They noted further that nutrient excess, including also protein and fat increase, lead to increased production of reactive oxygen species. Increased levels of reactive oxygen species can lead to mitochondrial DNA damage.

Reactive oxygen species and superoxide dismutases

Reactive oxygen species including superoxide O_2^- that contains one unpaired electron is produced primarily during mitochondrial metabolism. Production of excess superoxide molecules can be damaging. Superoxide dismutase enzymes play key roles in converting superoxide O_2^-, to O_2 and to H_2O_2 (hydrogen peroxide).

Two major superoxide dismutase enzyme forms occur, SOD1 soluble superoxide dismutase occurs in the cytoplasm. SOD2, mitochondrial superoxide dismutase occurs in mitochondria. Both enzymes are nuclear encoded. Tsang et al.[28] reported that SOD1 activity was linked in part to nutrient levels, and this was impacted by MTORC1 signaling. MTORC1 was found to impact phosphorylation of a specific amino acid T40 in SOD1. Under conditions of nutrient abundance SOD1 activity was suppressed through this specific phosphorylation. Under conditions when nutrients were deficient, SOD1 function was no longer suppressed. TORC1 therefore regulates redox homeostasis in response to nutrient conditions.

Nutrient sensing that impacts mitochondrial function

Liu et al.[29] reported that a specific protein Sirtuin 3 (SIRT3) acts as a sensor of nutrient deprivation. The SIRT3 gene encodes an NAD dependent deacetylase. Under conditions of fasting SIRT3 activity is induced. SIRT3 acts to deacetylate enzymes involved in synthesis of fatty acids and processes of fatty acid oxidation and ketogenesis are facilitated. Liu et al., reported that SIRT3 actively promotes a specific transcription program within mitochondria that enhances oxidative phosphorylation. SIRT3 actively promotes oxidative phosphorylation on fasting through deacetylation of a specific protein LRP130. This deacetylation increases the transcription promoting program of LRP130.

LRP130 is currently designated as LRPPRC leucine rich pentatricopeptide. This protein is deficient in a specific inborn error of metabolism associated with elevated levels of lactic acid due to impaired mitochondrial function.[30] LRP130 was identified more recently as a mitochondrial transcription factor and was specifically shown to increase oxidative phosphorylation. Deacetylation of a specific lysine residue in LPR130 was shown to lead to its inactivation.

Liu et al. reported that basal mitochondrial transcription factors include TFB2M (transcription factor B2 mitochondria) and POLRMT (mitochondrial RNA polymerase).

NRF2 nuclear regulatory factor, with human gene designation *NFE2L2*, has been shown to play a key role in the maintenance of redox homeostasis in mitochondria. Under basal and homeostatic conditions NRF is present in the cytoplasm bound to a protein designated KEAP1 (kelch like ECH associated protein 1). Under conditions when homeostasis is threatened and increased levels of oxidant substances are present, NRF2 is released from KEAP. Released NRF2 was reported by Ryoo and Kwak[31] to increase mitochondrial biogenesis through upregulation of specific genes. NRF2 was shown to bind to specific anti-oxidant response elements located in the promoter regions of specific genes.

NRF2 binding to promoters increased expression of glutathione biosynthetic enzyme and also leads to increased expression of glutathione peroxidase GPX1. NRF2 expression also increased synthesis of enzymes that increased NADPH production.

Mitochondrial biogenesis is increased by both NRF1 NRF2 and TFAM transcription factors.

Mitochondria and endoplasmic reticulum metabolism and nutrient sensing

Special sites of interaction between mitochondria and endoplasmic reticulum have been identified. These sites have been designated as mitochondria associated membranes (MAMs). Theurey and Rieusset[32] reviewed processes that occur at MAMs in response to alterations in nutrient and energy availability. They also noted that structural changes may occur in the MAMs as part of adaptations to nutrient availability.

Theurey and Rieusset emphasized the importance of nutrient sensors in mediating adaptations to differences in nutrient state. Specific dynamic processes that occur in mitochondria in response to nutrient state include fusion between mitochondria and formation of new mitochondrial through mitochondrial division and also through mitochondrial biogenesis. Adaptation also involves the destruction and catabolism of damaged mitochondria.

They noted that under conditions of nutrient deprivation there is increased production of specific transcription factors including peroxisomal proliferation receptor PPARA and its coactivator PPARG (PPARGC1A), and FOXO1.

When excess proteins accumulate in the endoplasmic reticulum, proteins may fail to assume the correct folded state and they are then functionally impaired. The accumulated unfolded proteins trigger a specific response, the unfolded protein response. Theurey and Rieusset noted the hypoglycemia and fatty acid accumulation can also trigger the unfolded protein response. Furthermore, under conditions of metabolic stress changes in the endoplasmic reticulum stimulate calcium release from the endoplasmic reticulum and also calcium release from mitochondria.

Metabolic flexibility and metabolic inflexibility

Theurey and Rieusset[32] reported that reduced capacity to adapt to different nutrient sates are key factors in the causation of metabolic disease, obesity and type 2 diabetes. These disorders are characterized by reduced insulin sensitivity and lipid accumulation. They emphasized the importance of nutrient sensors in mediating adaptations to different nutrient states.

Houtkooper and co-workers (see Smith et al.[33]) defined metabolic flexibility as:

"The ability to efficiently adapt metabolism by substrate sensing, trafficking, storage and utilization, depending on availability and requirement."

They noted that liver, adipose tissue and muscle play key roles in metabolic flexibility. In addition, metabolic flexibility is dependent on metabolic enzymes and pathways, transcription factors and on specific organelles. Metabolic flexibility is highly dependent on optimal mitochondrial function. Key features of metabolic flexibility include dramatic differences between metabolic processes in situations with metabolic abundance and situations when nutrient intakes are reduced.

They emphasized that human physiological adaptations primarily evolved during epochs when there were great fluctuations in the food and energy supply, depending on seasons and environmental conditions. Under such variable conditions metabolic flexibility was key to survival. However, constancy of food supply is the predominant current norm.

Smith et al., noted that carbohydrate rich intake leads to high levels of glucose that stimulate insulin secretion and increased glucose uptake into tissues and increased glucose breakdown and pyruvate oxidation with suppression of fatty acid oxidation. Fatty acids present are used to synthesize triglycerides and for fat storage. Smith et al., noted that under conditions of caloric excess the mitochondrial electron transfer system assumes greater activity. This leads to increased mitochondria membrane potential and to increased generation of reactive oxygen species. High nutrient activity alters the NADH/NAD ratio and the activity of Sirtuin and this reduces mitochondrial biogenesis. Under conditions of nutrient excess mitochondrial function ultimately becomes impaired.

Nutrient abundance also stimulates MTOR activity and increased protein synthesis and increased cell proliferation.

Key features of fasting involve activation of sensors of energy stress, such as AMPK. Entry of fatty acids into mitochondria is stimulated by activity of carnitine palmitoyl transferases and fatty acid oxidation in mitochondria is increased. During fasting glucose uptake into cells is reduced and glucose breakdown is slowed. Since glucose levels must ideally not fall below a certain level, when a hypoglycemic state is threatened gluconeogenesis is stimulated and glucose can be generated from other metabolites and molecules. Smith et al., noted that during prolonged fasting breakdown of amino acids can occur. Particularly important in this process is breakdown of branched chain amino acids through activity of branched chain keto-acid dehydrogenase. Under conditions of prolonged fasting autophagy and tissue breakdown occurs.

Smith et al., reviewed aspects of physical exercise and its impact on metabolism. They emphasized two types of skeletal muscle fibers. In type 1 oxidative fibers mitochondrial are abundant and oxidative phosphorylation and ATP production predominant. The type 1 fibers also carry out fatty acid oxidation. Type II skeletal muscle fibers are primarily glycolytic and carry out glycogen breakdown to produce ATP. Specific metabolic processes are carried out under low-intensity exercise and additional metabolic processes are initiated during high intensity exercise.

Smith et al., noted the importance of certain circulating factors involved in metabolic flexibility. These include factors released from the gut, e.g. GLP1 (glucagon like peptide 1) and ghrelin. In the fed state GLP1 levels are increased and there is reduced glucose production in the liver. Ghrelin functions as a neuropeptide it is secreted when nutrient levels are low and acts to stimulate appetite.

Other circulating factors that play important roles in metabolic flexibility include adipokines derived from fat tissue, myokines and hepatokines.

Smith et al., noted that diet and exercise also impact epigenetic modifications of DNA and histone. Epigenetic modifications impact gene regulation and play important roles in metabolic flexibility.

Adipose tissue

In a review of adipose tissue Esteve Ràfols[34] noted that is has long been known that white fat releases fatty acids during periods of negative energy balance. However, in recent decades white fat has also been found to release multiple regulatory factors, designated adipokines. These include factors that impact metabolism, factors that impact carbohydrate and energy metabolism and adipokines that impact immune response. One adipokine that has been intensely studied is leptin. Leptin receptors occur on cells in the peripheral tissue and in the central nervous system, Efeyan et al.[9] reported that binding of leptin to leptin receptors in hypothalamic neurons suppress appetite, in part through suppression of neuropeptides and neurotransmitters that function to stimulate appetite. Leptin receptor defects have been found in certain individuals with morbid obesity.

Adiponectin ADIPQ is a hormone produced by adipocytes and Efeyan et al.[9] noted that this adiponectin promotes insulin sensitivity and energy expenditure. Adiponectin mutations have been reported in some individuals with obesity and type 2 diabetes.

Metalloregulators factors

Bird[35] reviewed factors that control levels of metals, including iron, copper and zinc in humans. Metals that are essential for human health include iron, copper, manganese, cobalt and zinc. Metals function as co-factors for enzyme reactions. There is also evidence that they stabilize specific protein interactions and structures.

Bird noted that metal homeostasis is controlled by expression of genes that impact metal uptake, metal transport and metal storage. Zinc transport involves proteins encoded by ZNT and ZIP genes. There are 9 genes in the ZNT category currently they are classified as solute carriers in the SLC30A category. ZIP genes are now classified in as solute carriers in the SLC39A category.

Bird noted that when zinc levels are low specific ZIP genes are expressed. In contrast when Zinc levels are high specific ZNT genes are expressed. Another important factor that impacts transcription of genes when zinc levels are high is MTF1 metal regulatory transcription factor 1. It particularly induces transcription of the metal regulatory binding genes, the metallothioneins. MTF1 specifically binds to promoters of genes that contain a metal-responsive element.

Maret[36] reported that in humans approximately 3000 zinc metallo proteins occur and that in many of these proteins Zinc has a co-factor or a regulatory function. He classified ZIP genes as zinc importers that promote import of zinc into the cytoplasm and ZNTs as zinc exporters and are involved in zinc efflux. He noted that zinc ions can be stored inside cellular vesicles.

Metallothioneins are cysteine rich proteins that can bind metals, particularly copper and zinc and can serve to store metals and they can donate metals. More than 23 different genes encode metallothioneins in humans.

Bird[35] reported that when metal levels are high MTF1 binds to metal responsive elements and promote gene expression. Bird also reported that when iron levels are high the HAMP gene that encodes hepcidin, is expressed in liver and released into the circulation. In the circulation hepcidin undergoes modification and is activated. Active hepcidin limits intestinal absorption of iron. Active hepcidin also blocks iron release from storage sites. Bird emphasized that hepcidin plays the key role in iron homeostasis. The precise mechanisms through which iron triggers hepcidin release remain to be completely elucidated.

Bird reported that post-translational mechanisms also control activity of specific transcripts of genes involved in maintaining metal homeostasis. IRE's (iron responsive elements are found in the untranslated regions of several genes that encode products involved in iron homeostasis. The IREs can bind specific proteins IRPs. When iron levels are high, IREs bind IRPs and this limits the translation of the gene transcripts.

There is an IRE in the untranslated region of the ferritin encoding genes and this ensures that the ferritin gene transcript is only translated when the iron levels are high. Ferritin protein then promotes iron storage.

Bird noted that there are also mechanisms that impact mRNA stability and factors that promote mRNA stability have been shown to come into play when zinc levels are low.

Post-translational mechanisms also come into play in determining metal homeostasis. One such mechanism impacts the intra-cellular location of protein. Bird reported that under conditions when copper levels are low, the copper transported ATPases ATP7A and ATP7B are located in the Golgi network apparatus of the cell and copper is transported to copper requiring enzymes. When copper levels are high these enzymes move to the outer membrane of the cell where they can facilitate copper export. A specific protein COMMD1 also plays an important role in copper homeostasis and influences activity and location of ATP7A and ATP7B in response to copper levels.

Harmful damaging factors, endogenous or exogenous

Oxidative stress and mitochondria

Chance and co-workers (see Boveris et al.[37]) and Jensen[38] separately published evidence that mitochondria produce hydrogen peroxide (H_2O_2). Subsequently studies demonstrated that the mitochondrial derived H_2O_2 was derived from O_2^- superoxide. Superoxide is an oxygen molecule with one unpaired electron, and is classified as a reactive oxygen species molecule.

Murphy[39] reviewed production of reactive oxygen species in mitochondria. Murray reported that that superoxide radicals are derived primarily from mitochondrial complex 1 and are particularly present when mitochondria are not making ATP and protons are accumulating in the space between the inner and outer mitochondrial membranes. Superoxide radicals also accumulate when there is a high ratio of reduced nicotinamide adenine dinucleotide (NADH) to unreduced nicotinamide adenine dinucleotide (NAD) in the mitochondrial matrix.

Murphy documented the consequences of increased accumulations of reactive oxygen species in mitochondria. Reactive oxygen species can damage proteins and reduce the capacity of mitochondria to produce ATP. Reactive oxygen species (ROS) can also damage molecules in mitochondrial membranes through lipid peroxidation. ROS can increase the permeability of mitochondrial membranes and promote release of molecules and substances from the mitochondria into the cytoplasm and increase cell death apoptosis.

Collectively then the accumulation of reactive oxygen species in the mitochondria promotes mitochondrial dysfunction, cellular apoptosis and necrosis, disease and aging.

Within the mitochondria the enzyme superoxide dismutase (SOD2) acts to transform superoxide H_2O^- to hydrogen peroxide (H_2O_2). Efflux of hydrogen peroxide from mitochondria can occur. Murray noted that in addition, H_2O_2 can be degraded within mitochondria through the activity of specific enzymes, e.g. glutathione peroxidase, in the presence of the anti-oxidant glutathione. Thioredoxin and periredoxin can also degrade H_2O_2 when NADH is present.

Sies et al.[40] comprehensively reviewed oxidative stress. They emphasize that it is important to define the type of oxidative stress and the intensity of the oxidative stress and the physiological responses to stress. In addition, it is important to consider both the O_2^-, H_2O_2 and also other reactive species including reactive nitrogen species, reactive sulfur species, reactive carbonyl species and reactive selenium species.

In addition to damage to proteins and lipids, Sies et al., also documented evidence of oxidative damage to carbohydrates. Free carbohydrates undergo oxidation to yield reactive carbonyls.

Sies et al., reviewed biological responses to stress signaling. These include heat shock response, unfolded protein response, apoptosis, autophagy, inflammation and damage signals that impact cell proliferation.

They emphasized the importance of specific transcription factors that act as master regulators of responses to oxidative stress. These include the NRF2, KEAP1 and NFKB transcription factors.

Oxidative damage can lead to hydrolysis of DNA and can impact DNA methylation, Sies et al., noted that guanine is the DNA base most susceptible to oxidation. Oxidative damage leads to generation of 8-oxo-guanne or 8-hydroxyguanine. These converted forms of guanine pair with adenine rather than cytosine thus leading to transversion mutations. There are also reports that 8-oxyguanine can impact mitochondrial DNA.

Oxidative damage to molecules

Cadet and Davies[41] noted that oxygen is highly reactive with metals and with biologic molecules. Biologic systems are highly dependent for energy on oxidation processes in the mitochondria. Nevertheless, mitochondria also generate reactive oxygen species including hydrogen peroxide (H_2O_2) and the superoxide radical O_2^-. In addition to these damaging forms, nitro-oxygen species generated in the organism can be damaging. Oxidative damage can impact DNA, proteins and lipids.

Cadet and Davies noted that anti-oxidants, including vitamins C and E and polyphenols are found in foodstuffs. In addition, a large number of enzymes and cofactors with anti-oxidant properties are present in biological systems. Anti-oxidant enzymes present in humans include superoxide dismutases, glutathione peroxidase, peroxiredoxins (PRDX), glutaredoxins (GLRX). Coenzymes for antioxidants include glutathione and thioredoxin.

Systems to repair or remove damaged components

Cadet and Davies noted that damaged proteins and polypeptides can be removed through activities of the proteasome system and through specific proteases. Damage phospholipids can be degraded through phospholipases. Damaged proteins and lipids and damaged organelles can be removed through the autophagy processes.

There is also evidence that potentially damaging levels of oxidants can induce increased expression of genes that encode proteins or enzymes that can have anti-oxidant effects.

Cadet and Davies referred to the process of adaptive homeostasis and noted that an important factor in this process was the NRF2 transcription factor.

Oxidative DNA damage

Cadet and Davies[41] noted that that progress had been made in methods to detect oxidative DNA damage. Lindahl[42] reported that modified nitrogen base in DNA could be followed by removal of the modified base through the action of uracil DNA glycosylase. Cadet and Davies noted that additional specific DNA glycosylases utilized in base excision repair of DNA had been identified. These glycosylases include nei-like glycosylases NEIL1,2 and 3.

Cadet and Davies reported that oxidation reaction also occur on methylated cytosine. These oxidation reactions generate 5-hydroxy methylcytosine (5hmc), 5-formyl cytosine and 5-carboxy cytosine. TET enzymes play roles in these reactions and in the demethylation of cytosine.

Environmental agents and oxidative stress

Bisphenol A

Gassman[43] reported that Bisphenol A (BPA) production has steadily increased and that more than10 billion pounds of this substance are produced each year and human exposure to BPA has steadily increased. However, development of clear understanding of precise molecular actions of BPA in the human have lagged.

Gassman reported that evidence has emerged that indicates that BPA accumulation in the body results in increases in reactive oxygen species in part through reactions that impact peroxidase, NADPH and CYP450. Increases in oxidative stress have been demonstrated in several cell and tissue types.

Gassman noted that ROS generated as a result of BPA exposure may exceed the capacity of anti-oxidant systems. These anti-oxidant systems include reduced glutathione, ascorbic acid, thioredoxin and enzymes including superoxide dismutase, catalase and glutathione peroxidase.

Excess reactive species can lead to DNA damage including strand breaks, nucleotide changes, DNA cross links. Gassman reported further that BPA can alter chromatin folding and compaction and can down regulate DNA damage repair.

Early studies revealed that BPA accumulates in mitochondria through interaction with lipids in mitochondrial membranes.

Damaging metabolites

Specific inborn errors of metabolism can lead to the generation of abnormal metabolites or to abnormal levels of physiological substances. Both of these changes can lead to cellular damage. In addition, cellular damage can be caused by deficiency of certain molecular substances derived during the course of normal metabolism or by metabolic defects that result in energy deficiency. Gropman[44] reviewed patterns of injury that are generated through inborn errors of metabolism.

Excess levels of specific physiological substrates, referred to as substrate intoxication injury, can arise due to defects in amino acid metabolism or in organic acid metabolic defects. They can also arise as a result of urea cycle defects or in porphyrin generation defects. Specific inborn errors that lead to increased oxidative stress can result in cellular damage.

Phenylketonuria represents an example of a disorder where a block in the metabolism of the aminoacid phenylalanine leads to deficiency in the synthesis of the essential amino-acid tyrosine. Tyrosine is necessary for the generation of dopamine, a neurotransmitter. In addition, the block in the conversion of phenylalanine to tyrosine leads to generation of excess phenylketones. Gropman noted evidence of extensive brain white matter damage in untreated phenylketonuria.

Specific inborn errors of metabolism have been discovered to be due to cellular deficiency of certain essential compounds or due to deficiency or defective function of certain transporter proteins that are required to deliver substances into cells. One examples of such a disorder is cellular creatine deficiency due to defects in the creatine transporter SLC6A8. Gropman noted that the creatine deficiency disorder can also result from defective biosynthesis of creatine in the body.

Energy deficiency disorders can in some cases arise due to specific defects in mitochondrial function.[45,46]

DNA damage detection, consequences, and repair

It is important to take into account components in the organism's internal environment and factors in the external environment that can lead to DNA damage. In addition, throughout evolution mechanisms have evolved through which the organism can repair DNA damage.

Different mechanisms of repair are utilized depending on the sites within DNA where changes (mutations) have arisen: mutation may arise in the nitrogen bases in DNA, changes may impact the nucleotides that include the nitrogen bases and linked sugar, deoxyribose. As new strands of DNA are synthesized in replication processes inappropriate bases may be inserted these are referred to as mismatches, additional bases may be inserted or bases may be deleted.

In 2015 Nobel prizes were awarded to Tomas Lindahl, Aziz Sancar and Paul Modrich for their contributions to understanding of DNA repair mechanisms. Their contributions were presented in their Nobel lectures and Kunkel[47] reviewed their individual contributions.

Lindahl et al.[48] determined that processes involved in DNA damage repair included oxidative deamination or alkylation of the nucleotide bases and that damaged bases can be released from the DNA strand through cleavage of the glycosidic (deoxyribose) bond. Lindahl also identified base excision repair that involved generation of a modified nitrogen base followed by removal of the modified base through the action of glycosylase. The lesion in the DNA that resulted from the removal of the base was repaired through synthesis activity of DNA polymerase followed by ligase activity.

Aziz Sancar's early work concentrated on DNA damaged by ultra-violet light that generates photo products. He demonstrated that that the UV alterations could be reversed through an enzyme photolyase. In addition. Nucleotide excision repair and specific enzymes for nucleotide excision were identified. Sancar in his Nobel lecture[49] emphasized that an important action of ultra-violet light was to convert adjacent thymine to thymine dimers in which additional linkages were established between the bases. It was later determined that UV light could induce unusual linkages between pyrimidines these were referred to as cyclobutane pyrimidine dimers. The specific photoreactive enzyme Sancar and coworkers discovered, photolyase could break these abnormal linkages. Following cloning of the photolyase enzyme it was discovered that the enzyme utilized blue light absorbing cofactors methylene tetrahydrofolate and reduced flavine adenine dinucleotide. The purified photolysase was shown to bind to the cyclobutane dimers.

Later studies revealed that the pyrimidine dimers could be removed by excision and the DNA could then be repaired. Excision repair in *E. coli* was shown to be due to three different enzymes, designated UvrA, UvrB and UvrC. Following excision of the damaged nucleotides, DNA polymerase filled the gap with synthesis of new nucleotides and DNA ligase reunited the two ends of the DNA strand. Subsequent studies revealed that in humans 6 different genes and 16 different proteins were required for excision repair. Of particular significance was the discovery that defects in excision repair in human occurred in the condition known as Xeroderma pigmentosum.

Gene products involved in excision repair

XPA	XPA, DNA damage recognition and repair factor
RPA,	Heterotrimeric Replication Protein A (RPA) complex
TFIIH	TFIIH core complex helicase subunits ERCC2, ERCC3
XPC	XPC complex subunit, DNA damage recognition and repair factor
XPF	ERCC4 excision repair 4, endonuclease catalytic subunit
ERCC1	ERCC excision repair 1, endonuclease non-catalytic subunit
XPG	ERCC excision repair 5, endonuclease

Mismatch repair

Mismatched base pairs arise during DNA replication and recombination. Modrich[50] noted that a repair system must recognize the mismatched base pair and must also recognize the DNA strand on which the mismatch occurred. Early studies revealed that the newly synthesized strand could be recognized because it was transiently hypomethylated.

Mismatches that were unmethylated, were first reported by Glickman and Radman in 1980.[51] In *E. coli* this recognition involved four different enzymes mutH, mutL mutD and UvrD. Studies by Modrich's group led to isolation and purification of these enzymes. UvrD was shown to function as a helicase that unwound DNA. Mismatch recognition required mutS. The enzyme mutH was shown to slowly incise unmethylated DNA Other necessary components include the exonuclease and the single stranded DNA binding protein ssb. DNA polymerase II and DNA Ligase were required to complete repair.

Levinson et al.[52] reported that *E. coli* DNA mismatch repair mutants frequently led to nucleotide tandem repeats at a number of sites in the genome. These were referred to as microsatellite repeats.

Modrich noted that great stimulation to the identification of human proteins involved in mismatch repair were reports of high frequency of mononucleotide and dinucleotide repeat sequences in patients with the cancer predisposition syndrome Lynch syndrome and in patients with hereditary predisposition to colon cancer. Isolation and sequencing of the human homologs of *E. coli* mismatch repair enzymes was carried out and has led to identification of genes that are mutated in patients with Lynch syndrome and in patients with colon cancer, *MLH1*, *MSH2*, *MSH6*, and *PMS2*.[53] Mutations in these genes and in the *EPCAM* gene have been implicated in microsatellite instability and Lynch syndrome.[54] *EPCAM* encodes a molecule that also serves as an epithelial cell adhesion molecule.

Mismatch DNA repair genes in humans

It is important to note that mismatch repair defects occur not only in hereditary cancer predisposition to colon cancer but also in sporadic colon cancers, Li and Martin[55] reviewed colon cancer in context of mismatch repair deficiency but also in the context of dietary factors and studies on gut microbiome populations. They noted that large scale genomic sequencing transcriptome studies proteomic analyses and microbiome studies have been carried out in colon cancer. Ito et al.[56] reported that the micro-organism *Fusobacterium nucleatum* in associated with colorectal cancer. Loss of function mutations in mismatch repair gene encoded products were reported to occur in 15–20% of cases of sporadic cancer.

Pathogenic mutations in the following genes were reported in these cases of sporadic colon cancer.

MSH2	mutS homolog 2
MSH1	mutL homolog 1
MSH6	mutS homolog 6
PMS2	PMS1 homolog 2, mismatch repair system component
MSH3	mutS homolog 3
EXO1	exonuclease 1

Mismatch repair gene defects have also been reported to play roles in endometrial cancers and hematopoietic cancers

Altered CpG island DNA methylation patterns have been associated with some cases of colon cancer. In this phenotype, sometimes referred to as the CIMP phenotype, hypermethylation occurs in CpG islands in the promoter regions of specific genes, including frequently tumor suppressor genes, leading to decreased expression of these genes. Toyota in 1999[57] reported that in certain colon cancers hypermethylation occurred in the promoter region of *MLH1*. The CIMP colon cancers have other specific features including mutations of the *BRAF* gene (B-Raf proto-oncogene, serine/threonine kinase). CIMP tumors also often occur more frequently in the right colon.[58]

Peña-Diaz and Rasmussen[59] noted that inactivating germline gene mutation in mismatch repair genes were initially described in hereditary colon cancer and later non- hereditary somatic mutations were reported in colon cancer. In addition, in some cases of colon cancer epigenetic changes leading to decreased expression of DNA mismatch repair genes have been found. It is also important to note that deletions or insertion of nucleotides can also lead to decreased production of the products of these genes. These authors also reported that in some cases of colon cancer results of DNA sequence analysis revealed variants of unknown significance in mismatch repair genes.

Laboratory studies on tumors to search for DNA mismatch repair gene defects often involve immunochemical studies with antibodies specific to the mismatch repair gene encoded proteins.

With respect to DNA sequencing on blood samples of patients with suspected germline mutations in mismatch repair gene Peña-Diaz and Rasmussen noted that in 25–30% of these patients sequencing studies fail to reveal mutations. Some of these patients may have epigenetic changes that alter expression of the genes. They also noted evidence that appropriate expression of mismatch repair gene proteins has been shown in some cases to be due to aberrant microRNAs.

Signaling DNA damage

Ciccia and Elledge[60] reviewed the signaling system that is triggered when DNA is damaged and that triggers cellular response to this damage. They emphasized that enzymes and proteins involved in repairing DNA damage must be specific to the type of DNA damage and must be targeted to act at the specific sites of damage.

In addition to DNA damage repair the DNA damage response can impact the cell cycle and significant DNA damage may induce cell cycle arrest and even cell death.

Double stranded DNA damage

Double stranded DNA breaks that involve both strands of DNA at a particular site in the genome are considered to be the most damaging form of genomic disruption. These breaks activate checkpoint responses and cell cycle arrest in the cell.[61]

Exonucleases act at the site of breaks to remove damaged DNA and repair is then initiated. Repair mechanisms include non-homologous end joining and homology directed repair. Non-homologous end joining can lead to deletion in the DNA. Homologous DNA repair provides more accurate repair of double stranded breaks.[62]

Key genes involved in double stranded DNA repair include:

BRCA1	BRCA1, DNA repair associated (BRCA Breast cancer)
BRCA2	BRCA2, DNA repair associated (BRCA Breast cancer)
PALB2	partner and localizer of BRCA2
RAD51C	member of the RAD51 family.
RAD51D	member of the RAD51 family.
ATM	ATM serine/threonine kinase (ATM ataxia telangiectasia mutated)

Mutations in these genes have been found to lead to increased risk for breast cancer.

It is interesting to note that the BRCA1 and BRCA2 genes are not highly conserved in evolution.[62]

The BRCA2 protein was reported to play important roles in cell cycle checkpoint control.

Chen et al.[62] noted that cells that have low levels of BRCA1 or BRCA2 expression (e.g. due to germline mutations) are particular sensitive to specific environmental and metabolism generated chemicals, including aldehydes. Acetaldehyde and formaldehyde are generated during metabolism. Aldehydes are also generated from ingested alcohol and formaldehyde is present in tobacco smoke.

Interstrand cross-linking of DNA and the Fanconi pathway

Genes in the Fanconi pathway encode products that play particularly important roles in repairing the form of DNA damage that leads to cross links between DNA strands. Interstrand cross links can result from chemical damage induced by compounds including aldehydes and platinum.

Proteins that form the Fanconi complex are involved in recognizing interstrand cross-links, carrying out nucleotide incision, removing damaged material and then facilitating new DNA synthesis and lesion healing though homologous recombination.[63]

Sumpter and Levine[64] noted that in addition to their roles in DNA damage repair (DDR), the Fanconi pathway proteins have cytoprotective functions. They demonstrated that Fanconi complex associated (FA) proteins play roles in specific autophagy processes and contribute to removal of damaged cellular products and viruses.

The Fanconi pathway was initially discovered in investigations of a disease known as Fanconi anemia. This disease was found to be due to homozygous or compound heterozygous defects in any one of a number of genes that encode proteins that together constitute the Fanconi complex. The clinical manifestations of Fanconi syndrome include not only anemia due to bone marrow failure but also increased susceptibility to leukemia and to cancers of the upper gastro-intestinal tract and the urogenital tract.

Sumpter and Levine reported that there is evidence that somatic mutations and epigenetic variants that lead to decreased expression of the Fanconi protein genes also play roles in breast and ovarian cancers. They listed 23 different genes that encode proteins that participate in the Fanconi pathway functions.

Several of these proteins bind ubiquitin. Ubiquitylated FANC1 and FANCD2 proteins were found to bind to the forked structure that result during DNA replication and this binding protects the replication forks from degradation. Specific FA proteins were reported to suppress non-homologous joining of DNA segments separated by single strand breaks.

Cytoprotective functions include protecting cell components against reactive oxygen species and against pro-inflammatory cytokines. FANC protein was shown to protect against cytotoxic stress in part through its interaction with other proteins including activators of transcription, and through interactions with heat shock proteins. The FANC protein was also shown to activate the enzyme glutathione-S-transferase P1 (GSTP1) that protects against reactive oxygen species.

Sumpter and Levine determined that specific Fanconi proteins FANCC, FANCF and FANCL play roles in autophagy. They noted that FANCC protein was particularly involved in the process through which damaged mitochondria are removed.

Sumpter and Levine noted that removal of damaged mitochondria was important since they are an important source of reactive oxygen species. In addition, reactive oxygen species play roles in the generation of DNA interstrand cross links.

Exogenous and endogenous agents leading to DNA damage

Exogenous damaging agents can lead to several different forms of DNA damage including binding of chemical adducts to DNA, single and double stranded DNA breaks, strand cross-links and oxidative lesions.

Ciccia and Elledge[60] documented different physical and chemical agents that induced DNA damage. Physical agents including ionizing radiation and UV light can lead to pyrimidine dimers. Radiation including medical treatments and diagnostics can lead to base damage, to single stranded and double stranded DNA breaks. Chemical agents used in therapy can lead to a number of different forms of DNA damage including aberrant inter-strand crosslinks and to double or single stranded DNA breaks.

Barnes et al.[65] reviewed exogenous agents defined as carcinogenic, that can induce DNA damage. They defined two categories of DNA damaging carcinogens, activation dependent carcinogens that require metabolic processing prior to initiating changes in DNA and carcinogens that can directly initiate DNA changes. They noted electrophilic groups present in direct damaging agents interact with nitrogen and oxygen atoms present in DNA and other cellular components.

Indirect damaging agents undergo bioactivation in the organism to generate damaging compounds. The specific activating metabolic processes frequently involve cytochrome P450 enzymes, Other activating reactions can be carried out by enzymes such as sulfotransferases, *N*-acetyltransferase. Barnes et al., listed the following examples of indirectly acting carcinogens: polycyclic aromatic hydrocarbons, heterocyclic aromatic amines, N-Nitrosamines, mycotoxins and aristolochic acid present in certain plants.

Direct DNA damaging agents listed by Barnes et al., included ultra-violet light particularly UVB that induces cyclobutene dimers between thymidine and cytosine and also cytosine-cytosine dimers, thymidine-thymidine dimers were less impacted. UVA was noted to be poorly absorbed by human skin and its mutagenic effects were less clearly defined.

Damage through DNA strand cross linking was noted to be induced by specific chemotherapeutic agents, including cisplatin. Barnes et al., noted that DNA cross linking can also be due to specific endogenous compounds generated in the host, including acetaldehyde and malonaldehyde.

Tobacco

The combustion of tobacco that occurs during smoking was found to generate thousands of potentially damaging agents.[66] Tobacco smoking was reported to be associated with 90% of cases of lung cancer.[67] It was also reported to play key roles in cancer at other sites.[68]

Barnes et al.[65] noted that the key carcinogenic agents in tobacco smoke include polycyclic aromatic hydrocarbons and tobacco specific nitrosamines. The polycyclic hydrocarbons undergo bioactivation through the cytochrome P450 oxidase system. This activation leads to the generation of electrophilic compounds that can bind to DNA and other macromolecules leading to the formation of chemical adducts. In DNA tobacco derived molecules lead particularly to G-T transversion mutations.

Population differences in levels of expression of specific cytochrome P450 enzymes can potentially impact sensitivity to certain DNA damaging agents.

Barnes et al. reported that oxidative stress was responsible for the generation of endogenous hydroxyl (OH) radicals. The radicals that arise in the organisms interact with DNA causing disruption of phosphodiester bonds leading to DNA breaks.

DNA damage response and the cell cycle

During time periods when the DNA damage response is activated to repair DNA damage, it is important that cell cycle arrest is initiated to ensure that damaged DNA is not transmitted to subsequent generations of cells.

Transition from the stationary stage (S) of the cell cycle to the mitotic stage M is dependent on the activity of cyclins and activity of cyclin dependent kinase (CDK1).[69] Processes that inactivate CDK1 can lead to blocking of cell cycle transition.

Specific kinases play roles in regulating cell cycle transitions and are referred to as checkpoint kinases. These kinases include ATM (mutated in ataxia telangiecstasia), ATR serine threonine kinase and phosphoinositol-3 kinase like enzyme (DNAPK). Delia and Mizutani reported that although these kinases have similar functions in that the act as checkpoint kinases, they differ with respect to the types of DNA damage that trigger their activities. They noted that ATM kinase activity is primarily triggered by double stranded DNA breaks, and ATM impacts several steps in the cell cycle. ATR activity is primarily activated by single stranded DNA breaks and by stalled replication forks; in addition, it acts primarily to block transition from S phase. DNAPK was noted to be activated by double stranded DNA breaks.

Delia and Mizutani[70] reported that a specific complex the MRN complex senses double stranded DNA breaks. The MRN complex is composed of 3 protein MRE11-RAD50 and NBS1. This complex binds to double stranded DNA breaks and the MRN complex then binds ATM protein. This binding autophosphorylates ATM on serine 1981. The phosphorylated ATM then phosphorylates other proteins, including BRCA1 to facilitate DNA repair. The checkpoint kinase CHK2 is also targeted by ATM phosphorylation and the downstream effect of this is to prevent activity of the CDK1 and thereby lead to blocking of the G2/M cell cycle transition.

Heavy metal exposure

Park and Jeong[71] reviewed cellular responses in response to excess exposure to heavy metals. They noted that specific metals are essential for functions of enzymes and proteins, these metals include particularly, Iron (Fe), copper (Cu) and zinc (ZN). It is important to note that other specific metals are also utilized in physiological processes and in cofactors for enzyme reactions; these include magnesium (Mg), molybdenum (Mo) and even cobalt (Co). However physiological processes utilize very low quantities of these metals.

However excess levels of metals are damaging in part through their formation of abnormal protein linkages and through generation of reactive oxygen species. Specific proteins in the body have high metal affinities and are involved in binding and transport of metals. These proteins include metallothioneins. Park and Jeong noted the importance of a specific transcription factor MTF1, metal regulatory transcription factor 1 that induces transcription of metallothionein genes in response to the presence of excess metals. MTF1 is localized primarily in the cytoplasm. However, in the presence of excess metals it moves to the nucleus and stimulates expression of metallothionein encoding genes.

Park and Jeong noted that excess concentration of heavy metals in cells can also lead to increased expression of the heat shock factor HSF1 that promotes expression of the heat shock proteins HSP70 and HSP90. These heat shock proteins act as chaperones that help to refold and stabilize proteins.

Removal of damaged cellular material

Steady advances have been made in understanding the mechanisms and roles of protein modification through ubiquitination since its first discovery by Ciechanover and others.[72] For many years the primary roles of ubiquitination were thought to be to promote modifications of damaged proteins and to lead to their degradation in proteasomes. In recent years studies have revealed that modification of proteins by ubiquitin or ubiquitin-like molecules can promote interactions between proteins and ubiquitination can alter the function of a specific protein. In addition, there is now evidence for cross-talk and interaction between the ubiquitin-proteasome system and the autophagy system that includes phagosomes, endosomes and lysosomes and that can also lead to degradation of damaged molecules and organelles in the cell.[73]

The first step in ubiquitination involves the linkage of a 76 amino-acid protein ubiquitin to a targeted protein. Ubiquitination utilizes three enzymes, ubiquitin activating enzyme E1, ubiquitin transferring enzyme E2 and ubiquitin ligating enzyme E3, Activation of the ubiquitin protein involves processing at its C terminal. This processing requires ATP. Activated ubiquitin is transferred to the E2 enzyme and subsequently to the E3 enzyme that functions to ligate ubiquitin to the substrate protein Ubiquitin proteins can also be joined to each other to forms ubiquitin chains. Ubiquitin ligases predominantly transfer ubiquitin to lysine residues but can sometimes also transfer ubiquitin to serine or threonine residues.

Kwon and Ciechanover[73] reported that mammalian cells have more than 800 E3 ubiquitin ligase. E3 ubiquitin ligases are classified into 2main groups HECT ligase and Ring ligases. In addition, the ring ligase can form multi-subunit complexes with cullin scaffold proteins to generate cullin-ring ligases.

Kwon and Ciechanover reported that monoubiquitination of proteins primarily impacts protein interactions and protein localization. Proteins destined for degradation usually bind several or multiple ubiquitins. Polyubiquitination can involves ubiquitin and ubiquitin like modifiers SUMO (small ubiquitin modifier), NEDD8, ISG15. These ubiquitin-like modifiers usually bind at specific lysine residues.

Ubiquitin chains attached to proteins can bind adaptors. These are referred to as ubiquitin binding domains. Specific adaptors include RPN10 and RPN13 (proteasome regulatory particles) can link ubiquitinated substrates to proteasomes.

Ji and Kwon[74] reported that specific adapters can link ubiquitinate substrates to autophagic vacuoles. These adaptors include P62 (sequestosome 1), NBR1 (autophagy cargo receptor).

Ubiquitin chains can also be linked to organelles or to invading pathogens and through the action of adaptors can be passed to autophagosomes.

Different types of lysine linkages are associated with polyubiquitin chains. The lysine linkage system is described as a degron and it impacts the subsequent destination of the ubiquitinated substrate.

Lysine 48 linkage and to some degree lysine 11 linkages were reported to be primarily present when substrates were to bound to RPN10 and RPN13 adaptors and degraded in proteasomes. Lysine 48 and lysine 63 linkages on larger protein aggregates interacted with p62 (sequestosome and NBR1 and entered phagosomes for autophagy.

Cellular stress and the unfolded protein response

McMahon et al.[75] reviewed aspects of the Unfolded protein response (UPR) and the integrated stress response (ISR). They noted both of these responses can lead to altered transcription and facilitate expression of gene products that facilitate adaptation to stress. However, in the presence of significant stress apoptosis and cell death may be triggered.

McMahon et al. noted that the endoplasmic reticulum plays key roles in the maintenance of cellular homeostasis in part through facilitating protein folding and export. Under conditions of increased synthesis of proteins in the cell, misfolded or unfolded protein can accumulate and induce endoplasmic reticulum stress. The unfolded protein response is initiated to mitigate the endoplasmic reticulum stress. This response involved the activation of 3 key sensors, including ATF6 (activating transcription factor 6), PERK a kinase that phosphorylates the alpha subunit of eukaryotic translation-initiation factor 2, and to its inactivation. This reduces initiation of translation and results in repression of global protein synthesis. The third gene that exhibits increased expression in response to endoplasmic reticulum stress is IRE1 an inositol requiring kinase. This protein specifically impacts splicing and increases expression of specific factors that bind transcription factors. New symbols have been assigned, so that PERK is now designated as EIF2AK3 and IRE1 is now ERN1.

McMahon noted that control of RNA levels also constitutes part of the stress response. They proposed that non-protein coding RNAs play key roles in control of the stress response.

Telomere shortening

Turner et al.[76] reviewed telomere biology and they considered telomere length in the context of aging. They noted that inducers of biological aging include extrinsic factors and also intrinsic factors such as sub-optimal mitochondrial function, increased production of reactive oxygen species and inflammation. Telomere nucleic acid sequences are susceptible to oxidative damage.

Turner et al., documented evidence of telomere damage in specific disorders that are characterized by premature aging. Shortened telomeres were found in Hutchinson Gilford Progeria, and Werner syndrome. In addition. Shortened telomere or accelerated telomere shortening was documented in certain DNA breakage and genomic instability syndromes, including Nijmegen DNA breakage syndrome and ataxia telangiectasia.

Telomere length has also been found to be negatively correlated with number of units of alcohol consumers per year and with number of cigarettes smoked per year.

Tissue homeostasis and cell death

Green and Llambi[77] noted that death of potentially harmful cells is critical to the maintenance of tissue homeostasis. They reviewed three types of cell death, apoptosis, autophagy and necrosis and the signaling pathways associated with cell death.

Specific morphologic features characterize each of the three different forms of cell death. In apoptosis cells shrinks, blebs appear in the cell membrane and chromatin becomes condensed leading to pyknosis. Green and Llambi noted that the major signaling pathways involved in apoptosis include the intrinsic pathway that involve increased permeability of mitochondrial membranes and release of pre-apoptotic proteins such as BCL2 apoptosis regulator. The extrinsic cell death pathway is also involved in apoptosis. This involves binding of specific ligands to cell death receptors in the tumor necrosis factor receptor pathways.

They noted that cellular stress, including DNA damage and endoplasmic reticulum stress can lead to apoptosis. Green and Llambi also reported that apoptotic cell death can also occur in certain cell types that require availability for specific growth factors for survival. Death results if the growth factors are no longer available in sufficient quantities.

Morphologic features of autophagic cell death were reported to include development of enlarged cellular vesicles due to autophagy of degraded materials within the cells. Green and Llambi note that autophagic death can result from nutrient and or energy deficiency. Another key process in autophagy includes inhibition of mTORC1 and activation of a complex that contain ULK kinase (autophagy related kinase). This complex initiate activity of the family of ATG family of proteins that form the autophagosome. Autophagosomes subsequently fuse with lysosomes where phagocytosed products can be digested.

Necrosis was defined as the process found on microscopy to involve swelling of cells and rupture of the plasma membrane.

Key signaling processes in cells death also include activation of caspase enzymes, Caspases can also be activated by release of BCL from mitochondria.

Green and Llambi reported that key signaling factors in necrosis include RIP3 receptor interacting serine threonine kinase and TNFRs members if the family of tumor necrosis factor receptors. Toll receptor activation. Especially TLR4 activation, can also be involved in generating cell death following pathogen recognition and cytokine production.

Molecular mechanisms to counteract and detoxify xenobiotics

Xenobiotics are defined as chemical substances not produced in the human body. Specific systems exist in the body to counteract effects of these and detoxify them. These systems include enzymes and specific transcription activators to enhance expression of genes to promote xenobiotic destruction. Key enzyme systems that destroy Xenobiotics include the cytochrome P450 family of enzyme, flavin containing monooxygenase.

In addition, specific molecular systems promote extrusion of harmful chemicals from the body. These include members of the family of ATP binding cassette transporters. P-glycoproteins is an important eliminator of toxic substances.[78]

Pharmacologic agents as xenobiotics and pharmacogenetics

Bishop[79] reviewed pharmacogenetics and noted that genetic variation has relevance for dosing, side -effect occurrence and hyper-sensitivity reactions to medical treatments. He noted that many drugs require biotransformation before they become active this occurs primarily in the liver. Important enzymes involved in this process include cytochrome P450 enzymes, CYP2C9, CYP2C19 and CYP2D6. Additional steps in biotransformation of drugs include UDP glucuronyl-transferase and catechol aminotransferase. Bishop also noted the importance of P-glycoprotein in drug transport.

CYP2D6 is reported to metabolize 20% of marketed medications. The genetic variants that most commonly lead to altered CYP2D6 activity are single nucleotide polymorphisms. In addition, structural variants that alter activity also occur. CYP2D6 variation can lead to poor metabolism in some cases and to ultra-rapid metabolism in others

Important variants that impact drug metabolism also occur in CYP2C9. A specific enzyme Thiopurine methyltransferase (TPMT) is required for metabolism of thiopurine and related medications. These medications are used in treatment of cancer and in diseases associated with abnormal immune response. Bishop noted that 40 different TPMT variants have been described and that many of these lead to decreased enzyme activity and altered responses to standard doses of medication.

In addition to genetic variants that impact drug metabolism, there are genetic variants that lead certain individuals to manifest hypersensitivity reaction to medication. Hypersensitivity reactions include the Stevens-Johnson syndrome and also epidermal necrolysis, both are serious conditions. The presence of the allele HLAB*1502 predisposes to hypersensitivity reactions. The HLAB*1502 allele occurs with frequencies of 10–15% in individuals from East Asia and can lead to severe hypersensitivity reactions in response to treatment of epilepsy. Other specific HLA alleles are also sometime causative of hypersensitivity reaction, these include HLAA 1301, HLADRB 1602.

Arylhydrocarbon receptor (AHR)

This receptor is activated by specific ligands that include damaging exogenous and endogenous substances. On activation of the AHR by ligand binding, the receptor acts as a transcription factor, Petriello et al.[80] noted that dioxins were among the first identified ligands that bind to AHR. Unactivated AHR occurs in the cytoplasm as part of a complex that also contains heat shock proteins and a co-chaperone XAP2 (AIP aryl hydrocarbon interacting protein). Following binding to ligand this complex moves to the nucleus where it binds to AHR nuclear translocator and can then bind to DNA sequence elements defined as xenobiotic response elements (XREs). This then leads to expression of genes involved in xenobiotic destruction.

References

1. Bernard C. *An introduction to the study of experimental medicine.* trans. H.C. Green, New York: Dover Publications; 1865. [1865]1957.
2. Cannon WB. *The wisdom of the body.* New York, NY: W.W. Norton & Co; 1932.
3. Heitman J, Movva NR, Hall MN. Targets for cell cycle arrest by the immunosuppressant rapamycin in yeast. *Science* 1991;**253**(5022):905–9.
4. Laplante M, Sabatini DM. mTOR signaling in growth control and disease. *Cell* 2012;**149**(2):274–93. https://doi.org/10.1016/j.cell.2012.03.017. Review, 22500797.
5. Saxton RA, Sabatini DM. mTOR signaling in growth, metabolism, and disease. *Cell* 2017;**168**(6):960–76. https://doi.org/10.1016/j.cell.2017.02.004. Review. Erratum in: Cell. 2017;169(2):361–371, 28283069.
6. Efeyan A, Sabatini DM. Nutrients and growth factors in mTORC1 activation. *Biochem Soc Trans* 2013;**41**(4):902–5. https://doi.org/10.1042/BST20130063. 23863153.

7. Giguère V. Canonical signaling and nuclear activity of mTOR-a teamwork effort to regulate metabolism and cell growth. *FEBS J* 2018;**285**(9):1572–88. https://doi.org/10.1111/febs.14384. Epub 2018 Jan 31. Review, 29337437.

8. Huang K, Fingar DC. Growing knowledge of the mTOR signaling network. *Semin Cell Dev Biol* 2014;**36**:79–90. https://doi.org/10.1016/j.semcdb.2014.09.011.

9. Efeyan A, Comb WC, Sabatini DM. Nutrient-sensing mechanisms and pathways. *Nature* 2015;**517**(7534):302–10. https://doi.org/10.1038/nature14190. Review, 25592535.

10. Demetriades C, Plescher M, Teleman AA. Lysosomal recruitment of TSC2 is a universal response to cellular stress. *Nat Commun* 2016;**7**:10662. https://doi.org/10.1038/ncomms10662. 26868506.

11. Mihaylova MM, Shaw RJ. The AMPK signalling pathway coordinates cell growth, autophagy and metabolism. *Nat Cell Biol* 2011;**13**(9):1016–23. https://doi.org/10.1038/ncb2329. 21892142.

12. Sabatini DM. Twenty-five years of mTOR: uncovering the link from nutrients to growth. *Proc Natl Acad Sci USA* 2017;**114**(45):11818–25. https://doi.org/10.1073/pnas.1716173114. 29078414.

13. Menon S, Dibble CC, Talbott G, Hoxhaj G, Valvezan AJ, et al. Spatial control of the TSC complex integrates insulin and nutrient regulation of mTORC1 at the lysosome. *Cell* 2014;**156**(4):771–85. https://doi.org/10.1016/j.cell.2013.11.049. 24529379.

14. Appenzeller-Herzog C, Hall MN. Bidirectional crosstalk between endoplasmic reticulum stress and mTOR signaling. *Trends Cell Biol* 2012;**22**(5):274–82. https://doi.org/10.1016/j.tcb.2012.02.006. 22444729.

15. Ebrahimi-Fakhari D, Saffari A, Wahlster L, Sahin M. Using tuberous sclerosis complex to understand the impact of MTORC1 signaling on mitochondrial dynamics and mitophagy in neurons. *Autophagy* 2017;**13**(4):754–6. https://doi.org/10.1080/15548627.2016.1277310. 28121223.

16. Metallo CM, Vander Heiden MG. Understanding metabolic regulation and its influence on cell physiology. *Mol Cell* 2013;**49**(3):388–98. https://doi.org/10.1016/j.molcel.2013.01.018. Review, 23395269.

17. Harden A, Young WJ. The alcoholic ferment of yeast-juice. *Proc R Soc Lond Ser B Biol Sci* 1906;**77**(519):405–20. https://doi.org/10.1098/rspb.1906.0029.

18. Goldberger J. The etiology of pellagra. 1914. *Public Health Rep* 2006;**121**(Suppl 1):77–9. discussion 76, 16550768.

19. Warburg O, Christian W. *Biochem Z* 1935;**275**:464.

20. Cantó C, Menzies KJ, Auwerx J. NAD(+) metabolism and the control of energy homeostasis: a balancing act between mitochondria and the nucleus. *Cell Metab* 2015;**22**(1):31–53. https://doi.org/10.1016/j.cmet.2015.05.023. Review, 26118927.

21. Brown MS, Radhakrishnan A, Goldstein JL. Retrospective on cholesterol homeostasis: the central role of Scap. *Annu Rev Biochem* 2018;**87**:783–807. https://doi.org/10.1146/annurev-biochem-062917-011852. 28841344.

22. Südhof TC, Goldstein JL, Brown MS, Russell DW. The LDL receptor gene: a mosaic of exons shared with different proteins. *Science* 1985;**228**(4701):815–22. 2988123.

23. Hara T, Ichimura A, Hirasawa A. Therapeutic role and ligands of medium- to long-chain Fatty Acid receptors. *Front Endocrinol (Lausanne)* 2014;**5**:83. https://doi.org/10.3389/fendo.2014.00083. eCollection 2014. Review, 24917851.

24. Ichimura A, Hara T, Hirasawa A. Regulation of Energy Homeostasis via GPR120. *Front Endocrinol (Lausanne)* 2014;**5**:111. https://doi.org/10.3389/fendo.2014.00111. eCollection 2014. Review, 25071726.

25. Im DS. FFA4 (GPR120) as a fatty acid sensor involved in appetite control, insulin sensitivity and inflammation regulation. *Mol Aspects Med* 2018;**64**:92–108. https://doi.org/10.1016/j.mam.2017.09.001. 28887275.

26. Sheng R, Yang L, Zhang Y, Xing E, Shi R. Discovery of novel selective GPR120 agonists with potent anti-diabetic activity by hybrid design. *Bioorg Med Chem Lett* 2018;**28**(15):2599–604. https://doi.org/10.1016/j.bmcl.2018.06.047. 29980358.

27. Ruegsegger GN, Creo AL, Cortes TM, Dasari S, Nair KS. Altered mitochondrial function in insulin-deficient and insulin-resistant states. *J Clin Invest* 2018;**128**(9):3671–81. https://doi.org/10.1172/JCI120843. 30168804.

28. Tsang CK, Chen M, Cheng X, Qi Y, Chen Y, et al. SOD1 phosphorylation by mTORC1 couples nutrient sensing and redox regulation. *Mol Cell* 2018;**70**(3). 502–515.e8. https://doi.org/10.1016/j.molcel.2018.03.029. 29727620.

29. Liu L, Nam M, Fan W, Akie TE, Hoaglin DC, et al. Nutrient sensing by the mitochondrial transcription machinery dictates oxidative phosphorylation. *J Clin Invest* 2014;**124**(2):768–84. https://doi.org/10.1172/JCI69413. 24430182.

30. Oláhová M, Hardy SA, Hall J, Yarham JW. Haack et al. LRPPRC mutations cause early-onset multisystem mitochondrial disease outside of the French-Canadian population. *Brain* 2015;**138**(Pt 12):3503–19. https://doi.org/10.1093/brain/awv291. 2651095.

31. Ryoo IG, Kwak MK. Regulatory crosstalk between the oxidative stress-related transcription factor Nfe2l2/Nrf2 and mitochondria. *Toxicol Appl Pharmacol* 2018;**359**:24–33. https://doi.org/10.1016/j.taap.2018.09.014. Review, 30236989.

32. Theurey P, Rieusset J. Mitochondria-associated membranes response to nutrient availability and role in metabolic diseases. *Trends Endocrinol Metab* 2017;**28**(1):32–45. https://doi.org/10.1016/j.tem.2016.09.002. Review, 27670636.

33. Smith RL, Soeters MR, Wüst RCI, Houtkooper RH. Metabolic flexibility as an adaptation to energy resources and requirements in health and disease. *Endocr Rev* 2018;**39**(4):489–517. https://doi.org/10.1210/er.2017-00211. 29697773.

34. Esteve RM. Adipose tissue: cell heterogeneity and functional diversity. *Endocrinol Nutr* 2014;**61**(2):100–12. https://doi.org/10.1016/j.endonu.2013.03.011. [23834768].

35. Bird AJ. Cellular sensing and transport of metal ions: implications in micronutrient homeostasis. *J Nutr Biochem* 2015;**26**(11):1103–15. https://doi.org/10.1016/j.jnutbio.2015.08.002. Review, 26342943.

36. Maret W. Zinc in Cellular Regulation: The Nature and Significance of "Zinc Signals". *Int J Mol Sci* 2017;**18**(11). pii: E2285. https://doi.org/10.3390/ijms18112285. Review, 29088067.

37. Boveris A, Oshino N, Chance B. The cellular production of hydrogen peroxide. *Biochem J* 1972;**128**(3):617–30 [4404507].

38. Jensen PK. Antimycin-insensitive oxidation of succinate and reduced nicotinamide-adenine dinucleotide in electron-transport particles. I. pH dependency and hydrogen peroxide formation. *Biochim Biophys Acta* 1966;**122**(2):157–66. 4749271.

39. Murphy MP. How mitochondria produce reactive oxygen species. *Biochem J* 2009;**417**(1):1–13. https://doi.org/10.1042/BJ20081386. Review, 19061483.

40. Sies H, Berndt C, Jones DP. Oxidative stress. *Annu Rev Biochem* 2017;**86**:715–48. https://doi.org/10.1146/annurev-biochem-061516-045037. 28441057.

41. Cadet J, Davies KJA. Oxidative DNA damage & repair: an introduction. *Free Radic Biol Med* 2017;**107**:2–12. https://doi.org/10.1016/j.freeradbiomed.2017.03.030. 28363603.

42. Lindahl T. My journey to DNA repair. *Genomics Proteomics Bioinformatics* 2013;**11**(1):2–7. https://doi.org/10.1016/j.gpb.2012.12.001. 23453014.

43. Gassman NR. Induction of oxidative stress by bisphenol A and its pleiotropic effects. *Environ Mol Mutagen* 2017;**58**(2):60–71. https://doi.org/10.1002/em.22072. 28181297.

44. Gropman AL. Patterns of brain injury in inborn errors of metabolism. *Semin Pediatr Neurol* 2012;**19**(4):203–10. https://doi.org/10.1016/j.spen.2012.09.007. Review 23245553.

45. DiMauro S, Hirano M. Pathogenesis and treatment of mitochondrial disorders. *Adv Exp Med Biol* 2009;**652**:139–70. https://doi.org/10.1007/978-90-481-2813-6_10. Review, 20225024.

46. Gorman GS, Chinnery PF, DiMauro S, Hirano M, Koga Y, et al. Mitochondrial diseases. *Nat Rev Dis Primers* 2016;**2**:16080. https://doi.org/10.1038/nrdp.2016.80. Review, 27775730.

47. Kunkel TA. Celebrating DNA's repair crew. *Cell* 2015;**163**(6):1301–3. https://doi.org/10.1016/j.cell.2015.11.028. 26638062.

48. Lindahl T, Modrich P, Sancar A. The 2015 Nobel prize in chemistry the discovery of essential mechanisms that repair DNA damage. *J Assoc Genet Technol* 2016;**42**(1):37–41. 27183258.

49. Sancar A. Mechanisms of DNA Repair by Photolyase and Excision Nuclease (Nobel Lecture). *Angew Chem Int Ed Engl* 2016;**55**(30):8502–27. https://doi.org/10.1002/anie.201601524. Review, 27337655.

50. Modrich P. Mechanisms in E. coli and human mismatch repair (Nobel lecture). *Angew Chem Int Ed Engl* 2016;**55**(30):8490–501. https://doi.org/10.1002/anie.201601412.

51. Glickman BW, Radman M. Escherichia coli mutator mutants deficient in methylation-instructed DNA mismatch correction. *Proc Natl Acad Sci U S A* 1980;**77**(2):1063–7. 6987663.

52. Levinson G, Gutman GA. High frequencies of short frameshifts in poly-CA/TG tandem repeats borne by bacteriophage M13 in Escherichia coli K-12. *Nucleic Acids Res* 1987;**15**(13):5323–38. 3299269.

53. Lagerstedt Robinson K, Liu T, Vandrovcova J, Halvarsson B, et al. Lynch syndrome (hereditary nonpolyposis colorectal cancer) diagnostics. *J Natl Cancer Inst* 2007;**99**(4):291–9. 17312306.

54. Latham A, Srinivasan P, Kemel Y, Shia J, Bandlamudi C, et al. Microsatellite instability is associated with the presence of lynch syndrome pan-cancer. *J Clin Oncol* 2019;**37**(4):286–95. https://doi.org/10.1200/JCO.18.00283.

55. Li SKH, Martin A. Mismatch repair and colon cancer: mechanisms and therapies explored. *Trends Mol Med* 2016;**22**(4):274–89. https://doi.org/10.1016/j.molmed.2016.02.003. 26970951.

56. Ito M, Kanno S, Nosho K, Sukawa Y, Mitsuhashi K, et al. Association of Fusobacterium nucleatum with clinical and molecular features in colorectal serrated pathway. *Int J Cancer* 2015;**137**(6):1258–68. https://doi.org/10.1002/ijc.29488. 25703934.

57. Toyota M, Ahuja N, Ohe-Toyota M, Herman JG, Baylin SB, Issa JP. CpG island methylator phenotype in colorectal cancer. *Proc Natl Acad Sci U S A* 1999;**96**(15):8681–6. 10411935.

58. Gallois C, Pernot S, Zaanan A, Taieb J. Colorectal cancer: why does side matter? *Drugs* 2018;**78**(8):789–98. https://doi.org/10.1007/s40265-018-0921-7. 29790124.

59. Peña-Diaz J, Rasmussen LJ. Approaches to diagnose DNA mismatch repair gene defects in cancer. *DNA Repair (Amst)* 2016;**38**:147–54. https://doi.org/10.1016/j.dnarep.2015.11.022. 26708048.

60. Ciccia A, Elledge SJ. The DNA damage response: making it safe to play with knives. *Mol Cell* 2010;**40**(2):179–204. https://doi.org/10.1016/j.molcel.2010.09.019. Review. [20965415].

61. Dasika GK, Lin SC, Zhao S, Sung P, et al. DNA damage-induced cell cycle checkpoints and DNA strand break repair in development and tumorigenesis. *Oncogene* 1999;**18**(55):7883–99. Review, 10630641.

62. Chen CC, Feng W, Lim PX, Kass EM, Jasin M. Homology-directed repair and the role of BRCA1, BRCA2, and related proteins in genome integrity and cancer. *Annu Rev Cancer Biol* 2018;**2**:313–36. https://doi.org/10.1146/annurev-cancerbio-030617-050502. [30345412].

63. Walden H, Deans AJ. The Fanconi anemia DNA repair pathway: structural and functional insights into a complex disorder. *Annu Rev Biophys* 2014;**43**:257–78. https://doi.org/10.1146/annurev-biophys-051013-022737. 24773018.

64. Sumpter Jr R, Levine B. Emerging functions of the Fanconi anemia pathway at a glance. *J Cell Sci* 2017;**130**(16):2657–62. https://doi.org/10.1242/jcs.204909. 28811338.

65. Barnes JL, Zubair M, John K, Poirier MC, Martin FL. Carcinogens and DNA damage. *Biochem Soc Trans* 2018;**46**(5):1213–24. https://doi.org/10.1042/BST20180519. 3028751.

66. International agency for research on cancer IARC Working Group on the Evaluation of Carcinogenic Risks to Humans. Tobacco smoke and involuntary smoking. *IARC Monogr Eval Carcinog Risks Hum* 2004;**83**:1–1438. 15285078.

67. Vargas AJ, Harris CC. Biomarker development in the precision medicine era: lung cancer as a case study. *Nat Rev Cancer* 2016;**16**(8):525–37. https://doi.org/10.1038/nrc.2016.56. 27388699.

68. Sasco AJ, Secretan MB, Straif K. Tobacco smoking and cancer: a brief review of recent epidemiological evidence. *Lung Cancer* 2004;**45**(Suppl 2):S3–9. https://doi.org/10.1016/j.lungcan.2004.07.998. 15552776.

69. Swaffer MP, Jones AW, Flynn HR, Snijders AP, Nurse P. CDK substrate phosphorylation and ordering the cell cycle. *Cell* 2016;**167**(7):1750–1761.e16. https://doi.org/10.1016/j.cell.2016.11.034. 27984725.

70. Delia D, Mizutani S. The DNA damage response pathway in normal hematopoiesis and malignancies. *Int J Hematol* 2017;**106**(3):328–34. https://doi.org/10.1007/s12185-017-2300-7. Epub 2017 Jul 13. Review, 28707218.

71. Park C, Jeong J. Synergistic cellular responses to heavy metal exposure: a minireview. *Biochim Biophys Acta Gen Subj* 2018;**1862**(7):1584–91. https://doi.org/10.1016/j.bbagen.2018.04.003. 29631058.

72. Ciechanover A, Elias S, Heller H, Ferber S, Hershko A. Characterization of the heat-stable polypeptide of the ATP-dependent proteolytic system from reticulocytes. *J Biol Chem* 1980;**255**(16):7525–8. 6249802.

73. Kwon YT, Ciechanover A. The ubiquitin code in the ubiquitin-proteasome system and autophagy. *Trends Biochem Sci* 2017;**42**(11):873–86. https://doi.org/10.1016/j.tibs.2017.09.002. Review. [28947091].

74. Ji CH, Kwon YT. Crosstalk and interplay between the ubiquitin-proteasome system and autophagy. *Mol Cells* 2017;**40**(7):441–9. https://doi.org/10.14348/molcells.2017.0115. 28743182.

75. McMahon M, Samali A, Chevet E. Regulation of the unfolded protein response by noncoding RNA. *Am J Physiol Cell Physiol* 2017;**313**(3):C243–54. https://doi.org/10.1152/ajpcell.00293.2016. Review, 28637678.

76. Turner KJ, Vasu V, Griffin DK. Telomere biology and human phenotype. *Cells* 2019;**8**(1) pii: E73. https://doi.org/10.3390/cells8010073. Review, 30669451.

77. Green DR, Llambi F. Cell death signaling. *Cold Spring Harb Perspect Biol* 2015;**7**(12) pii: a006080. https://doi.org/10.1101/cshperspect.a006080. Review, 26626938.

78. Efferth T, Volm M. Multiple resistance to carcinogens and xenobiotics: P-glycoproteins as universal detoxifiers. *Arch Toxicol* 2017;**91**(7):2515–38. https://doi.org/10.1007/s00204-017-1938-5. 28175954.

79. Bishop JR. Pharmacogenetics. *Handb Clin Neurol* 2018;**147**:59–73. https://doi.org/10.1016/B978-0-444-63233-3.00006-3. [29325628].

80. Petriello MC, Hoffman JB, Morris AJ, Hennig B. Emerging roles of xenobiotic detoxification enzymes in metabolic diseases. *Rev Environ Health* 2017;**32**(1–2):105–10. https://doi.org/10.1515/reveh-2016-0050. 27837601.

7

MICROORGANISMS AND MICROBIOME

Microorganisms in soil

In a review in 2017 Fierer[1] noted the most soil microorganisms remained undescribed. Key questions remain including how the soil microbiome is impacted by human activities and by environmental and climate changes.

Fierer emphasized the key roles of soil in nutrient cycling and survival of plants and animals. In addition, soil microorganisms have in the past and they continue in the present to constitute important sources of antibiotics.

Fierer noted that the composition of soil microorganisms varies in different environments. Key factors of soil that influence microbiome composition include moisture level, pH, soil oxygen content, presence of sources of nitrogen and phosphorus, temperature, soil texture, plant species present and animal predation. Fierer noted that abiotic soil regions also occur.

Recent studies in countries and regions across the globe have provided information on the global distribution of microbial biomass, including bacteria, fungi, archaea, protists and viruses. Protists are defined as eukaryotes that are not plants, fungi or animals. They are predominantly unicellular. Protists in the ocean include plankton. Free living protists are considered to be important in ecosystem stability.[2]

Soils that have a higher water content are richer in biomass. Therefore, forested regions of the globe have higher biomass content and desert regions have the lowest biomass.

Nucleotide sequencing is used for classification of microorganisms, however Fierer emphasized that precise classification of many microorganisms has not yet been achieved. Sequence in 16SRNA is used for classification of bacteria and Archaea. For classification of fungi, specific sequence in internal transcribed spacer (ITS1) is used. For protist identification 18SRNA sequence is used.

Soils are rich in viruses with 10^7 to 10^9 viral particles present per gram of soil. Freier noted that recent advances in viral metagenomics will lead to deeper of the viral characterizations of soils.

Gene Environment Interactions. https://doi.org/10.1016/B978-0-12-819613-7.00007-4

Horizontal gene transfer between microorganisms

Fierer emphasized that within soil there are high rates of horizontal gene transfer that occurs through transduction or through conjugation. Particular characteristics reported to occur as a result of horizontal gene transfer include transfer of antibiotic resistant genes and also transfer of arsenic detoxification mechanisms.

Efforts to restore degraded soils

In 2015 de Vrieze[3] authored a report related to efforts to restore soils in a region where farming practices had destroyed soil vitality and rendered barren an area, previously described as a bread-basket region.

There is evidence that diverse soil microbiome result in fewer plant diseases. In addition, specific microbes have been shown to enable plants to survive extreme conditions. A specific bacterium Stenotrophomonas rhizophilia was reported to increase resistance to drought in specific crops. This bacterium secreted molecules referred to as osmoprotectants. These molecules prevent outflow of water out of plants in salty conditions.

There is particular interest in rhizobiome organisms that inhabit plant roots. There is also growing concepts regarding the interactions of plant root systems with their environment, referred to as the rhizosphere and evidence that plant roots not only take up nutrients from soil but also transmit exudates into soil. Mommer et al.[4] reviewed the rhizosphere frame work. They noted that plants take up water from soil and 15 essential nutrients including copper, zinc nitrogen and phosphorus. Nitrogen utilized by plants is mostly present in organic materials such as ammonia and is also present as NO_3 (nitrate). The nitrogen forms in soil therefore differ from the predominant nitrogen form in air.

Symbiotic relationships exist between plant roots and microorganisms in the root environments to provide necessary elements.

Plant root exudates

Mommer et al. reported that many different substances are exuded from plant roots including proteins, fatty acids, sterols, organic acids, amino acids and sugars. The substances exuded from roots ultimately influence nutrient availability for the plant. Root exudates also stimulate mineralization of organically linked molecules. The amount and types of exudates produced by the plant roots are influenced by microorganisms present in the root systems.

In addition, there is evidence that plant root development is integrated with available nutrients and that plant exudates and the rhizobiome collectively interact in development of the plant root system.

Climate change and microorganisms

Climate change has different consequences in different environments. Hutchins et al.[5] noted that increased temperatures and decreased pH shift the ocean locations of plankton and would lead to loss of particular plankton forms. They noted that specific species of plankton that decline includes coccolithophores, that currently absorb carbon including fossil fuel derived carbon. Loss of this microorganism will represent loss of the ecosystem system function it provides.

Hutchins et al. also reported that changes in soil moisture content impact cycling of nutrients in soil. They concluded that climate change is impacting the geographic range and the diversity of microbial species and that the precise consequences of these changes remain unknown.

Agroecosystems

Toju et al.[6] emphasized the importance of studying agroecosystems. Agroecosytem is defined as "a spatially and functionally coherent unit of agricultural activity, the living and nonliving components involved in that unit and their interactions".

Toju et al. note that studies of microbiome functions in ecosystems are particularly important in processes of degradation of waste. They emphasized that characterization of the core microorganisms was particularly enhanced through the development of new techniques including microfluidics and informatics. They also emphasized the complex webs of interactions that microorganisms have with each other and with the endosphere in which they exist.

They considered that key imperiling factors for microbial communities important in agriculture included chemical fertilizers and climate change. Toju et al. concluded that studies on the effects on microorganisms of agricultural practices and specific environmental changes continue to be of great importance.

Antibiotic resistance

Resistance of pathogenic organisms to therapeutically administered antibiotics has become a problem of increasing concern throughout the world in recent decades.

Antibiotic resistance is not a new problem, evidence for resistance to penicillin was documented a few years of it initial clinical therapeutic use.

Watkins and Bonomo[7] reviewed current concepts and future directions in addressing antibiotic resistance. They emphasized that there were two key problems in that the occurrence of increased frequencies of antibiotic resistance coincides with decades when very few new antibiotics are being developed. The misuse of antibiotics contributes to emergence of resistance.

The increased frequencies of antibiotic resistance increase the death rates of patients and also contribute to increases in medical costs through prolongations of hospital stays.

Mechanisms by which microorganisms develop antibiotic resistance

New mutations in bacterial DNA may lead to antibiotic resistance. Acquired resistance may arise through lateral transfer of genes by means of plasmid or phage. Alterations in expression of specific bacterial genes may facilitate antibiotic resistance. Examples include altered expression of enzymes such as lactamase, methylase or lipopolysaccharidase.[8]

Watkins and Bonomo[7] noted that particular environments facilitate the development of antibiotic resistant organisms. These include environments where high levels of bacteria occur in proximity to low levels of antibiotics. Examples of such environments include waste water facilities.

Antibiotic resistance in Gram negative organisms

Gram negative organisms have emerged as particularly likely to develop antibiotic resistance. This results in part from the low permeability of the Gram- negative organisms to antibiotics. Cama et al.[9] reviewed tools that are available for advance testing of antibiotic uptake into Gram negative organisms.

They reported that the outer membrane of Gram-negative organisms is a complex double membrane envelope with a lipopolysaccharide polymer mesh on the outer membrane surface. Coma et al. emphasized that the lipopolysaccharide mesh prevents uptake of large molecules into the bacteria. Small molecules can transfer into the bacterial through porin structures.

Cama et al. noted that Gram- negative organisms also have efficient means of expelling molecules, including antibiotics, through efflux pumps.

Research in measuring antibiotic accumulation within microorganisms is a current focus of research efforts. Various techniques used in these studies include microfluidic assays, single cell microscopy and mass spectrometry.

Antibiotic tolerant persisters and microfilms

A particular problem that arises is the occurrence of persister bacterial cells that are slow growing and antibiotic tolerant. Lewis[10] note that currently available antibiotics are particular active against rapidly growing bacteria and do not effectively work against slow growing organisms. Persister cells are particularly a problem if they occur in biofilms. Biofilms tend to occur on indwelling medical devices, e.g. catheters and prostheses.

Antibiotic use in animal food production

Numerous studies have revealed the occurrence of antibiotic resistant organisms in industrial farming settings. Watkins and Bonomo noted that use of antibiotics in livestock production was usually for growth promotion and disease prevention rather than for disease treatment. In addition, antibiotics were administered in sub-therapeutic doses.

The Pew Charitable Trust compiled a bibliography of articles that reported the presence of antibiotic resistant organisms in animal production settings https://grist.files.wordpress.com/2011/04/pew_abr_bibliography.pdf.

Antibiotic stewardship programs

The World Health Organization and governmental agencies in different countries, including the Center for Disease Control in the USA have outlined and sought to implement antibiotic stewardship programs, https://www.cdc.gov/hicpac/recommendations/antibiotic-stewardship-statement.html.

Key precepts of these programs include:

Do not use antibiotics when antibiotics will not help, e.g. in disorders likely viral in origin.

Use the right antibiotic at the right dose and for the right duration.

Development of rapid test methods to identify pathogenic microorganisms and their likely antibiotic sensitivity will be important in achieving goals.

In addition to antibiotic stewardship measures, Monnier et al.[11] noted the WHO introduction of measures for equitable availability of effective antibodies and engagement of patient communities in responsible use of antibiotics.

Monnier et al. also noted that barriers to the development of new antibiotics exist. These include insufficient financial incentives for companies and regulatory barriers.

Mycobacterium tuberculosis (M. Tb) and antibiotic resistance

MacNeil et al.[12] reported that in 2017 worldwide approximately 19 million cases of tuberculosis occurred; 70% of cases occurred in South-East Asia and Africa, and 27% of cases occurred in individuals with HIV infections. They emphasized that innovative approaches to case finding and new treatment regimens were required.

Point of care diagnostics

Kozel and Burnham-Marusich[13] reported that enzyme linked immunoassays (ELISA) and lateral flow immunoassay platforms constituted the major technologies used in point of care diagnostics. Currently such immunoassays provide the basis of tests for numerous infections. In addition to antibody-based assays designed to detect specific microbial antigens, point of care antibody assays are utilized to test for specific host protein that manifest altered levels in response to infections, e.g. C-reactive protein.

M. Tb drug susceptibility testing

The World Health Organization recommends pre-treatment drug sensitivity test in cases of tuberculosis. A comprehensive investigation of validity M. Tb DNA sequence data to identify evidence of mutation that lead to antibiotic resistance was undertaken by the CRyPTIC consortium[14] and results of their study were reported in 2018.

The CRyPTIC consortium analyzed 10,290 isolates of M. Tb. Their data confirmed that a specific mutation in the organism S315T was associated with resistance to the mainline tuberculosis medication Isoniazid. A S450L mutation in M. Tb. organism led to resistance to the medication Rifampin.

The samples studied by the CRyPTIC consortium revealed that 48.1% of the isolates from patients revealed that the Mtb. organism was susceptible to all 4 drugs Isoniazid, Rifampin, Ethambutol and Pyrazinamide.

In this study phenotypic drug susceptibility was also carried out in patients. These tests included analyses of drug effects on cultured specimens and microscopic observations. The authors concluded that DNA sequence analysis of molecular determinants of M. Tb antibiotic resistance can guide therapy of tuberculosis.

Diagnostic tests for tuberculosis

Different new diagnostic tests have been developed to screen for M. Tb. One new test is designed to tests for cell free M. Tb DNA in human plasma. Click et al.[15] noted that diagnostic testing for M. Tb in sputum is often difficult in children. In addition, some individuals have extra-pulmonary tuberculosis infection. A real-time polymerase chain reaction test allows for rapid tests of cell free DNA in plasma.

New techniques for identifying anti-microbial agents

Lewis[10] noted that following identification of Streptomycin, Actinomyces species were progressively mined for antibiotics. In addition, minor modifications of isolated antibiotics were carried out. However, production of new antibiotics slowed.

Diffusion chambers and I-chip techniques

Kaeberlein et al.[16] initiated the use of diffusion chambers that allowed organisms to be in contact with fluid and components present in the natural environments in which organism survived. In the diffusion chamber concept organisms were separated from surrounding fluids by membranes that permitted exchange of chemicals between the fluid and the microorganism cells. For example, to culture marine microorganisms present in sediments, the platform containing the sediments was covered with a membrane and placed in a chamber containing sea-water from the vicinity where the sediment was isolated. Organisms grew and several rounds of harvesting and replication were carried out in order to achieve pure cultures of organism. Organisms were then subjected to nucleic acid sequencing.

Kaeberlein et al. emphasized that micro-organisms require specific signals that are present in the environment in which they normally flourish.

The diffusion chamber techniques facilitated growth of organisms and use of serial dilution and plating, facilitated isolation and amplification of different types of organisms present in an environment.

The diffusion chamber method for isolation of individual microorganisms was subsequently modified to high throughput technologies and development of the I-chip. The I-chip is composed of multiple plastic plates with through holes. Holes are covered with membranes that allow passage of natural environment nutrients.

Ling et al.[17] used the I-chip technique to isolate soil organisms. They then investigated substances produced by pure cultures of organisms. Their research led to isolation of a new soil organism Eleuthera terrae that produced a compound with antibiotic activity. This compound was designated Teixobactin and was found to be an inhibitor of peptidoglycan synthesis in microorganism.

Teixobactin was active against *Clostridium difficile*, multiple resistant *Staphylococcus aureus* and multiple resistant *Mycobacterium tuberculosis*. Important Teixobactin was not found to be toxic to human cells.

Nucleotide analyses and identification of anti-bacterial products

New techniques emerged based on evidence that microorganisms produce antimicrobial peptides. Efforts have been initiated to identify nucleotide sequences that encode these peptides. This provides an avenue to develop compounds corresponding to the peptides.

Hover et al.[18] used nucleotide sequencing of bacterial genomes and bioinformatic analyses to identify gene with sequences that encoded biosynthetic peptides of specific types. A specific family of calcium dependent cyclic peptides with a conserved calcium binding motif Asp-X-Asp-Gly was found to have antibiotic activity.

Hover et al. noted that peptides containing these motifs were known to be synthesized by a specific peptide synthase NPPS nonribosomal peptide synthase. They therefore scanned sequences from a number of different soil organisms to identify presence of sequences that encoded the NRPS peptide. Further analyses of complete genes that encoded this peptide led to identification of a class of biosynthetic gene clusters that encoded the specific peptides with antibiotic properties. This study led to the discovery of Malacidins that have antibiotic properties against Gram positive organisms previously determined to be multi-drug resistant. Hover et al. emphasized that sequence guided metagenomic pipeline analyses of environmental organism permitted discovery of new antibiotics

The search for anti-microbial medicines in plants

In 2019 Dettweiler et al.[19] published an account of studies on plant specimens that were recorded to have been used as medication during the American Civil War (1861–1865) long before antibiotics were available. They studied three species in particular *Liriodendron tulipifera* (tulip tree) *Aralia spinosa* (devils walking stick), *Quercus alba* (white oak).

Extracts of leaves of Quercus species were shown to have inhibitory effects on microorganisms in biofilms. Extracts of leaves of *L. tulipifera* had low toxicity for keratinocytes and were shown to have inhibitory effects on microorganisms in biofilms. Extract of bark from *L. tulipifera* was shown to exhibit toxicity toward keratinocytes.

Extracts from *L. tulipifera*, *Aralia spinosa* and Quercus species were shown to inhibit the growth of *Staphylococcus aureus*.

Phage therapy

Increasing evidence for antibiotic resistance, leading to difficulties in curing serious infections, has led to growing interest in bacteriophage therapy for infections. Bacteriophage are small virus like particles that can infect bacteria. They are considered to be among the most numerous and diverse life forms on earth. Twort[20] described the characteristics of these organisms, that had been alluded to by other investigators.

Gordillo Altamirano and Barr[21] reviewed phage therapy. They noted that phage therapy was in use before antibiotics became available. It continued to be used in Eastern Europe, particularly in Poland and Georgia, after the introduction of antibiotics. They noted that that the development of multiple antibiotic resistant forms of *Staphylococcus aureus*, Klebsiella pneumonia. *Pseudomonas aeruginosa* and Enterobacter, Campylobacter, Salmonella and *Neisseria gonorrhoeae*, has stimulated intense efforts to identify new therapeutic agents and rekindled interest in phage therapy. This interest is facilitated by current improvements in understanding of phage biology.

Gordillo Altamirano and Barr noted that phage therapy procedures are becoming more standardized. They noted that therapeutic design must include evidence that a particular phage has activity against the pathogen to be treated and that information is available indicating that bacteria have the receptor to which that phage will bind. Following phage binding to bacterial receptors, phage inject material into bacteria. Phage that have entered bacteria then control the bacterial genome to promote phage growth and proliferation. Ultimately this process leads to lysis of bacteria.

Gordillo Altamirano and Barr emphasized that lytic phage must be used since they destroy bacteria. Phage of the lysogenic type exist in bacteria and do not destroy them and should not be used since gene transfer from bacteria to phage genomes can occur.

Gordillo Altamirano and Barr noted that randomized control trials of phage therapy are ongoing in a number of countries including the USA. Initial trials included use of phage for topical treatment of skin infections and otitis media. Oral administration has been investigated for treatment of severe infections including typhoid fever. Following approval from regulatory agencies for compassionate use, intravenous phage therapy has been used for treatment of severe non-antibiotic responsive septicemia.

Dedrick et al.[22] reported successful use of lytic phage therapy in treatment of a cystic fibrosis patient who had disseminated infection with drug resistant Mycobacterium abscessus.

Implication of bacteriophage therapies for clinical laboratories

In 2019 Caflisch and Patel[23] reported on implications of phage therapy for clinical microbiology facilities. They noted that previously microbiologists primarily used phage to classify strains of bacteria. However, they emphasized the expansion of phage therapy will requires additional knowledge and expertise regarding bacteriophage.

Phage to be used for therapy must be lytic phage with ability to bind to the specific bacteria causing infection. Lytic phage not only bind to bacterial membranes they also produce enzymes defined as holins that insert holes into bacterial membranes. They also produce enzymes that hydrolyze cell wall and bacterial proteins.

Caflisch and Patel noted that with introduction of phage therapy clinical laboratory facilities would need to become adept at phage susceptibility testing of the infectious bacteria causing disease in a particular patient. They noted that specific regulatory procedures had been put in place in particular countries. In Belgium master phage stocks are quality controlled by accredited laboratories.

Microbiome

Recent studies have drawn attention to the multiple microorganisms that inhabit various regions of the human body, collectively referred to as the microbiome. In a 2017 review Young[24] emphasized that in recent years clinicians have come to appreciate the crucial ecosystem services provided by the microbiome and that diseases can result from loss of microbiome function. Young defined the major categories of the indigenous microbiome as, catabolic and synthetic and emphasized the important microbiome host interactions.

The microbiome is involved in catabolism and bioconversion function of host derived substances or of dietary components. The microbiome is involved in synthesis of cofactors and signaling molecules. Host microbiome interaction particularly impact mucus production and immune response.

Young noted that although nucleotide sequencing constitutes a major focus of classification of organism in the microbiome, other forms of analysis are currently also important. These include, proteomics, metabolomics, functional genomics and single gene studies by means of PCR or fluorescent in situ hybridization.

Comprehensive microbiome analyses

Comprehensive analyses of the commensal human microbiome have revealed that it is composed of bacteria, viruses, fungi archaea and the virome consists of DNA viruses, RNA viruses and bacteriophage.

Recognition of the complexity of the microbiome in humans and other species was facilitate by advances in the development of culture independent techniques and advances in genetic methods of organism characterization. Sequencing of RNA in the small ribosomal subunit 16S RNA provided to provide important information for classification of microorganisms.[25]

Results of sequencing the 16S ribosomal genes on 8168 different bacterial culture isolates were reported by Janda and Abbott in 2007.[26] The 16S ribosome gene sequences provided genus information in greater than 90% of the different isolates. Species identification was however achieved in a lower percentage of cases.

In 2008 Roberts, and colleagues[27] confirmed that different phyla of microorganisms could be distinguished on the basis of sequences in 16S ribosomal RNA.

Subsequently more comprehensive sequencing was undertaken this was referred to as shotgun metagenomics and metatranscriptomics. Srinivasan et al.[28] analyzed 617 different clinical microbiology isolates. The individual isolates were cultured and 16SRNA was also sequenced. Of particular importance was the finding of 87.5% concordance in results achieved with sequencing and culture methodologies.

Tamburini et al.[29] drew attention to the fact that genome sequencing does not distinguish between live and dead micro-organisms. However, transcription analysis may circumvent this problem (Fig. 1).

Human Microbiome

Microbiome organism identification methods

qPCR

16S RNA sequencing

Single cell Genome sequencing

Cultivation

Metabolic assessment

Proteome Analyses

Predominant Organism Types in Healthy Human Gut Microbiome

Bacteroidetes, Firmicutes, Lactobacilli

Potential pathogenic organisms in unhealthy gut microbiome

Clostridium, Salmonella, Salmonella, Shigella, Pseudomonas

Factors that may lead to Gut Microbial Dysbiosis

Nutrient Poor Food, Enteropathogens, Inflammation

Consequences of Gut microbial Dysbiosis

Impaired weight gain, stunting in children

Abdominal discomfort

Microbiome in Healthy Lung

Prevotella, Veillonella

Microbiome in Asthma and in Chronic Obstructive Lung Disease

Proteobacteria, Hemophilus, Neisseria

Fig. 1 *Human microbiome.* A number of different techniques have been developed to characterize organisms present in the microbiome. Specific classes of organisms constitute the healthy microbiome present at different body sites, e.g. healthy gut and healthy lung. In specific disorders pathogenic organisms may displace the typical healthy microbiome organisms.

The microbiome in early post-natal life

Tamburini et al.[29] reviewed the microbiome in early life. In the post-partum period in infants delivered through vaginal delivery the predominant microbial species in the mouth and gut and in the skin were Lactobacillus species. Following Cesarean section delivery, the skin mouth and gut of infants were found to harbor Staphylococcus, Streptococcus and Propionibacteria. The differences between vaginally delivered and Cesarean section delivered species remained until 12–24 months. Antibiotic treatment in the early post-natal period enhanced the abundance of fungal species, including *Candida albicans*.

Breast feeding

Tamburini et al. noted that breast feeding is known to be beneficial to infant in part through the availability in breast milk of immunoglobulins, particularly IgA, and also lactoferrin and defensins. They noted also that the micro-organisms Bifidobacterium and Lactobacillus are abundant in breast milk and increased levels of these organisms result in breast- fed infants having more acidic intestinal contents. This increased acidity leads to higher levels of short chain fatty acids in the intestinal content and this is thought to offer increased protection against pathogenic organisms. Breast milk also contains organisms that are common on maternal skin, including Staphylococcus, Streptococcus, Lactobacillus and Bifidobacterium.

Microbiome, intestinal function and health

In a 2017 review Postler and Ghosh[30] emphasized that intestinal microbiome disruptions were associated with metabolic disease and inflammatory bowel disease. They cited evidence that the gut microbiome in human is composed of approximately 1150 species. Two phyla predominate in the gut microbiome Firmicutes that are predominantly Gram-positive organisms and Bacteroidetes that are classified as Gram-negative organisms.

Postler and Ghosh noted than an optimally functioning gut microbiome has considerable advantages for the host. It has been known for some years that microorganisms in the gut synthesize B vitamins and vitamin K. On ingestion of plant derived material species of the Bacteroidetes phylum can breakdown fibers leading to the generation of short chain fatty acids (SCFAs) that promote integrity of the gut mucosa.

When the gut mucosa integrity is compromised gut bacteria can pass through to the bloodstream. In addition, there is evidence that lipoprotein complexes can pass from the gut to the blood stream when the mucosa is damaged.

Short chain fatty acids

Adequate production of short chain fatty acids through bacterial digestion of plant fiber derived material has also been shown to be protective against inflammatory bowel disease in model organisms.[31]

Postler and Ghosh cited reports indicating that levels of short chain fatty acids were found to be low in patients with diabetes mellitus. The low levels of short chain fatty acids promoted the entrance of bacterial derived material including bacterial cell wall lipopolysaccharides from the mucosa into the bloodstream and the lymphatic system. These bacterial components promoted the release of inflammatory cytokines that impacted the pancreas and in adipose tissue[32] leading to decreased insulin release and to increased insulin resistance. There are also reports that the microbiome in patients with type 2 diabetes differs from that of non-diabetic individuals.

Species of the Bacteroidetes phylum were found to predominantly produce acetate and propionate while Firmicutes predominantly produced butyrate. Butyrate was reported to constitute an energy source for gut mucosa. Acetate and propionate were reported to more readily enter the circulation.

Short chain fatty acids were shown to promote intestinal epithelial integrity in part through stimulating the proliferation of goblet cells in the mucin and to stimulate the production of mucin by these cells.

Postler and Ghosh cited several reports that presented evidence that short chain fatty acids interacted with immune cells in the lamina propria of the colon and they limited secretion of pro-inflammatory cytokines from macrophages in the lamina propria. Short chain fatty acids and their transporter SLC25A8 were shown to limit the maturation of dendritic cells. Dendritic cells are considered to be key links between the innate and the adaptive immune system. Postler and Ghosh concluded that short chain fatty acids derived from the microbiome acted as regulators of the intestinal immune system.

Low density lipoproteins

Low density Lipoproteins can leak through damaged gut mucosa into the circulation. They can then accumulate in the intima of arteries. There they can undergo oxidation and the oxidized lipoprotein (OXLDL) have inflammatory effects and promote the infiltration of macrophages that generate foam cells and lead to plaque formation that can rupture and can lead to myocardial infarction and stroke.[33] Other lipid components that can leak through damaged gut mucosa into the circulation include trimethylamine-N-oxide. These oxides have also been shown to promote arteriosclerosis.

Amino acids

Microorganism in the gut were reported to breakdown tryptophan consumed in the diet to indoles that bind to the arylhydrocarbon receptor in the mucosal cells. The ligand bound arylhydrocarbon receptor acts as a transcription factor that promotes transcription of specific genes that encode products that protect the mucosa. Lamas et al.[34] reported that the altered microbiome in patients with inflammatory bowel syndrome produced lower amounts of indoles.

Postler and Ghosh[30] noted that polyamines derived from the metabolism of diet derived arginine by gut organisms was promoted to promote intestinal integrity and function.

Bile components secreted into the intestine

Components of the bile that is secreted into the intestine include primary bile acids cholic acid and chenodeoxycholic acid. These primary bile acids are partly reabsorbed from the lower intestine. Primary bile acids that remain in the lower intestine are converted are converted to secondary bile acids that have been reported by Levy et al.[35] to promote mucosal barrier function and barrier maintenance.

In addition to breaking down food components, micro-organisms in the gut also synthesize specific molecules including ATP and polysaccharide A. Polysaccharide A was shown to be protective against development of colon inflammatory reactions in part through inhibition of a specific pro-inflammatory cytokine IL8.[36]

In the conclusion of their review in 2017 Postler and Ghosh noted that much about the host microbiome relationship remained to be discovered.

Microbiota and the immune system development

Studies on rodents kept under germ-free and specific pathogen free conditions were carried out by Abrams in 1967[37] and Gustafsson in 1970[38] revealed that these animals had reduced intestinal mobility and they developed enlargements of the ceca and the cecum was filled with undegraded material.

Studies on germ free animals revealed that in spleen, thymus and lymph nodes lymphoid tissue was poorly developed.[39]

Studies by Forsthuber[40] revealed that neonatal immune cells, primarily T cells developed tolerance in part to antigens to which they were exposed. Defects in the development of mesenteric lymph nodes of germ-free rodents were reported by MacPherson.[41] El-Aidy et al.[42] revealed that there was a critical time-period during which intestinal immune system development occurred in response to colonization.

CD4+ and CD8+ T cells numbers were reduced in germ free animals. These T helper cells were reported to be a source of interleukin 17 that is important in maintenance of the integrity of the intestinal mucosal barrier. In addition, studies demonstrated an increased frequency of natural killer T cells in germ free animals. Studies on germ-free mice revealed that immunoglobulins IgA and IgG levels were decreased while levels of IgE were increased.[43]

In 2016 Gensollen et al.[44] reviewed information on colonization of mammals by commensal microbiota and the role of this in priming of the immune system development. They noted that several studies had revealed that children raised in farm environments are at decreased risk for development of allergies.

Antibiotic use during the first six months of life was reported by Mai et al.[45] to correlate with later susceptibility to allergic manifestations.

Gut microbiota, diet and health

In 2018 Gentile and Weir[46] reviewed the gut microbiota, the impact of diet and the relationship to health. They emphasized that the gut microorganisms are significantly impacted by diet. These authors placed particular emphasis on dietary content of microbiota accessible carbohydrate (MAC) as forms of carbohydrate that promote microbial diversity. There is evidence that diets low in MAC lead to decreases in the diversity of the gut microorganism types. In addition, low MAC intake is associated with reduction in production of short chain fatty acids in the intestine. Gentile and Weir emphasized that short chain fatty acids derived by microbial action are important in signaling via G-protein coupled receptors. This signaling impacts energy homeostasis and metabolism and suppresses inflammatory reactions.

Studies in mice have revealed that the gut microbiota are altered by high fat diets. Gentile and Weir noted that there are reports in human studies that high fat diets lead to movement of lipopolysaccharide from the bacterial cell walls into the circulation where activation of Toll like receptors induce an inflammatory reaction.

Microbes also alter bile acids in the intestine Gentile and Weir noted that the effect of microorganisms on bile acid components and their abundance likely constitute important factors in lipid absorption.

With respect to proteins ingested, metabolic transformation of specific amino acids by gut microorganism also contribute to the generation of short chain fatty acids and other metabolites. Gentile and Weir noted that studies on the physiological impact of microbiome derived amino-acids in humans required further study. They emphasized that bacteria produce certain vitamins especially B vitamins, and they utilize vitamins provided by the host dietary intake. Microorganisms also require certain minerals including zinc and iron. Deficient zinc intake is known to alter the gut microbiome and to lead to diarrhea.

Several studies have revealed that particular food additives impact the normal gut microbiome. These include emulsifiers.[47]

Microbiome related disorders and antibiotic usage

Dicks et al.[48] noted that in the healthy colon *Clostridium difficile* organisms are present in low numbers. However antibiotic treatment, especially for extensive periods, can lead to generation of mutant *C. difficile* strains that can cause severe disease. Treatment with broad-spectrum antibiotics, particularly when used long-term can lead to *Clostridium difficile* infection.

One particular type of *C. difficile* ribotype 027 has been reported to be contagious. Dicks et al. documented spread of this type of *C. difficile* through 14 European countries. Infection with pathogenic forms of *C. difficile* can lead to a range of manifestations. Infection may be asymptomatic in some cases while in other cases severe disease occurs associated with severe inflammatory reactions in the colon, with damage to the mucosa and pseudomembranous colitis leading to bleeding and in some cases to ulceration of the mucosa.

Dicks et al. reported that metabolism in normal, non-pathogenic forms of *C. difficile* leads to production of butyrate. Butyrate is reported to have a beneficial effect on the colon mucosa and to decrease permeability of the epithelial cells.

However pathogenic forms of *C. difficile* were shown to weaken the tight junctions between gut epithelial cells and to increase permeability of the mucosa so that infiltrations of the gut mucosa by organisms occurred more readily. Increased permeability and bacterial invasions trigger immune responses, cytokine activation and inflammatory reactions. There is also evidence that *C. difficile* organism that infiltrate into the colon tissue release specific toxins.

The microbiome in malnutrition

Kwashiorkor

In a report published in 1952 Trowell and Davies[49, 50] noted that the unique manifestations of the malnutrition syndrome kwashiorkor had been clearly described by Williams in 1933[51] and 1935[52] although previous studies had alluded to similar conditions in children in French Indo-China and Africa.

The condition designated kwashiorlor usually develops in infants following termination of breast feeding often before one year of age, followed by a period of inadequate nutrition. In Africa the hair of children with this condition changed in color from black to red and the skin became pale. Subsequently dermatitis developed and peeling of the skin occurred. Children with kwashiorkor have markedly protuberant abdomens, their growth becomes retarded, swelling of the limbs occurs and children become irritable. The condition may lead to death.

Studies carried out by Coward et al. in 1972[53] revealed that primates fed low protein high carbohydrate diets developed pathological features reminiscent of those found in kwashiorkor. Golden et al. in 1998[54] published data that revealed that micronutrient deficiency played an important role in causation of kwashiorkor manifestations.

Smith et al.[55] published results of studies on the gut microbiome in children with kwashiorkor. They studied twin pairs discordant for kwashiorkor manifestations and children with and without kwashiorkor from the same regions. Smith et al. analyzed material containing a combination of local diet components and gut microbiome containing material derived from kwashiorkor affected-children and material from unaffected children. They administered this material to gnotobiotic mice. They reported that material derived from kwashiorkor affected children produced marked weight loss and metabolic perturbations in mice.

Based on these results Smith et al. concluded that the gut microbiome was an important factor in the causation of kwashiorkor.

In 2017 Million et al.[56] reviewed gut microbiota and malnutrition. They noted that worldwide malnutrition was the leading cause of death in children under five years of age. They defined different categories of malnutrition, stunting: height more than 2 standard deviation below the mean; wasting: height and weight more than 2 standard deviations below the mean; underweight: low weight for height. They noted that wasting is associated with higher mortality.

Marasmus is malnutrition condition without all of the typical manifestations of kwashiorkor. Million et al. emphasized that breast feeding, food and water security were the major factors that prevented malnutrition and that a healthy gut microbiome also helped protect against malnutrition. They defined proliferation of bifidiobacteria in early life as important in establishing healthy gut microbiome. Bifidiobacteria longum was reported to usually be transferred from the mother to the infant. Depletion of this organism in the gut of the infant or child exacerbated energy and vitamin deficiency and reduced immunoprotection against pathogenic microbes.

Interaction of the human genome and the gut microbiome

Goodrich et al.[57] reported that higher levels of the normally beneficial gut microorganisms of the Bifidiobacteria type occurred in humans with the genotype that was associated with lactose intolerance.

Gut microbiome and colorectal cancer

Hale et al.[58] reported results of studies carried out on human colon tumors and adjacent normal mucosa to assess functioning of the DNA mismatch repair gene status. They also carried out studies on these tissue samples to determine microbiome status through ribosomal RNA analyses. In addition, they carried out metabolite analyses to quantify short chain amino acids and specific amino acid derived metabolites.

Specific gene mutations impact the function of DNA mismatch repair complexes. Key components of this complex are proteins encoded by the genes *MLH1*, *PMS2*, *MSH2* and *MSH6*. Impaired function. Impaired production of the products of these genes is often associated with hypermethylation.

Richman et al.[59] reported that 15% of colorectal cancers were associated with DNA mismatch repair (MMR) defects. Distinctions are made between MMR deficient and MMR proficient cancers. Sweetser et al.[60] reported MMR deficient colorectal cancers (dMMR) originate in sessile serrated adenomas while MMR proficient colorectal cancers (pMMR) arise in tubular adenomas. Colorectal cancers of the dMMR type were reported to arise more frequently in the right-side colon.

Hale et al. undertook studies to establish whether specific microbiome differences occurred in dMMR colorectal tumors versus pMMR colorectal tumors. Results of their study revealed the intact MMR function, (pMMR status) enhanced suppression of growth of *Bacteroides fragilis*. Their studies also revealed that deficient MMR status (dMMR) increased proliferation of *Bacteroides fragilis* and increased proliferation of the sulfidogenic produced bacterial specie *Fusobacterium nucleatum*.

Results of prior studies have revealed that increased levels of hydrogen sulfide in gut increase methylation of DNA of the mutation carrying mismatch repair genes and also increase methylation of the proto oncogene *BRAF*.

High levels of hydrogen sulfide were reported to promote proliferation of colon cancer cells by Hellmich et al. in 2015.[61]

Hale et al. noted that the finding that deficient mismatch repair colo-rectal cancer is associated with proliferation of specific bacterial species, *Bacteroides fragilis* and *Fusobacterium nucleatum* raise the potential for therapeutic manipulation of the microbiome in patients with deficient mismatch repair function.

Increased growth of Fuseobacterium nucleatum independent of CpG hypermethylation phenotype and independent of microsatellite instability was reported to be associated with shorter colon cancer survival by Mima et al.[62]

Microsatellite instability in colon cancer tumors can result from germline mutation in the mismatch repair genes. Inefficient mismatch repair can also result from epigenetic inactivation of the *MLH1* gene. Methylation changes in the *MSH1* gene also occurs in sporadic cancers.[63]

Gut/brain axis and the intestinal microbiome

Mayer et al.[64] reported that progress had been made in in characterizing connection between the gastrointestinal, the enteric nervous system and the central nervous system. In addition, information had been gathered on substances produced by the gut microbiota that could act in the central nervous system and modify behavior and pain sensitivity. They noted that studies had primarily been carried out in rodents and that research on humans was required.

The lung microbiome

In a review of the lung microbiome in health and disease Moffat and Cookson[65] defined the microbiome as including bacteria, Archaea, fungi and viruses. They defined microbiome metagenomics as the study of DNA, RNA, proteins and metabolites.

Moffat and Cookson referred to the work of Woese and Fox in 1977[25] who first reported the use of 16S RNA in classifying different micro-organisms. Additional studies have established the use of nucleotide sequencing of the 15S RNA hypervariable regions to distinguish different micro-organisms.[66]

Moffat and Cookson[65] noted that despite earlier studies that proposed that the lower airways were sterile, human bronchoscopy samples have revealed the presence of microorganism. The bacterial topography of the lower respiratory tract in healthy humans was reported by Dickson et al.[67] Phyla of organisms in the healthy lung include Firmicutes, Bacteroidetes, Proteobacteria, Fusobacteria, Actinobacteria. There is evidence that the oropharynx and intrathoracic airways carry similar organisms to those in the lung.

Respiratory diseases

Moffat and Cookson noted that microbiome studies of bronchial brushings from patients with asthma have revealed that the microbiome differs in those samples from the microbiome found in healthy lungs. In addition, potential pathogens identified in patients with asthma included Hemophilus and Neisseria.

Chronic obstructive lung disease (COPD) is associated with inflammatory reaction in small airways. Moffat and Cookson noted that COPD occurs most frequently in smokers, however it may also occur in non-smokers. During period of exacerbation of symptoms in cases of COPD with infection Proteobacteria and Hemophilus influenzae occurred in increased numbers.

Chronic suppurative lung disease occurs in patient with cystic fibrosis and in cases with bronchiectasis. Moffat and Cookson reported that complex microbiomes occur in bronchial samples from these patients. Key pathogens in these cases included *Pseudomonas aeruginosa* and Burkholderia cepacian. Cystic fibrosis arises due to homozygous mutations in the CFTR transmembrane conductance regulator.

Non-CF bronchiectasis was found in some cases to be associated with mutations in the human fucosyltransferase gene *FUT2*. Mutations in this gene were found to facilitate the growth of *Pseudomonas aeruginosa*.

Moffat and Cookson noted that the condition referred to as idiopathic pulmonary fibrosis arises due to pathogenic mutation in the Mucin 5B gene.

Skin microbiome

The normal (commensal) organisms on the skin are sometimes referred to resident organisms. In a 2013 review Zeeuwen et al.[68] noted that the diversity of microbiome on the skin differs in different body sites, depending on the degree of moisture, pH, and skin secretions (sebum). The organisms in the skin microbiota comprise predominantly members of 4 different phyla, Actinobacteria, Firmicutes, Proteobacteria and Bacteroidetes. Corynebacteria and Staphylococcus species predominate in moist areas and Propionobacteria predominate in areas with higher levels of sebaceous secretion, e.g. the forehead.

Zeeuwen et al. emphasized that organisms predominate is the superficial regions of the skin, particularly in the stratum corneum while the stratum granulosum is relatively organism free.

Allergic responses and gene environment interactions

Atopy the predisposition to develop hypersensitivity, allergic reactions, can lead to reaction in the skin such as dermatitis. Filaggrin is a protein that functions as a barrier and to prevent organism invasion into the deeper layers of the skin.

McAleer et al.[69] reported that specific mutations in the filaggrin gene *(FLG)* that encodes intermediate filament-associated protein that aggregates keratin intermediate filaments in mammalian epidermis.

Thomsen et al.[70] reported that loss of function mutations in filaggrin *FLG* (R501X, 2282del4 and R2447X) were particularly common in Northern European populations and these mutations predisposed to atopic dermatitis.

Paller et al.[71] reported that in atopic dermatitis there was a decrease in microbiome diversity in the skin and colonization with pathogenic forms of *Staphylococcus aureus*. They emphasized the importance of cross-talk between commensal organisms and the immune system and that commensal organisms impact development of the innate and the adaptive immune systems.

The skin disease psoriasis

Psoriasis has been found to be more common in individuals who have genomic copy number variants in the beta defensin genes.[72] Beta defensins have been shown to have anti-microbial properties.

Zeeuwen et al.[68] noted that several studies have revealed that diverse organisms occur in the skin lesions in psoriasis and that Firmicutes species were particularly abundant. They documented important factors involved in skin homeostasis. These included an intact epithelial barrier and production of anti-bacterial peptides in the epidermis and also adequate innate and adaptive immune responses.

In 2014 Kiatsurayanon et al.[73] reported that human beta defensins also play roles in keratinocyte proliferation and migration. They also reported that beta-defensins were over-expressed in psoriasis. Smithrithee et al.[74] reported that beta defensins increase expression of the interleukin IL37. IL37 normally acts as an inhibitor of inflammatory response and auto-immune reactions.

The late cornified epithelial proteins, LCE3B and LCE3C, proteins, have been shown to be reduced in some cases of psoriasis. Niehues et al.[75] reported that these proteins have anti-bacterial activity.

Microorganisms environmental influences. Host factors and auto-immune diseases

In 2012 Selmi et al.[76] presented information on the environmental influences on auto-immunity and noted that the leading hypothesis was that auto-immune responses involved the impact of specific environmental factors and individuals with a specific genetic constitution. Key environmental factors included infectious agents, chemicals, xenobiotics, adjuvant and certain physical agents include ultra-violet exposure.

Adjuvants are defined as non-antigenic agents that stimulate the immune response. Adjuvants are sometimes included with vaccines. They include lipoproteins, lipopeptides and lipoglycans from microorganisms. Also included in the adjuvant category are specific micro-organism proteins, e.g. flagellin, profilin and microorganism DNA and RNA.

Adjuvants were identified as specific components that led to nonantigenic specific adaptive immune responses and induced the release of pro-inflammatory cytokines and chemokines.[77]

Auto-immunity and molecular mimicry

Rojas et al.[78] reviewed auto-immunity and molecular mimicry. They noted that molecular mimicry refers to the situation where a similarity exists between specific foreign peptides and host peptides. They emphasized that induction of auto-immune responses seldom results from molecular mimicry alone but depends on other host factors.

The term molecular mimicry was used in 1964 by Damian[79] to define similarity between infectious agents and human factors. In 1966 Zabriski and Freimer[80] produced a publication entitled, "An immunological relationship between group A streptococcus and mammalian muscle "

Rojas et al. noted that subsequently other reports have been generated regarding similarities of specific pathogen components and specific human tissue components. These include homology of Campylobacter jejuni polysaccharides and human myelin carbohydrates. Epitopes in certain influenza vaccine components have been implicated in the etiology of the auto-immune disorder Guillain-Barre syndrome.

Rojas et al. noted that clearly in most patients, infectious disease or vaccine administration are not followed by auto-immune diseases. With respect to immunization some investigators have postulated that adjuvants that are added to vaccines may play a role in causation of auto-immune responses.

Host factors likely play roles in the etiology of auto-immune diseases. Xiao et al.[81] reported that individuals with the HLA DRB1 15.01 allele who were also infected with Epstein Barr virus were at greater risk for developing the auto-immune disease multiple sclerosis.

Studies have been carried out to define the specific immunological components that contribute to the auto-immune response; both B cells and the immunoglobulins they produce and aberrant T cells functions have been implicated.

Another concept that has gained attention given the fact that central nervous system components are damaged in several auto-immune diseases, is that the central nervous system (CNS) is not as previously postulated an immune privileged site but that there are significant connections but that there are significant connections between the CNS and the immune system.

References

1. Fierer N. Embracing the unknown: disentangling the complexities of the soil microbiome. *Nat Rev Microbiol* 2017;**15**(10):579–90. https://doi.org/10.1038/nrmicro.2017.87. 28824177.
2. Geisen S, Bonkowski M. Methodological advances to study the diversity of soil protists and their functioning in soil food webs. *Appl Soil Ecol* 2018;.
3. de Vrieze J. The littlest farmhands. *Science* 2015;**349**(6249):680–3. https://doi.org/10.1126/science.349.6249.680.

4. Mommer L, Kirkegard J, van Ruijven J. Root–root interactions: towards a rhizosphere framework. *Trends Plant Sci* 2016;**21**(3):209–17. https://doi.org/10.1016/j.tplants.2016.01.009.

5. Hutchins DA, Jansson JK, Remais JV, Rich VI, Singh BK, Trivedi P. Climate change microbiology—problems and perspectives. *Nat Rev Microbiol* 2019;**17**(6):391-6. https://doi.org/10.1038/s41579-019-0178-5. 31092905.

6. Toju H, Peay KG, Yamamichi M, Narisawa K, Hiruma K, et al. Core microbiomes for sustainable agroecosystems. *Nat Plants* 2018;**4**(5):247–57. https://doi.org/10.1038/s41477-018-0139-4. Review. Erratum in: Nat Plants. 2018 Sep;4(9):733, 29725101.

7. Watkins RR, Bonomo RA. Overview: global and local impact of antibiotic resistance. *Infect Dis Clin N Am* 2016;**30**(2):313–22. https://doi.org/10.1016/j.idc.2016.02.001. Review, 27208761.

8. Blair JM, Webber MA, Baylay AJ, Ogbolu DO, Piddock LJ. Molecular mechanisms of antibiotic resistance. *Nat Rev Microbiol* 2015;**13**(1):42–51. https://doi.org/10.1038/nrmicro3380.

9. Cama J, Henney AM, Winterhalter M. Breaching the barrier: quantifying antibiotic permeability across Gram-negative bacterial membranes. *J Mol Biol* 2019. pii: S0022-2836(19)30181-0. https://doi.org/10.1016/j.jmb.2019.03.031. 30959052.

10. Lewis K. New approaches to antimicrobial discovery. *Biochem Pharmacol* 2017;**134**:87–98. https://doi.org/10.1016/j.bcp.2016.11.002. Review. 2782396.

11. Monnier AA, Schouten J, Tebano G, Zanichellie V, Huttner BD, et al. Ensuring antibiotic development, equitable availability and responsible use of effective antibiotics: recommendations for multisectoral action. *Clin Infect Dis* 2019;**68**(11):1952–9. https://doi.org/10.1093/cid/ciy824.

12. MacNeil A, Glaziou P, Sismanidis C, Maloney S, Floyd K. Global epidemiology of tuberculosis and progress toward achieving global targets—2017. *MMWR Morb Mortal Wkly Rep* 2019;**68**(11):263–6. https://doi.org/10.15585/mmwr.mm6811a3. 30897077.

13. Kozel TR, Burnham-Marusich AR. Point-of-care testing for infectious diseases: past, present, and future. *J Clin Microbiol* 2017;**55**(8):2313–20. https://doi.org/10.1128/JCM.00476-17. 28539345.

14. CRyPTIC Consortium and the 100,000 Genomes Project, Allix-Béguec C, Arandjelovic I, et al. Prediction of susceptibility to first-line tuberculosis drugs by DNA sequencing. *N Engl J Med* 2018;**379**(15):1403–15. https://doi.org/10.1056/NEJMoa1800474. Epub 2018 Sep 26, 30280646.

15. Click ES, Murithi W, Ouma GS, McCarthy K, Willby M, et al. Detection of apparent cell-free M. tuberculosis DNA from plasma. *Sci Rep* 2018;**8**(1):645. https://doi.org/10.1038/s41598-017-17683-6. 29330384.

16. Kaeberlein T, Lewis K, Epstein SS. Isolating "uncultivable" microorganisms in pure culture in a simulated natural environment. *Science* 2002;**296**(5570):1127–9. https://doi.org/10.1126/science.1070633. 12004133.

17. Ling LL, Schneider T, Peoples AJ, Spoering AL, Engels I, et al. A new antibiotic kills pathogens without detectable resistance. *Nature* 2015;**517**(7535):455–9. https://doi.org/10.1038/nature14098. 25561178.

18. Hover BM, Kim SH, Katz M, Charlop-Powers Z, Owen JG, et al. Culture-independent discovery of the malacidins as calcium-dependent antibiotics with activity against multidrug-resistant Gram-positive pathogens. *Nat Microbiol* 2018;**3**(4):415–22. https://doi.org/10.1038/s41564-018-0110-1.

19. Dettweiler M, Lyles JT, Nelson K, Dale B, Reddinger RM, et al. American Civil War plant medicines inhibit growth, biofilm formation, and quorum sensing by multidrug-resistant bacteria. *Sci Rep* 2019;**9**(1):7692. https://doi.org/10.1038/s41598-019-44242-y. 31118466.

20. Twort FW. An investigation on the nature of ultra-microscopic viruses. *Lancet* 1915;**186**(4814):1241–3. https://doi.org/10.1016/S0140-6736(01)20383-3.

21. Gordillo Altamirano FL, Barr JJ. Phage therapy in the post-antibiotic era. *Clin Microbiol Rev* 2019;**32**. e00066–18.

22. Dedrick RM, Guerrero-Bustamante CA, Garlena RA, Russell DA, Ford K, et al. Engineered bacteriophages for treatment of a patient with a disseminated drug-resistant Mycobacterium abscessus. *Nat Med* 2019;**25**(5):730–3. https://doi.org/10.1038/s41591-019-0437-z. 31068712.

23. Caflisch KM, Patel R. Implications of bacteriophage- and bacteriophage component-based therapies for the clinical microbiology laboratory. *J Clin Microbiol* 2019. pii: JCM.00229-19. https://doi.org/10.1128/JCM.00229-19. 31092596.

24. Young VB. The role of the microbiome in human health and disease: an introduction for clinicians. *BMJ* 2017;**356**:j831. https://doi.org/10.1136/bmj.j831. Review. 28298355.

25. Woese CR, Fox GE. Phylogenetic structure of the prokaryotic domain: the primary kingdoms. *Proc Natl Acad Sci U S A* 1977;**74**(11):5088–90. 270744.

26. Janda JM, Abbott SL. 16S rRNA gene sequencing for bacterial identification in the diagnostic laboratory: pluses, perils and pitfalls. *J Clin Microbiol* 2007;**45**(9):2761–4. 17626177.

27. Roberts E, Sethi A, Montoya J, Woese CR, Luthey-Schulten Z. Molecular signatures of ribosomal evolution. *Proc Natl Acad Sci U S A* 2008;**105**(37):13953–8. https://doi.org/10.1073/pnas.0804861105. 18768810.

28. Srinivasan R, Karaoz U, Volegova M, MacKichan J, et al. Use of 16S rRNA gene for identification of a broad range of clinically relevant bacterial pathogens. *PLoS ONE* 2015;**10**(2):e0117617. https://doi.org/10.1371/journal.pone.0117617. eCollection 2015, 25658760.

29. Tamburini S, Shen N, Wu HC, Clemente JC. The microbiome in early life: implications for health outcomes. *Nat Med* 2016;**22**(7):713–22. https://doi.org/10.1038/nm.4142. Review. 27387886.

30. Postler TS, Ghosh S. Understanding the holobiont: how microbial metabolites affect human health and shape the immune system. *Cell Metab* 2017;**26**(1):110–30. https://doi.org/10.1016/j.cmet.2017.05.008. Review. 28625867.

31. Smith PM, Howitt MR, Panikov N, Michaud M, et al. The microbial metabolites, short-chain fatty acids, regulate colonic Treg cell homeostasis. *Science* 2013;**341**(6145):569–73. https://doi.org/10.1126/science.1241165. 23828891.

32. Donath MY, Dalmas É, Sauter NS, Böni-Schnetzler M. Inflammation in obesity and diabetes: islet dysfunction and therapeutic opportunity. *Cell Metab* 2013;**17**(6):860–72. https://doi.org/10.1016/j.cmet.2013.05.001. 23747245.

33. Brophy ML, Dong Y, Wu H, Rahman HN, et al. Eating the dead to keep atherosclerosis at Bay. *Front Cardiovasc Med* 2017;**4**:2. https://doi.org/10.3389/fcvm.2017.00002. eCollection 2017, 28194400.

34. Lamas B, Richard ML, Leducq V, Pham HP, et al. CARD9 impacts colitis by altering gut microbiota metabolism of tryptophan into aryl hydrocarbon receptor ligands. *Nat Med* 2016;**22**(6):598–605. https://doi.org/10.1038/nm.4102.

35. Levy M, Thaiss CA, Elinav E. Metabolites: messengers between the microbiota and the immune system. *Genes Dev* 2016;**30**(14):1589–97. https://doi.org/10.1101/gad.284091.116. 27474437.

36. Jiang F, Meng D, Weng M, Zhu W, et al. The symbiotic bacterial surface factor polysaccharide A on *Bacteroides fragilis* inhibits IL-1β-induced inflammation in human fetal enterocytes via toll receptors 2 and 4. *PLoS ONE* 2017;**12**(3):e0172738. https://doi.org/10.1371/journal.pone.0172738. eCollection 2017, 28278201.

37. Abrams GD, Bishop JE. Effect of the normal microbial flora on gastrointestinal motility. *Proc Soc Exp Biol Med* 1967;**126**(1):301–4. 6066182.

38. Gustafsson BE, Midtvedt T, Strandberg K. Effects of microbial contamination on the cecum enlargement of germfree rats. *Scand J Gastroenterol* 1970;**5**(4):309–14. 5429896.

39. Gordon HA, Bruckner-Kardoss E, Wostmann BS. Aging in germ-free mice: life tables and lesions observed at natural death. *J Gerontol* 1966;**21**(3):380–7. 5944800.

40. Forsthuber T, Yip HC, Lehmann PV. Induction of TH1 and TH2 immunity in neonatal mice. *Science* 1996;**271**(5256):1728–30. 8596934.

41. Macpherson AJ, Harris NL. Interactions between commensal intestinal bacteria and the immune system. *Nat Rev Immunol* 2004;**4**(6):478–85. 15173836.

42. El Aidy S, Hooiveld G, Tremaroli V, Bäckhed F, Kleerebezem M. The gut microbiota and mucosal homeostasis: colonized at birth or at adulthood, does it matter? *Gut Microbes* 2013;**4**(2):118–24. https://doi.org/10.4161/gmic.23362. 23333858.

43. Cahenzli J, Köller Y, Wyss M, Geuking MB, McCoy KD. Intestinal microbial diversity during early-life colonization shapes long-term IgE levels. *Cell Host Microbe* 2013;**14**(5):559–70. https://doi.org/10.1016/j.chom.2013.10.004. 24237701.

44. Gensollen T, Iyer SS, Kasper DL, Blumberg RS. How colonization by microbiota in early life shapes the immune system. *Science* 2016;**352**(6285):539–44. https://doi.org/10.1126/science.aad9378. 27126036.

45. Mai XM, Kull I, Wickman M, Bergström A. Antibiotic use in early life and development of allergic diseases: respiratory infection as the explanation. *Clin Exp Allergy* 2010;**40**(8):1230–7. https://doi.org/10.1111/j.1365-2222.2010.03532.x. 20545711.

46. Gentile CL, Weir TL. The gut microbiota at the intersection of diet and human health. *Science* 2018;**362**(6416):776–80. https://doi.org/10.1126/science.aau5812. Review. 30442802.

47. Chassaing B, Koren O, Goodrich JK, Poole AC, Srinivasan S, et al. Dietary emulsifiers impact the mouse gut microbiota promoting colitis and metabolic syndrome. *Nature* 2015;**519**(7541):92–6. https://doi.org/10.1038/nature14232. 25731162.

48. Dicks LMT, Mikkelsen LS, Brandsborg E, Marcotte H. Clostridium difficile, the Difficult "Kloster" Fuelled by Antibiotics. *Curr Microbiol* 2018. https://doi.org/10.1007/s00284-018-1543-8. 30084095.

49. Trowell HC, Davies JN. Kwashiorkor. I. Nutritional background history, distribution, and incidence. *Br Med J* 1952;**2**(4788):796–8. 12978334.

50. Trowell HC, Davies JN, Dean JF. Kwashiorkor. II. Clinical picture, pathology, and differential diagnosis. *Br Med J* 1952;**2**(4788):798–801. 12978335.

51. Williams CD. A nutritional disease of childhood associated with a maize diet. *Arch Dis Child* 1933;**8**(48):423–33. 21031941.

52. Williams CD. Kwashiorkor. A nutritional disease of children associated with a maize diet by Cicely D Williams from the Lancet. *Nutr Rev* 1935;1151.

53. Coward DG, Whitehead RG. Experimental protein-energy malnutrition in baby baboons. Attempts to reproduce the pathological features of kwashiorkor as seen in Uganda. *Br J Nutr* 1972;**28**(2):223–37. 462724.

54. Golden MH. Oedematous malnutrition. *Br Med Bull* 1998;**54**(2):433–44. 9830208.

55. Smith MI Yatsunenko T, Manary MJ, Trehan I, et al. Gut microbiomes of Malawian twin pairs discordant for kwashiorkor. *Science* 2013;**339**(6119):548–54. https://doi.org/10.1126/science.1229000.

56. Million M, Diallo A, Raoult D. Gut microbiota and malnutrition. *Microb Pathog* 2017;**106**:127–38. https://doi.org/10.1016/j.micpath.2016.02.003. Review. 26853753.

57. Goodrich JK, Davenport ER, Clark AG, Ley RE. The relationship between the human genome and microbiome comes into view. *Annu Rev Genet* 2017;**51**:413–33. https://doi.org/10.1146/annurev-genet-110711-155532. 28934590.

58. Hale VL, Jeraldo P, Chen J, Mundy M, et al. Distinct microbes, metabolites, and ecologies define the microbiome in deficient and proficient mismatch repair colorectal cancers. *Genome Med* 2018;**10**(1):78. https://doi.org/10.1186/s13073-018-0586-6. 3037688.

59. Richman S. Deficient mismatch repair: read all about it (Review). *Int J Oncol* 2015;**47**(4):1189–202. https://doi.org/10.3892/ijo.2015.3119. 26315971.

60. Sweetser S, Jones A, Smyrk TC, Sinicrope FA. Sessile serrated polyps are precursors of colon carcinomas with deficient DNA mismatch repair. *Clin Gastroenterol Hepatol* 2016;**14**(7):1056–9. https://doi.org/10.1016/j.cgh.2016.01.021. 26898652.

61. Hellmich MR, Szabo C. Hydrogen sulfide and cancer. *Handb Exp Pharmacol* 2015;**230**:233–41. https://doi.org/10.1007/978-3-319-18144-8_12. Review. 26162838.

62. Mima K, Nishihara R, Qian ZR, Cao Y, Sukawa Y, et al. Fusobacterium nucleatum in colorectal carcinoma tissue and patient prognosis. *Gut* 2016;**65**(12):1973–80. https://doi.org/10.1136/gutjnl-2015-310101. 26311717.

63. Gelsomino F, Barbolini M, Spallanzani A, Pugliese G, Cascinu S. The evolving role of microsatellite instability in colorectal cancer: a review. *Cancer Treat Rev* 2016;**51**:19–26. https://doi.org/10.1016/j.ctrv.2016.10.005. 27838401.

64. Mayer EA, Tillisch K, Gupta A. Gut/brain axis and the microbiota. *J Clin Invest* 2015;**125**(3):926–38. https://doi.org/10.1172/JCI76304. 25689247.

65. Moffatt MF, Cookson WO. The lung microbiome in health and disease. *Clin Med (Lond)* 2017;**17**(6):525–9. https://doi.org/10.7861/clinmedicine.17-6-525. 29196353.

66. Klindworth A, Pruesse E, Schweer T, Peplies J, et al. Evaluation of general 16S ribosomal RNA gene PCR primers for classical and next-generation sequencing-based diversity studies. *Nucleic Acids Res* 2013;**41**(1):e1. https://doi.org/10.1093/nar/gks808. 22933715.

67. Dickson RP, Erb-Downward JR, Freeman CM, McCloskey L, et al. Bacterial topography of the healthy human lower respiratory tract. *MBio* 2017;**8**(1). pii: e02287-16. https://doi.org/10.1128/mBio.02287-16. 28196961.

68. Zeeuwen PL, Kleerebezem M, Timmerman HM, Schalkwijk J. Microbiome and skin diseases. *Curr Opin Allergy Clin Immunol* 2013;**13**(5):514–20. https://doi.org/10.1097/ACI.0b013e328364ebeb. Review, 23974680.

69. McAleer MA, Irvine AD. The multifunctional role of filaggrin in allergic skin disease. *J Allergy Clin Immunol* 2013;**131**(2):280–91. https://doi.org/10.1016/j.jaci.2012.12.668. Review. 23374260.

70. Thomsen SF, Elmose C, Szecsi PB, Stender S, et al. Filaggrin gene loss-of-function mutations explain discordance of atopic dermatitis within dizygotic twin pairs. *Int J Dermatol* 2016 Dec;**55**(12):1341–4. https://doi.org/10.1111/ijd.13401. 27653621.

71. Paller AS, Kong HH, Seed P, Naik S, et al. The microbiome in atopic dermatitis. *J Allergy Clin Immunol* 2018. pii: S0091-6749(18)31664-6. https://doi.org/10.1016/j.jaci.2018.11.015. 30476499.

72. Hollox EJ, Huffmeier U, Zeeuwen PL, Palla R, Lascorz J, et al. Psoriasis is associated with increased beta-defensin genomic copy number. *Nat Genet* 2008;**40**(1):23–5. https://doi.org/10.1038/ng.2007.48. 18059266.

73. Kiatsurayanon C, Niyonsaba F, Smithrithee R, Akiyama T, et al. Host defense (Antimicrobial) peptide, human β-defensin-3, improves the function of the epithelial tight-junction barrier in human keratinocytes. *J Invest Dermatol* 2014;**134**(8):2163–73. https://doi.org/10.1038/jid.2014.143. 24633129.

74. Smithrithee R, Niyonsaba F, Kiatsurayanon C, Ushio H, et al. Human β-defensin-3 increases the expression of interleukin-37 through CCR6 in human keratinocytes. *J Dermatol Sci* 2015;**77**(1):46–53. https://doi.org/10.1016/j.jdermsci.2014.12.001. 25541254.

75. Niehues H, Tsoi LC, van der Krieken DA, Jansen PAM, et al. Psoriasis-associated Late Cornified Envelope (LCE) proteins have antibacterial activity. *J Invest Dermatol* 2017;**137**(11):2380–8. https://doi.org/10.1016/j.jid.2017.06.003. 28634035.

76. Selmi C, Leung PS, Sherr DH, Diaz M, et al. Mechanisms of environmental influence on human autoimmunity: a National Institute of Environmental Health Sciences expert panel workshop. *J Autoimmun* 2012;**39**(4):272–84. https://doi.org/10.1016/j.jaut.2012.05.007. Review, 22749494.

77. Meroni PL. Autoimmune or auto-inflammatory syndrome induced by adjuvants (ASIA): old truths and a new syndrome? *J Autoimmun* 2011;**36**(1):1–3. https://doi.org/10.1016/j.jaut.2010.10.004. 21051205.

78. Rojas M, Restrepo-Jiménez P, Monsalve DM, Pacheco Y, et al. Molecular mimicry and autoimmunity. *J Autoimmun* 2018;**95**:100–23. https://doi.org/10.1016/j.jaut.2018.10.012. Epub 2018 Oct 26. Review, 30509385.

79. Damian RT. Molecular mimicry: antigen sharing by parasite and host and consequences. *Am Nat* 1964;**98**:129–49.

80. Zabriskie JB, Freimer EH. An immunological relationship between the group. A streptococcus and mammalian muscle. *J Exp Med* 1966;**124**(4):661–78. 5922288.

81. Xiao D, Ye X, Zhang N, Ou M, et al. A meta-analysis of interaction between Epstein-Barr virus and HLA-DRB1*1501 on risk of multiple sclerosis. *Sci Rep* 2015;**5**:18083. https://doi.org/10.1038/srep18083. 26656273.

8

GENOMIC CHANGES AND ENVIRONMENTAL FACTORS IN CAUSATION OF BIRTH DEFECTS AND NEURODEVELOPMENTAL DISORDERS

Birth defects

Introduction

Kraus and Hong[1] noted that compelling information from several sources revealed that causation of birth defects was not adequately explained by genetic factors and that environmental factors were important. It is also important to note that for many specific birth defects, mutations in several different genes have been shown to be causative. Several examples of such birth defects will be discussed below.

Mosaicism

Efforts to identify causative genomic and molecular defects in birth defects may be complicated by the presence of mosaicism in a particular individual. Mosaicism refers to the presence of genotypic abnormalities present in some cells but not in others. The possibility of mosaicism is important to take into account in genetic counseling since the phenotype can be greatly influenced by the degree of mosaicism.

Keppler-Noreuil et al.[2] reported that heterogenous segmental somatic overgrowth that impacted skeletal system and vascular elements, occurred as a result of somatic activating mutations in the PIK3CA gene. Individuals with segmental overgrowth were found to be mosaics. The somatic activation mutations were found to be present in cells of the tissues in the specific aberrant segments and not in cells of tissue with normal growth.

Gene Environment Interactions. https://doi.org/10.1016/B978-0-12-819613-7.00008-6

Holoprosencephaly

This disorder is characterized by incomplete cleavage of the forebrain during development. This incomplete cleavage leads to loss of midline brain structures, abnormal fusion of ventricles of the brain.

Holoprosencephaly may occur as a malformation in cases that have abnormalities in systems beyond the brain and face, these cases are sometimes referred to as secondary holoprosencephaly. In some cases, the holoprosencephaly is present in the brain and may be associated with impaired development of midline structure of the face that may lead to defects of the eyes, nose and mouth, but more wide spread manifestations are not present, this form is referred to as primary holoprosencephaly.

In the National Organization for Rare diseases (NORD) database the incidence of holoprosencephaly in embryos is 1 in 250, the incidence in live births is 1 in 16,000, https://rarediseases.org/rare-diseases/holoprosencephaly/.

Roessler et al.[3] reported that cytogenetic abnormalities that led to deletion or disruption of the Sonic Hedgehog gene caused primary holoprosencephaly. The sonic hedgehog gene product SHH is now know to be a member of a signaling pathway that plays a key role in development processes.

Kruszka et al.[4] reported that most syndromic forms of holoprosencephaly were due to trisomy of chromosome 13, or 18, or 22. In addition, holoprosencephaly was reported to be present in Smith Lemli Opitz syndrome due to deficiency of 7-dihydro cholesterol reductase (DHCR7) and in Hartsfeld syndrome, due to specific mutations or deletions in the fibroblast growth factor 1 encoding gene (*FGFR1*). This syndrome is also associated with limb abnormalities.

In a 2018 review of holoprosencephaly Roessler et al.[5] documented both major and minor genes that harbored pathogenic mutations in holoprosencephaly. They also noted whether genes were involved in primary and or secondary holoprosencephaly and genes that were modifiers of the holoprosencephaly (HPE) mutations in other genes. Products of these genes are listed are listed below.

SIX3	Major factor	Six homeobox protein, a transcription factor
SHH	Major factor	Sonic Hedgehog signaling molecule
ZIC2	Major factor	Member of the ZIC family of C2H2-type zinc finger proteins
TGIF1	Minor factor	TGFB induced factor homeobox 1
CDON	Minor factor	Cell adhesion associated protein
STIL	Rare	Centriolar associated protein involved in autosomal recessive primary HPE

Continued

DLL1	Potential driver	Delta like NOTCH ligand
GLI2	Syndromic HPE	Member of the C2H2-type zinc finger protein subclass of the Gli family
FGFR1	Syndromic HPE	Fibroblast growth factor receptor 1
FGF8	Syndromic HPE	Fibroblast growth factor 8
PTCH1	Likely modifier	Patched 1 protein member of the sonic hedgehog family
BOC	Likely modifier	Cell adhesion associated protein with fibronectin type 3 repeats
DISP1	Likely modifier	Dispatched RND transporter, plays a role in embryonic patterning

Roessler et al. noted that even within a specific family with a particular mutation a spectrum of phenotypes could occur from mild to severe. They postulated that this variable expressivity was likely due to genetic or environmental modifiers.

Ethanol as holoprosencephaly inducing teratogen

In studies on a mouse model of holoprosencephaly Hong and Krauss[6] reported that the sonic hedgehog coregulator CDON undergoes mutations that synergize with transient in utero exposure to ethanol to induce holoprosencephaly.

Dubourg et al.[7] reported that the gene abnormalities encountered were sometimes inherited and sometimes occurred as de novo events. They also noted that gene defects had only been identified in 25% of cases.

Holoprosencephaly in cases where definitively classified pathogenic mutations SHH pathways genes were not identified

Kim et al.[8] undertook a comprehensive study of 26 families where holoprosencephaly had occurred and where no gene defects had been reported. For this study they utilized next generation exome sequence information and also included analyses of 822 ancestrally matched trios. Importantly, genes analyzed intensely included genes that were identified in developmental databases as being involved in forebrain development and genes that had been identified as having abnormalities in mouse studies where phenotypic abnormalities in mice resembled the phenotypic abnormalities that occurred in holoprosencephaly. Results were analyzed under an oligogenic model of disease.

The authors emphasized that in identifying significant variants in modifier genes the ACMG guidelines for variant calling were not helpful in variant classification. In establishing their variant calling algorithm the investigators used clinician generated lists of phenotypes present in holoprosencephaly and in similar conditions, using data from phenotype database resources, OMIM, literature reports and databases with variant information. They also used information on genes involved in generation of holoprosencephaly and similar phenotypes in the mouse studies. In addition, they also used transcriptome analysis data from the human developmental biology resource (HDBR) that included information on the expression of genes at different developmental stages.

The authors reported that they identified oligogenic variants in significant genes at much higher frequency in 10 of the holoprosencephaly families than in the ancestry matched families. The oligogenic events occurred in genes in the sonic hedgehog pathway and in the ciliary pathway. Cilia are reported to sense extra-cellular generated signals and to convey these to the interior of the cells to trigger appropriate responses. Defects in ciliary function can lead to a range of phenotypic abnormalities, including developmental defects in the brain

Extensive details on phenotypic abnormalities in brain and face were documented and correlated with gene variants identified. The products and functions genes found to have significant variants are documented below.

SHH	Sonic hedgehog signaling molecule
FAT1	FAT atypical cadherin 1 involved in cell proliferation
NDST1	*N*-deacetylase and N-sulfotransferase 1
COL2A1	Alpha-1 chain of type II collagen
PTCH1	Patched 1 component of the hedgehog signaling pathway.
LRP2	Low density lipoprotein-related protein 2 (LRP2)
BOC	Cell adhesion associated
SCUBE2	Signal peptide, CUB domain and EGF like domain containing 2
HIC1	HIC ZBTB transcriptional repressor 1, growth regulator
STK36	Serine/threonine kinase 36
WNT4	WNT secreted signaling protein
B9D1	B9 domain-containing protein, one of several that are involved in ciliogenesis
CELSR	Cadherin EGF LAG seven-pass G-type receptor 1
MKS1	Required for formation of the primary cilium in ciliated epithelial cells
IFT172	Necessary for ciliary assembly and maintenance
PRICKLE1	Nuclear receptor involved in transcription factor passage
SIX3	Transcription factor
TCTN	Member of the tectonic gene family which functions in Hedgehog signal transduction
TULP3	Member of the tubby gene family of bipartite transcription factors

Kim et al., noted that FAT1 mutations in mice were reported to lead to midline deformities in mice. They noted that the patients in their series who had variants in the ciliary pathway genes also manifested polydactyly that is a well- known feature in ciliopathy syndromes. Patient in their series with variants in SCUBE and BOC had midline brain defects and ocular hypotelorism and single nostril defects.

Cleft lip and palate: Gene variants and environmental pathways

In a review published in 2011 Dixon et al.[9] emphasized that clefts of the lip and/or palate (CLP) could be syndromic or non-syndromic in origin. The frequency of CLP on average is 1 in 700 births, however the frequency differs in various parts of the world. They noted that twin studies and family studies revealed that genetic factors played a role in many cases. However, many cases of CLP are sporadic and occur without a known family history.

Dixon et al. reviewed development of the tissue planes that led to structural positioning and integrity of the nose, upper lip, soft and hard palate. The also noted that in addition to cleft of the lip and palate there were other related minor abnormalities that included lip pits and dental anomalies.

Dixon also noted that environmental factors play roles in CLP and that interactions between environmental factors and particular genetic variants, were likely causative in some cases. They emphasized that studies on environmental factors and gene environment interactions required that large cohorts be studied and that information was beginning to emerge from such studies. Maternal smoking during pregnancy was related to an increased incidence of facial clefting and there was some evidence that variants in specific genes interacted with smoking to increase risk. The specific gene with implicated variants encoded for glutathione-S-transferase (GSST1) and nitric oxide synthase 3 (NOS3). There are reports that alcohol consumption during pregnancy increase the risk of facial clefts.[10]

There are reports indicating that low vitamin A intake of the mother in early pregnancy increases risk of clefting.[11] Jahanbin et al.[12] reported that maternal multi-vitamin supplementation significantly reduced the risk of non-syndromic clefting.

Studies on gene defects that give rise to syndromes in which orofacial clefts occur, have provided information on genes that encode products involved in orofacial development. One example is Van Der Woude syndrome. This disorder is associated with cleft lip, cleft palate, lip pits and teeth abnormalities, and is inherited as an autosomal dominant condition. Mutations in the IRF6 (interferon regulatory factor 6 (IRF6) were shown to lead to this condition.[13]

Leslie et al.[14] published information on the best supported loci associated with non-syndromic cleft lip and palate and noted the function of the products of these genes had biological plausibility. The products and functions of associated genes they identified are listed below.

IRF6	Interferon regulatory transcription factor 6
MAFB	MAF bZIP transcription factor B
ARHGAP29	Rho GTPase activating protein 29, signaling
VAX1	Ventral anterior homeobox regulation of body development and morphogenesis.
PAX7	Paired box 7 transcription factor

Studies in multi-generational families each with multiple cases of cleft lip and palate

Cox et al.[15] analyzed results of exome sequencing in 209 individuals in 72 multi-generation families in which multiple cases of cleft-lip and palate occurred. In these families, cleft-lip and palate abnormality was likely inherited as an autosomal dominant trait with variable penetrance.

Based on exome sequencing results Cox et al., identified 5 genes in which pathogenic variant occurred in affected individuals. The products of these genes are listed below

CTNND1	Catenin delta 1, which function in adhesion between cells and signal transduction
PLEKHA5	Pleckstrin homology domain containing A5 junctional marker of epithelial cells
PLEKHA7	Pleckstrin homology domain containing A7 junctional marker of epithelial cells
ESRP2	Epithelial cell-type-specific splicing regulator
CDH1	Cadherin a calcium-dependent cell-cell adhesion protein

Cox et al., then carried out sequencing analyses on 497 individuals including members of small families and individual cases of cleft lip/palate and they identified additional individuals with pathogenic variants in CTNND1, PLEKHA7, ESRP2 and CDH1.

Cox et al., also carried out analyses on palate epithelium from human tissues and mouse tissues. These studies revealed that PLEKHA4, PLEKHA7 and ESRP2 and p120 an adherens junction protein colocalized in the developing palate epithelium. They also demonstrated that knockout of Ctnnd1 in mice led to overt clefting in the homozygous mice.

Based on these studies Cox et al., concluded that defects in regulation of adhesion of epithelial cells contributed significantly to cleft lip and palate.

Funato and Nakamura[16] reported that more than 350 different genes had been implicated in the etiology of cleft lip and palate. They generated separate functional categories for the products of the implicated genes and they proposed that different genes were implicated in causation of cleft lip only, other genes were implicated in cleft palate only and defects in some genes led to both cleft lip and palate. Their study involved comprehensive analyses of literature information and of database information.

Funato and Nakamura noted that the main categories of genes involved in cleft lip and palate included transcription factors, signaling molecules and extra-cellular matrix proteins. In cleft palate only they reported that transcription factors were involved.

These authors also categorized environmental toxins and impact that led to cleft lip and palate. They reported that transretinoin equally induced cleft lip and palate or cleft palate only while tetrachlorobenzodioxin exposure particularly led to cleft palate.

Neurodevelopmental disorders: Genetic and environmental factors

Introduction

In a 2015[17] review Boivin et al. defined neurodevelopment as: "The dynamic inter-relationship between genetic, cognitive, emotional and behavioral processes across the developmental life."

With respect to neurodevelopmental disorders they emphasized the importance of continued research in three major categories: Gene and Brain, Environment and Brain, Environment and Gene. Boivin et al., noted also the importance of continued cross-disciplinary research in different countries including low income and middle-income countries (LMIC).

Key environmental factors included infections, maternal and child malnutrition, trauma in women particularly in the pre-natal period. With respect to infections, Boivin et al., noted that adequate treatment of diseases such as malaria and HIV are important since brain damage can occur in these diseases. In addition, following control of infection, children may require cognitive rehabilitation.

Trauma and depression in pregnant women were reported to be associated with DNA methylation differences in offspring. Differences in methylation of *NR3C1* the gene that encodes nuclear receptor subfamily 3 group C member 1, a glucocorticoid receptor and in methylation of the *BDNF* gene that encodes brain derived neurotrophic factor, were reported by Braithwaite et al.[18] Murgatroyd et al.[19] reported that prenatal depression impacted glucocorticoid receptor gene (*NR3C1*) 1-F promoter methylation in infants.

Childhood malnutrition, including protein-caloric deficiencies and micronutrient deficiencies were reported to impair neurodevelopment in infants and children.[20]

Boivin et al. presented information on an unusual condition referred to as Nodding syndrome. Episodes of this condition were reported to occur in South Sudan, Northern Uganda and Southern Tanzania. The syndrome is apparently a form of seizure disorder, atonic seizure. A number of different etiologies are under consideration including nematode infections and also malnutrition.

Food contamination and nutritional malabsorption have also been found to cause defects in neurocognitive development. Of particular interest in this regard, is a condition referred to as Konzo. This occurs when women and children are displaced because of violence and when droughts occur. Under such conditions in Africa, specific varieties of Cassava are the main source of food. Inadequate breakdown of Cassava leads to production of cyanide CN^- that can induce neurological damage and memory impairment.[21]

Hydrocephalus is more common in low and middle- income countries than in high income countries; Kiwanuka et al.[22] reported that neonatal infections are an important contributory cause.

Genomic and genetic factors in neurodevelopment

Boivin et al.[17] defined six classes of genetic related forms of neuro-disability. These included:

1. Genomic disorders.
2. Monogenic disorders.
3. Disorders due to mitochondrial functional defects.
4. Common disorders associated with behavioral and specific learning defects (autism and attention deficit disorders).
5. Disorders that were triggered by genetic risk and environmental interactions.
6. Alterations in the epigenome.

They emphasized that disorders in categories 4–6 were multifactorial in origin.

Concepts relevant to neurodevelopmental disorders

Homberg et al.[23] in a review in 2016 describe neurodevelopmental disorders as "a group of heterogeneous conditions in which defects occurred in affect, social, cognitive, motor and language functions". They considered six main categories of neurodevelopmental disorders: autism spectrum disorders, intellectual disability, communication or speech disabilities, motor/tic disorders, attention deficit hyperactivity disorder.

Homberg et al. emphasized that genetic, epigenetic and environmental factors played roles in these disorders. In addition, they emphasized that many of these disorders were polygenic in origin. They listed processes and cellular functions that were found in some neurodevelopmental disorders:

Neuroprogenitor proliferation defects.

Cortical migration and patterning defect.

Defects in mRNA translation.

Abnormal proteasomal degradation.

Impaired synaptic scaffolding.

Impaired signal transduction.

Impaired synaptic function.

Impaired neurotransmission.

DNA methylation defects.

Impaired immune functions.

Neuroprogenitor cell defects may in some cases be due to impaired centriole and centrosome defects.

With respect to environmental factors involved in causation of neurodevelopmental defects, they emphasized poor prenatal nutrition, prenatal exposure to harmful substances, e.g. alcohol, post conceptional maternal stress.

Homberg et al., concluded that effective treatment strategies for these disorders remained unclear. They emphasize that it is important to not only consider defects in these children but also to augment their strengths and to promote their well-being.

Infant and child development, environment, genes and their interactions

"The debate between Nature and Nurture as determinant of early childhood development is over. Today we understand that the two are inextricably linked.[24]"

In considering cognitive development in infants and children it is important to consider prenatal and post-natal factors. Prenatal brain developmental is followed by post-natal processes to expand connections.

Developmental trajectories of neuronal circuits and networks

In 2017 Keunen et al.[25] reviewed brain development and the emergence of functional neuronal architecture. They noted that the complex adult type brain gyrification pattern is in place in the neonatal brain cortex.

In describing the early brain development Keunen et al. cited evidence that neurogenesis is largely completed by mid-gestation. During the second trimester of pregnancy there is in the fetal brain abundant dendritic branching and synaptogenesis and establishment of connections involving the cortex and thalamus develop between 24 and 32 weeks. Subsequently, neuronal activity, documented as spontaneous transient activity, occurs. During the third trimester the cortical surface area expands significantly.

Processes that occur during the third trimester and that continue in the post-natal period, include dendritic and synaptic growth and connection and maturation of function of oligodendrocytes that enable myelination of axons. Keunen et al. cited work by Huttenlocher et al.[26] that provided information on post-natal processes that included explosive synaptogenesis, dendritic arborization and the generation of increased connectivity. The synaptic and dendritic over-production of early infancy was followed by periods of pruning of synapses that occurred during childhood and during adolescence.[27]

Keunen et al.[25] reviewed progress in analysis of functional neuronal connectivity. They noted that further methodological improvements will be required for optimal studies. Results of available studies revealed that in preterm birth infants, connectivity impairments were present. These included reduced thalamo-cortical connectivity, reduced inter and intra-hemispheric connectivity. Impaired connectivity during development significantly impacts communication efficiency. They noted further that there is an urgent need for studies that investigate gene-environment interactions and relationships to connectivity development.

Key factors in child development

The Lancet has published several editions of comprehensive assessment of important factors in child development, starting with a landmark series in 2007. A report in this series in 2017 by Black et al.[28] was entitled "Early childhood development coming of age: science through the life course." A key finding in this report was that in low and middle- income countries 43% of children under the age of 5 years were at risk to not achieve their developmental potential.

The report documented processes and factors that begin before conception, and pre-natal and post-natal processes that can disrupt normal brain development. Black et al., emphasized that it is important to consider child development as the "ordered progression of perceptual, motor, cognitive, language, socio-emotional and self-regulation skills". They emphasized the key importance of nurturing care, opportunities for play and exploration and safety of the environment.

Nurturing includes nutritious food, adequacy of micronutrients for prospective mothers, pregnant women, infants and children. Black et al., cited evidence that the key period when adequate nutrition. Was necessary for brain development included the 1000 days between conception and the age of 2 years.

Autism

Kanner[29] described a condition in children characterized by: "autistic disturbances of affective contact" and "obsessive desire for maintenance of sameness".

The DSM 5 (Diagnostic and statistical manual for mental disorders) criteria for autism include the following:

A. Deficits in social emotional reciprocity
1. Deficits in non-verbal communication behaviors used for social interaction
2. Deficits in developing, maintaining, understanding relations

B. Restrictive repetitive patterns of behavior, interests or activities.
1. Stereo repetitive motor movements or activities
2. Insistence on sameness

It is important to note that autistic behaviors are now considered to be present in a number of different disorders, referred to as autism spectrum disorders. Children with autism may have high IQ, average IQ or they manifest cognitive impairments. In some children language is not affected while in other children with autism there may be significant language impairments.

Autism that occurs in association with physical abnormalities is referred to as syndromic autism. However, most cases of autism are non-syndromic. Non-syndromic autism is more common in boys than in girls. Family studies and twin studies indicate significant heritability of non-syndromic autism.

Syndromic autism

Sztainberg and Zoghbi[30] reviewed syndromic autism. Examples include occurrence of autism that occurs in children with chromosome abnormalities, including segmental genomic duplications of deletions, also defined as copy number variants (CNVs), CNVs that have been reported as associated with autism include deletions or duplications particularly those that occur in chromosome 15q 11-q13, 16p11.2, 22q11.2 or 7q11.23.

They noted that in addition there are specific monogenic diseases in which autism frequently occurs. They documented important disorders, the estimated autism prevalence in patients with the disorder and the specific gene products impacted in the disorders.

Disorder	Autism prevalence	Gene product and function
Fragile X syndrome	30–60%	FMR1 mRNA trafficking from nucleus
Rett syndrome	61%	MECP2 methyl-CpG binding protein 2
Angelman syndrome	34%	UBE3A ubiquitin ligase, protein decay
Phelan-McDermid syndrome	75%	SHANK3 postsynaptic scaffold proteins
Timothy syndrome	60%	CACNA1C calcium channel
Neurofibromatosis	18%	NF1 negative regulator of Ras signal

Autism epidemiology

Lyall et al.[31] reviewed epidemiology of autism spectrum disorders. They noted changes in the reported prevalence of autism and that in 2017 the estimated frequency in developed countries was 1.5% and that this reflected an increase in prevalence over the years. The increased prevalence was primarily due to increases in the frequency of cases of autism without intellectual disability. In 2017 30% of cases of autism were reported to have intellectual disability and 30–40% of cases were reported to have attention deficit disorders. Sensory sensitivity deficits were noted to be relatively common in individuals with autism. Other manifestations that occurred in some cases included sleep disorders, gastro-intestinal problems and impaired immunity.

Heritability estimates in autism, based on twin studies and family studies ranged between 76 and 95%.[32-34]

Lyall et al.[31] noted evidence for multifactorial etiology of autism with genetic and environmental factors both playing roles. Questions also arise concerning gene environmental interactions and their potential roles in autism.

Possible causative environmental factors in autism

Lyall et al. noted that sub-optimal conditions during pregnancy were considered to be important in causation of autism. These conditions include deficiency in intake of key vitamins, exposure to air pollutants, exposure to endocrine disruptive chemicals including organic phosphates, organo-chlorine pesticides, phthalates, polychlorinated bisphenols (PCB) and/or heavy metals. They also noted however that many of the environmental studies had small sample sizes and results of different studies were sometimes conflicting.

Other environmental factors apparently important in autism included advanced biparental age and advanced paternal age. Short interpregnancy interval was reported to be important in some studies.

Lyall et al., noted that there are several studies linking maternal bacterial and viral infections during pregnancy as important in autism.[35,36] The occurrence of autoimmune disease in the family was reported to be associated with an increased incidence of autism.[37]

Evidence for increased autism risk in infants born by Caesarean section was reported in some studies but presence of confounding factors could not be ruled out.[38]

Epigenetics and autism

In a 2017 review Siu and Weksberg[39] noted that physiological processes and metabolic processes were disrupted in some cases of autism. These disruptions led to immunological impairments, oxidative stress and mitochondrial dysfunction. Aspects of impacted immune function included alterations in cytokines, interleukins and microglia. There are also reports of microbiome dysbiosis.

Siu and Weksberg reported that 10–15% of cases of autism represent syndromic autism spectrum disorders (ASD) and in several of these cases the altered genes take part in significant epigenetic processes. Specific syndromes in which autism occurs, the genes impacted and their function are listed below.

Syndrome	Gene product	Gene product function
Rett syndrome	MECP2	Methyl-CpG binding protein 2
Sotos syndrome	NSD1	Histone methyl transferase writer
Kabuki syndrome	KMT2D	Lysine methyltransferase
CHARGE syndrome	CHD7	Chromodomain helicase DNA binding protein 7
Autism susceptibility	CHD8	Chromodomain helicase DNA binding protein 7

In addition, there are 3 syndromes due to defects in epigenetically imprinted regions, Angelman syndrome (15q11.2), Prader Willi syndrome (15q11.2) and Turner syndrome (Xp22.33). The prevalence of autism in these syndromes varied. Autism was reported to occur in 3% of cases of Turner syndrome, in Charge syndrome autism was reported to occur in 85% of cases.

Non-syndrome autism and common risk variants

In 2019 Grove[40] and members of the Autism Special Working Group of the Psychiatric Genomics Consortium reported analyses of sequencing studies on 18,381 cases and 27,969 controls. They reported that variant alleles that occurred commonly in the general population were frequently found in the autism population and that specific alleles at these loci constituted risk alleles. It is also important to note that the common variant alleles contributed to polygenic autism. Grove et al., noted that the ASD associated variants occurred commonly in regions of the genome enriched in sequence elements that regulated gene expression.

Based on the locations of significantly associated risk variants, Grove et al., identified genetic several loci that could potentially have altered expression as a result of the risk alleles in autism. The Product of the genes with altered expression are listed below

Gene product	Function
NEGR1	Developmentally expressed in hippocampus, hypothalamus adhesion molecule that modulates synapse formation
PTPB2	Splicing regulatory factor particularly active in neurogenesis and neuron differentiation
CADPS	Calcium binding protein involved in neurotransmitter and neuro-peptide exocytosis
KCNN2	Voltage independent calcium activated potassium channel modulated neuronal excitation
KMT2E	Histone lysine methyl transferase involved in chromatin regulation
MACROD2	Nuclear enzyme involved in mono-adenosine phosphate ribosylation

Genome-wide analyses of de novo variants in autism

An et al.[41] reported results of whole genome sequencing carried out on 1902 quartet families, in each family one child was diagnosed with autism spectrum disorder. Their study revealed an excess of de novo (non-inherited) nucleotide mutations in protein coding genomic region and a weaker signal of rare nucleotide variants in the non-protein coding genome. They noted that the variant signal in the non-protein coding regions were located in the promoter regions of genes. They defined promoter regions as the genome sequence that extended 2000 nucleotides upstream of the gene transcription start site. An et al., noted further that the variants occurred most frequently in transcription binding sites.

Functional magnetic neuroimaging in autism

Holiga et al.[42] used resting state functional magnetic resonance imaging of brain for studies in four ASD cohorts. They noted that resting state functional magnetic neuroimaging (rsFMRI) measurements are due to fluctuations in oxygen demands associated with spontaneous neuronal activity.

They noted that there had been a number of prior reports indicating altered resting state neuronal activity in autism, however different reports varied with respect to the anatomical sites involved and the directional of the altered activity.

Holiga et al., assembled data from 841 autistic individuals and 984 controls. The autism related data collected included assessment of social and communication skills (ADOS and ADI assessments), Vineland adaptive behavior measures, daily living skills and also Vineland assessments of social and communications skills.

They reported that their study revealed reproducible evidence for both hyperactivity and hypoactivity alterations in ASD individuals. Hypoactivity was primarily restricted to the motor sensory regions and hyperactivity hubs occurred mainly in the prefrontal and parietal cortices. They noted that cortex connectivity changes were observed and that there was no evidence for altered sub-cortical or altered cerebellar connectivity. They concluded that these changes indicated that autistic individuals may have deficits in mental switching and cognitive flexibility. However, their studies across cohorts revealed differences in the range of effect sizes.

Holiga et al. concluded there is a growing consensus across studies that certain cortical networks are characterized by hyperactivity while other are characterized by hypoactivity. They considered the key drivers of ASD to be connectivity alterations. They noted that their studies were consistent with diffusion tension imaging studies reported by Solso et al.[43] that reported evidence of axonal overconnectivity on the frontal lobes of children with autism.

Holiga noted that the findings of their study were also consistent with those of Geschwind and Levitt[44] who reported spatially distinct alterations of network connections in autism with regions of hyperconnectivity and regions of hypoconnectivity.

They emphasized that there was only a moderate link between the level of connectivity changes and clinical symptoms.

In a 2019 report Nunes et al.[45] also noted that intrinsic connectivity differences were observed on fMRI studies they conducted in ASD individuals. In their study they analyzed different brain networks including the visual network (VN) the somatosensory network (SN), the dorsal attention network (DAN) the ventral attention network(VA) and the fronto-parietal network (FPN), the default mode network (DMN) and the limbic system network.

Their study revealed idiosyncrasies in the DMN and SMN networks in autistic subjects. Furthermore, they demonstrated that the extent of idiosyncrasies in the network were correlated with degree of autism symptoms.

Nunes et al., emphasized that the default mode network (DMN) was linked to social cognitive processes, ability to have insight into the mental states of others and it was also linked to emotional processing and ability to understand narrative.

Nunes et al. noted that the SMN network is related to sensory and sensory motor interactions. They conclude the intracranial network alterations were idiosyncratic in extent, differing in different ASD individuals. However, the specific network alterations were correlated with the ASD manifestations.

Epilepsy

Epileptic encephalopathies can lead to significant neurodevelopmental impairments. In earlier decades much emphasis was placed on structural brain malformations as causes of child epilepsy. In recent decades there have been considerable efforts applied elucidation of genetic factors and biological factors the play roles in causation of epilepsy.

Epilepsy due to rare genomic changes of major effect

Myers and Mefford[46] reported that epileptic encephalopathies are largely due to de novo genomic changes. They noted that deleterious genomic copy number variants were detected in 5–10% of epileptic encephalopathy. DNA sequencing methodologies have revealed pathogenic mutations in a number of cases with this disorder. The proteins impacted by these mutations have expanded insight into biological pathways involved in epilepsy.

Orsini et al. reported that genetic abnormalities have been identified in 30% of cases of epilepsy syndromes. They noted that epilepsy syndromes particularly in early childhood are often classified on the bases of electroclinical information, examples include West syndrome, Dravet syndrome, Lennox Gestault syndrome. Mutations in specific ion channels have turned out to be important causes of epilepsy diagnosed in childhood. In addition, abnormalities in solute transporters and in neurotransmitters have been frequently involved and mutations each of these proteins have been identified across the different electroclinical syndromes.

Orsini et al.[47] provided information on key genes identified in specific epileptic encephalopathies of childhood. Important gene products in their listing and functions of these are listed below.

Ion channel proteins	
Sodium ion channel subunits	SCN2A, SCN1A, SCN8A
Potassium ion channel subunits	KCNQ2, KCNA2, KCNT1
Calcium channel subunit	CACNA1A
<u>Solute carrier proteins</u>	SLC25A22, SLC12A5

Neurotransmitter related	
STXBP1	Syntaxin-binding protein, plays a role in release of neurotransmitters
GRIN2A	Glutamate ionotropic receptor NMDA type subunit 2A

Signal transduction/other	
PIGA	Phosphatidylinositol glycan anchor biosynthesis class A
CDKL5	Cyclin dependent kinase like 5 (defects in West syndrome)
ERBB2	Tyrosine kinase, binds to and is activated by neuregulins
CHD2	Chromohelicase domain protein modifies chromatin structure

Common epilepsy (idiopathic generalized epilepsy)

It is important to note that there is no evidence that monogenic defects of major effect lead to common epilepsy. In a 2018 review Koeleman[48] noted that common epilepsy occurs in 1 in 200 individuals in the population. There is evidence for heritability of this form of epilepsy, however be noted that precise causative molecular factors have not been identified.

Mefford[49] emphasized that results from monogenic forms of epilepsy have not translated to generalized epilepsy. Generalized epilepsy does show evidence of heritability and it is likely that this heritability may be due to polygenic effects. Mefford also noted that few insights into genetic factors have been obtained from genome wide association studies.

May et al.[50] carried out an exome sequencing case control study on an initial panel of 152 cases of familial cases defined as generalized epilepsy and 549 ethnically matched controls. Results from this initial discovery cohort revealed enrichment in the cases in rare missense variants in genes that encode subunits of the GABA$_A$ neurotransmitter receptor. May et al., subsequently analyzed data from a validation cohort comprised of 724 epilepsy affected individuals from different countries in Europe. This validation cohort included individuals with familial or sporadic cases of generalized epilepsy.

May et al. reported an overall enrichment for rare missense variants in 15 different genes that encode subunits of the GABA$_A$ neurotransmitter receptor. The nonsynonymous variants increased epilepsy risk with

OR between 1.4 and 4.10. They carried out functional studies and reported loss of function in rare non-synonymous variants in the *GABRB2* and *GABRA5* genes that encode subunits encode GABA$_A$ subunits.

The GABA$_A$ neurotransmitter receptor is an ionotropic receptor. The ligand for this receptor is gamma aminobutyric acid. GABA$_A$ receptors constitute the major inhibitory receptors in the central nervous system. These receptors are each composed of 19 different subunits each subunit is encoded by different genes.

This study illustrates that relatively uncommon coding variants may also increase the risk of common disorders.

Genomic and genetic mosaicism in epilepsy

Examples of the impact of mosaicism in epilepsy were reported by de Lange et al.[51] in 2018 following studies of Dravet syndrome. Dravet syndrome is frequently associated with severe epilepsy due to heterozygosity for loss of function mutations in a neuronal sodium channel gene SCN1A. These mutations are frequently de novo in origin. However, all cases with loss of functions pathogenic mutations in SCN1A do not manifest severe epilepsy.

In their study de Lange et al., obtained DNA from different cell types in 128 individuals with epilepsy reported to be due to de novo SCN1A mutations. The cells used included including buccal cells, blood cells, and DNA was also isolated from saliva. They used next generation sequencing and methods to analyze the pathogenic and normal alleles in SCN1A. Results obtained were confirmed by quantitative polymerase chain reaction.

The de Lange study revealed mosaicism for SNC1A variants in 7.5% of patients, and they noted that disease was less severe in patients with mosaicism.

References

1. Krauss RS, Hong M. Gene-environment interactions and the etiology of birth defects. *Curr Top Dev Biol* 2016;**116**:569–80. https://doi.org/10.1016/bs.ctdb.2015.12.010. Review, 26970642.
2. Keppler-Noreuil KM, Sapp JC, Lindhurst MJ, Parker VE, Blumhorst C, et al. Clinical delineation and natural history of the PIK3CA-related overgrowth spectrum. *Am J Med Genet A* 2014;**164A**(7):1713–33. https://doi.org/10.1002/ajmg.a.36552. Epub 2014 Apr 29, 24782230.
3. Roessler E, Belloni E, Gaudenz K, Jay P, Berta P, et al. Mutations in the human sonic hedgehog gene cause holoprosencephaly. *Nat Genet* 1996;**14**(3):357–60. 8896572.
4. Kruszka P, Martinez AF, Muenke M. Molecular testing in holoprosencephaly. *Am J Med Genet C Semin Med Genet* 2018;**178**(2):187–93. https://doi.org/10.1002/ajmg.c.31617. 29771000.

5. Roessler E, Hu P, Muenke M. Holoprosencephaly in the genomics era. *Am J Med Genet C Semin Med Genet* 2018;**178**(2):165–74. https://doi.org/10.1002/ajmg.c.31615. [29770992].

6. Hong M, Krauss RS. Cdon mutation and fetal ethanol exposure synergize to produce midline signaling defects and holoprosencephaly spectrum disorders in mice. *PLoS Genet* 2012;**8**(10):e1002999. https://doi.org/10.1371/journal.pgen.1002999. 23071453.

7. Dubourg C, Kim A, Watrin E, de Tayrac M, Odent S, et al. Recent advances in understanding inheritance of holoprosencephaly. *Am J Med Genet C Semin Med Genet* 2018;**178**(2):258–69. https://doi.org/10.1002/ajmg.c.31619. 29785796.

8. Kim A, Savary C, Dubourg C, Carré W, Mouden C, et al. Integrated clinical and omics approach to rare diseases: novel genes and oligogenic inheritance in holoprosencephaly. *Brain* 2019;**142**(1):35–49. https://doi.org/10.1093/brain/awy290. 30508070.

9. Dixon MJ, Marazita ML, Beaty TH, Murray JC. Cleft lip and palate: understanding genetic and environmental influences. *Nat Rev Genet* 2011;**12**(3):167–78. https://doi.org/10.1038/nrg2933. Review, 21331089.

10. DeRoo LA, Wilcox AJ, Drevon CA, Lie RT. First-trimester maternal alcohol consumption and the risk of infant oral clefts in Norway: a population-based case-control study. *Am J Epidemiol* 2008;**168**(6):638–46. https://doi.org/10.1093/aje/kwn186. 18667525.

11. Boyles AL, Wilcox AJ, Taylor JA, Shi M, Weinberg CR, et al. Oral facial clefts and gene polymorphisms in metabolism of folate/one-carbon and vitamin A: a pathway-wide association study. *Genet Epidemiol* 2009;**33**(3):247–55. https://doi.org/10.1002/gepi.20376. 19048631.

12. Jahanbin A, Shadkam E, Miri HH, Shirazi AS, Abtahi M. Maternal folic acid supplementation and the risk of oral clefts in offspring. *J Craniofac Surg* 2018;**29**(6):e534–41. https://doi.org/10.1097/SCS.0000000000004488. Review, 29762322.

13. Kondo S, Schutte BC, Richardson RJ, Bjork BC, Knight AS, et al. Mutations in IRF6 cause Van der Woude and popliteal pterygium syndromes. *Nat Genet* 2002;**32**(2):285–9. 12219090.

14. Leslie EJ, Carlson JC, Shaffer JR, Feingold E, Wehby G, et al. A multi-ethnic genome-wide association study identifies novel loci for non-syndromic cleft lip with or without cleft palate on 2p24.2, 17q23 and 19q13. *Hum Mol Genet* 2016;**25**(13):2862–72. 27033726.

15. Cox LL, Cox TC, Moreno Uribe LM, Zhu Y, Richter CT, et al. Mutations in the epithelial cadherin-p120-catenin complex cause Mendelian non-syndromic cleft lip with or without cleft palate. *Am J Hum Genet* 2018;**102**(6):1143–57. https://doi.org/10.1016/j.ajhg.2018.04.009.

16. Funato N, Nakamura M. Identification of shared and unique gene families associated with oral clefts. *Int J Oral Sci* 2017;**9**(2):104–9. https://doi.org/10.1038/ijos.2016.56. 28106045.

17. Boivin MJ, Kakooza AM, Warf BC, Davidson LL, Grigorenko EL. Reducing neurodevelopmental disorders and disability through research and interventions. *Nature* 2015;**527**(7578):S155–60. https://doi.org/10.1038/nature16029. Review, 26580321.

18. Braithwaite EC, Kundakovic M, Ramchandani PG, Murphy SE, Champagne FA. Maternal prenatal depressive symptoms predict infant NR3C1 1F and BDNF IV DNA methylation. *Epigenetics* 2015;**10**(5):408–17. https://doi.org/10.1080/15592294.2015.1039221. 25875334.

19. Murgatroyd C, Quinn JP, Sharp HM, Pickles A, Hill J. Effects of prenatal and postnatal depression, and maternal stroking, at the glucocorticoid receptor gene. *Transl Psychiatry* 2015;**5**:e560. https://doi.org/10.1038/tp.2014.140. 25942041.

20. Abubakar A, Holding P, Newton CR, van Baar A, van de Vijver FJ. The role of weight for age and disease stage in poor psychomotor outcome of HIV-infected children in Kilifi, Kenya. *Dev Med Child Neurol* 2009;**51**(12):968–73. https://doi.org/10.1111/j.1469-8749.2009.03333.x. 19486107.

21. Boivin MJ, Okitundu D, Makila-Mabe Bumoko G, Sombo MT, Mumba D, et al. Neuropsychological effects of konzo: a neuromotor disease associated with poorly processed cassava. *Pediatrics* 2013;**131**(4):e1231–9. https://doi.org/10.1542/peds.2012-3011. 23530166.

22. Kiwanuka J, Bazira J, Mwanga J, Tumusiime D, Nyesigire E, et al. The microbial spectrum of neonatal sepsis in Uganda: recovery of culturable bacteria in mother-infant pairs. *PLoS ONE* 2013;**8**(8):e72775. https://doi.org/10.1371/journal.pone.0072775. eCollection 2013, 24013829.

23. Homberg JR, Kyzar EJ, Scattoni ML, Norton WH, Pittman J, et al. Genetic and environmental modulation of neurodevelopmental disorders: Translational insights from labs to beds. *Brain Res Bull* 2016;**125**:79–91. https://doi.org/10.1016/j.brainresbull.2016.04.015. Review, 27113433.

24. Lake A, Chan M. Putting science into practice for early child development. *Lancet* 2015;**385**(9980):1816–7. https://doi.org/10.1016/S0140-6736(14)61680-9. 25245180.

25. Keunen K, Counsell SJ, Benders MJNL. The emergence of functional architecture during early brain development. *Neuroimage* 2017;**160**:2–14. https://doi.org/10.1016/j.neuroimage.2017.01.047. 28111188.

26. Huttenlocher PR, Dabholkar AS. Regional differences in synaptogenesis in human cerebral cortex. *J Comp Neurol* 1997;**387**(2):167–78. 9336221.

27. Innocenti GM, Price DJ. Exuberance in the development of cortical networks. *Nat Rev Neurosci* 2005;**6**(12):955–65. Review, 16288299.

28. Black MM, Walker SP, Fernald LCH, Andersen CT, DiGirolamo AM, Lancet Early Childhood Development Series Steering Committee, et al. Early childhood development coming of age: science through the life course. *Lancet* 2017;**389**(10064):77–90. https://doi.org/10.1016/S0140-6736(16)31389-7. Review, 27717614.

29. Kanner L. Autistic disturbances of affective contact. *Nervous Child* 1943;**2**:217–50.

30. Sztainberg Y, Zoghbi HY. Lessons learned from studying syndromic autism spectrum disorders. *Nat Neurosci* 2016;**19**(11):1408–17. https://doi.org/10.1038/nn.4420. Review, 27786181.

31. Lyall K, Croen L, Daniels J, Fallin MD, Ladd-Acosta C, et al. The changing epidemiology of autism Spectrum disorders. *Annu Rev Public Health* 2017;**38**:81–102. https://doi.org/10.1146/annurev-publhealth-031816-044318. 28068486.

32. Hallmayer J, Cleveland S, Torres A, Phillips J, Cohen B, et al. Genetic heritability and shared environmental factors among twin pairs with autism. *Arch Gen Psychiatry* 2011;**68**(11):1095–102. https://doi.org/10.1001/archgenpsychiatry.2011.76. 21727249.

33. Colvert E, Tick B, McEwen F, Stewart C, Curran SR, et al. Heritability of autism spectrum disorder in a UK population-based twin sample. *JAMA Psychiatry* 2015;**72**(5):415–23. https://doi.org/10.1001/jamapsychiatry.2014.3028. 25738232.

34. Sandin S, Lichtenstein P, Kuja-Halkola R, Hultman C, Larsson H, Reichenberg A. The heritability of autism Spectrum disorder. *JAMA* 2017;**318**(12):1182–4. https://doi.org/10.1001/jama.2017.12141.

35. Lee BK, Magnusson C, Gardner RM, Blomström Å, Newschaffer CJ, et al. Maternal hospitalization with infection during pregnancy and risk of autismspectrum disorders. *Brain Behav Immun* 2015;**44**:100–5. https://doi.org/10.1016/j.bbi.2014.09.001. 25218900.

36. Zerbo O, Qian Y, Yoshida C, Grether JK, Van de Water J, Croen LA. Maternal infection during pregnancy and autism spectrum disorders. *J Autism Dev Disord* 2015;**45**(12):4015–25. https://doi.org/10.1007/s10803-013-2016-3. 24366406.

37. Keil A, Daniels JL, Forssen U, Hultman C, Cnattingius S, et al. Parental autoimmune diseases associated with autism spectrum disorders in offspring. *Epidemiology* 2010;**21**(6):805–8. https://doi.org/10.1097/EDE.0b013e3181f26e3f. 20798635.

38. Curran EA, Cryan JF, Kenny LC, Dinan TG, Kearney PM, Khashan AS. Obstetrical mode of delivery and childhood behavior and psychological development in a British cohort. *J Autism Dev Disord* 2016;**46**(2):603–14. https://doi.org/10.1007/s10803-015-2616-1. 26412364.

39. Siu MT, Weksberg R. Epigenetics of autism Spectrum disorder. *Adv Exp Med Biol* 2017;**978**:63–90. https://doi.org/10.1007/978-3-319-53889-1_4. Review, 28523541.

40. Grove J, Ripke S, Als TD, Mattheisen M, Walters RK, et al. Identification of common genetic risk variants for autism spectrum disorder. *Nat Genet* 2019;**51**(3):431–44. https://doi.org/10.1038/s41588-019-0344-8. 30804558.

41. An JY, Lin K, Zhu L, Werling DM, Dong S, et al. Genome-wide de novo risk score implicates promoter variation in autism spectrum disorder. *Science* 2018;**362**(6420). pii: eaat6576. https://doi.org/10.1126/science.aat6576. 30545852.

42. Holiga Š, Hipp JF, Chatham CH, Garces P, Spooren W, et al. Patients with autism spectrum disorders display reproducible functional connectivity alterations. *Biol Psychiatry* 2019. pii: S0006-3223(19)30147-7. https://doi.org/10.1016/j.biopsych.2019.02.019. 31010580.

43. Solso S, Xu R, Proudfoot J, Hagler Jr DJ, Campbell K, et al. Diffusion tensor imaging provides evidence of possible axonal overconnectivity in frontal lobes in autism Spectrum disorder toddlers. *Biol Psychiatry* 2016;**79**(8):676–84. https://doi.org/10.1016/j.biopsych.2015.06.029. 26300272.

44. Geschwind DH, Levitt P. Autism spectrum disorders: developmental disconnection syndromes. *Curr Opin Neurobiol* 2007;**17**(1):103–11. Review, 17275283.

45. Nunes AS, Peatfield N, Vakorin V, Doesburg SM. Idiosyncratic organization of cortical networks in autism spectrum disorder. *Neuroimage* 2019;**190**:182–90. https://doi.org/10.1016/j.neuroimage.2018.01.022. Review. [29355768].

46. Myers CT, Mefford HC. Genetic investigations of the epileptic encephalopathies: recent advances. *Prog Brain Res* 2016;**226**:35–60. https://doi.org/10.1016/bs.pbr.2016.04.006.Review.PMID:27323938.

47. Orsini A, Zara F, Striano P. Recent advances in epilepsy genetics. *Neurosci Lett* 2018;**667**:4–9. https://doi.org/10.1016/j.neulet.2017.05.014. Review, 28499889.

48. Koeleman BPC. What do genetic studies tell us about the heritable basis of common epilepsy? Polygenic or complex epilepsy? *Neurosci Lett* 2018;**667**:10–6. https://doi.org/10.1016/j.neulet.2017.03.042. Review, 28347857.

49. Mefford HC. Expanding role of GABA$_A$ receptors in generalised epilepsies. *Lancet Neurol* 2018;**17**(8):657–8. https://doi.org/10.1016/S1474-4422(18)30252-7. 30033051.

50. May P, Girard S, Harrer M, Bobbili DR, Schubert J, et al. Rare coding variants in genes encoding GABA$_A$ receptors in genetic generalised epilepsies: an exome-based case-control study. *Lancet Neurol* 2018;**17**(8):699–708. https://doi.org/10.1016/S1474-4422(18)30215-1. 30033060.

51. de Lange IM, Koudijs MJ, van 't Slot R, Gunning B, Sonsma ACM, et al. Mosaicism of de novo pathogenic SCN1A variants in epilepsy is a frequent phenomenon that correlates with variable phenotypes. *Epilepsia* 2018;**59**(3):690–703. https://doi.org/10.1111/epi.14021. 29460957.

PERSONALIZED PRECISION MEDICINE—PART A: CONCEPTS AND RELEVANCE IN MENDELIAN DISORDERS

Concepts

The concepts of personalized precision medicine were promoted initially by Hood and Flores.[1] Collins and Varmus[2] announced a new initiative on Precision Medicine. They wrote "What is needed now is a broad research program to build the evidence base needed to guide clinical practice".

Jameson and Longo[3] noted that in response, physicians emphasized that they had always practiced individualized, personalized medicine. However, inclusion of the term precision signaled inclusion of advances in genetics, genomics, proteomics, metabolomics and informatics. Jameson and Longo noted that advances in imaging technologies should be included as part of Precision Medicine. They also noted that interests of patients, physicians, payors, pharmaceutical industry and the health system were not fully aligned.

The precision medicine program and goals have been met with skepticism on the part of some individuals. Bayer and Galea[4] drew attention to the social determinants of health and to marked inequality in access to healthcare. They drew attention to a 2013 report from the National Research Council and the Institute of Medicine in the USA, entitled "Shorter lives, Poorer Health": that revealed that American citizens fared worse than citizens of other high-income countries. A key statement in this report was, "decades of research have demonstrated that health is determined by far more than health care". Bayer and Galea concluded that improvement in population health required that certain social realities needed to be addressed.

Davis and Shanley[5] noted that the scientific aspects of precision medicine were being promoted while less attention was being given to socio-behavioral and environmental factors that are major factors in human morbidity and mortality.

Gene Environment Interactions. https://doi.org/10.1016/B978-0-12-819613-7.00009-8

Precision medicine goals

Morgan et al.[6] outlined goals of precision medicine as the use of genomic data and electronic medical records to improve human health. They considered genomic data to be the inclusion of DNA sequence information, data on DNA methylation, data on metabolism and signaling and the relationship of all of these to physiology and disease. In addition, they noted that precision medicine also included information that defined disease drivers and differences in these drivers in different individuals. Collection of physical measurements and assessment of physical activity were also noted to be important factors in precision medicine.

Morgan et al., noted that the concept that records of patient care and outcomes could improve understanding of disease and promote improved patient care, can be traced back to Hippocrates. They noted that in addition to patient records disease registries provided useful information.

Whitsel et al.[7] emphasized that analyses of genes, environmental exposures, information on longitudinal behaviors, epigenetics and genome expression can potentially transform human health and well- being.

They noted that in parallel with these efforts it will be important for regulatory authorities to enforce standards related to air quality, food systems and to consider optimization of health care.

Whitsel et al. emphasized the important role of governments in promoting research and activities related to precision medicine. In addition, they emphasized that precision medicine initiatives and public health enterprises would need to promote trust in vulnerable populations and increase access to healthcare in these populations,

Ginsburg and Phillips[8] reviewed precision medicine and its potential values. They noted the important conceptual distinction between personalized medicine and precision medicine.

> *Personalized medicine refers to an approach to patients that considers their genetic make-up but with attention to their preferences, beliefs, attitudes, knowledge and social context whereas precision medicine describes a model for healthcare that relies heavily on data analytics and information.*

They noted that data capture in the precision medicine ecosystem includes information on family history, environmental exposures, phenotype and medical care. They emphasized that data and findings from clinical practice and from research can inform each other to improve healthcare. They emphasized the importance of generating knowledge from data gathered during healthcare delivery.

Ginsburg and Phillips noted the importance of taking genetic variants into account when prescribing therapies. Important evidence has been gathered on adverse responses to specific medication based on the presence of genetic variants in particular patients. Examples include adverse responses to Abacavir in individuals who have the HLAB *5701 variants, abnormal responses to the anti- tumor medication mercaptopurine in patients with variants in thiopurine *S*-methyltransferase (TMPT) and unusual responses to warfarin therapy in patients with specific sequence variants in cytochrome mono-oxygenase (CYP2C9) and vitamin K epoxide reductase complex subunit 1 (VKORC1).

Important definitions related to precision medicine studies

Genome wide association studies (GWAS) are defined as analyses of relatively common single nucleotide variants to determine their association with a specific phenotype or with a specific clinical condition.

Phewas studies aim to determine the different phenotypes associated with a specific genomic variant, or phenotypic data associated with specific alterations in gene expression or phenotypic manifestations associated with a specific disease. Phewas data utilizes the medical phenome database to search for association of specific clinical features with specific genetic variants as the input function. Although single nucleotide variants are most often used as the input function other input functions can be used including disease category, or therapeutic response (https://phewascatalog.org/).

Expression quantitative trait loci are defined as genomic loci that correlate with variations in expression of a specific gene or genes.

Transcription wide association studies (TWAS) are designed to identify genetic variants associated with alteration in gene expression. TWAS studies depend on data in a reference data base on specific genetic variants that have been shown to impact gene expression. If data on gene expression from a test individual is obtained it may be possible to use data in the reference data set to establish which genetic variants are likely involved. This use of data to infer connections is referred to as imputation. Alternatively, when data on the presence of a specific variant in the test individual is obtained, imputation of reference data may be applied to estimate likely expression changes in consequence of the presence of the specific SNP.

The UK Biobank: An example of connecting genomic and phenotypic data and health

In addition to collection of samples for genetic and metabonomic studies on approximately 500,000 individuals, the UK Biobank study involves collection of socio-demographic, lifestyle and health information. In addition, physical measurements are taken and parameters of heart and lung function are obtained. Data on cognitive function and data on bone and joints are obtained.[9]

The biological samples collected include blood, urine and saliva. Genomic analyses include whole genome sequencing, genotyping and array analyses of genomic variants.

Interim release of data on 15,000 participants in 2018 has facilitated identification of links between genetic variants and rare diseases, genetic variants and common diseases and associations of diseases with environmental and lifestyle factors.

Studies that have utilized data from UK Biobank

Genome wide association studies (GWAS) have revealed that altered blood pressure is a polygenic trait and note than 250 gene loci are associated with altered blood pressure.[10] Genetic variants have been identified that impact systolic and diastolic blood pressure levels and variants that impact pulse pressure. These variants have been predominantly identified in non-protein coding genomic regions. Specific studies have led to the identification of regulatory loci associated with altered blood pressure.

Giri et al.[11] carried out a trans-ethnic study that included 459,777 participants from the Million Veterans study program and 140,886 participants from the UK Biobank. They later carried out replication studies on 316,301 individuals enrolled from the International Consortium for Blood Pressure (ICBP) cohort.

Giri et al. revealed that in the single variant analyses, 505 independent loci were found to be associated with one or more parameters of blood pressure including systolic, diastolic blood pressure or pulse pressure. Giri et al. reported that both common and rare variants were associated with these blood pressure parameters and they integrated their genetic studies with studies on phenotypes.

Giri et al. reported that genes in the transforming growth factor beta (TGFB) and Notch signaling pathways were important in determining systolic blood pressure. Particularly important were genes that encoded soluble guanylate cyclases, GUCY1A, GUCY1B2 and FURIN. Soluble guanylate cyclases are heterodimeric proteins that catalyze the conversion of GTP to $3',5'$-cyclic GMP and pyrophosphate. Giri noted that guanyl cyclase plays important roles in nitric oxide function in the vascular wall. FURIN (PCSK3) encodes a protease, (proprotein convertase) and has been implicated in atherosclerosis.[12]

Elevated pulse pressure is noted to be a marker for arterial stiffness. Giri et al.[11] reported pulse pressure showed correlation with variants in TCF7L2 a transcription factor, and CDH13 (cadherin 13) that is reported to protect vascular endothelial cells from apoptosis due to oxidative stress. Altered pulse pressure was also found to be associated with resistance to atherosclerosis and with gene variants in PHACTR1 (phosphatase and actin regulator 1) that impacts the actin cytoskeleton and MTHFR (methylene tetrahydrofolate reductase). Variants in these genes have been previously implicated in risk for cardiovascular disease.

Giri et al. reported that blood pressure levels were also correlated with variants in RXFP2 that encodes a receptor for the hormone relaxin, and correlations with variants ADK adenosine kinase and in PDE3A a protein that mediates platelet aggregation and also plays important roles in cardiovascular function by regulating vascular smooth muscle contraction and relaxation.

UK Biobank data and calcific aortic stenosis

Thériault et al.[13] carried out genome wide association studies (GWAS) and eQTL (expression quantitative trait loci) analyses and transcriptome studies to identify genes involved in calcific aortic stenosis using data from cases and controls, including data from the UK Biobank.

Thériault et al. reported that calcific aortic stenosis occurs in approximately 2% of individuals older than 65 years. The calcifications lead to damage of the valve, to valve narrowing and eventually to heart failure. Statins and other medications are not effective in treating this disorder.

There is evidence that genetic factors are involved in the etiology of calcific aortic stenosis. Thériault et al. carried out their studies in cases of calcific aortic valvular stenosis (CAVS) confirmed during surgery. They also carried out studies on ethnically matched controls without valve disease and also on cases that had heart disease without valve disease.

Genome wide association studies included single nucleotide polymorphism (SNP) analyses using the Illumina bead chip on DNA derived from white blood cells. SNPs with populations frequencies <1% were studied and significant different between patients and controls emerged. A lead SNP (rs6702619) was identified that that was significantly associated calcific aortic stenosis.

In subsequent studies cases where individuals with this significant SNP variant occurred ultra-sound measurements of the aortic root diameter were measured.

Transcriptome analyses were carried out on 240 stenotic aortic valves. Quantitative analyses of RNA from aortic valve leaflets were carried out. In addition, eQTL (expression quantitative trait loci) studies were carried out in cases where genotype information and gene expression analyses were available. A specific program Matrix eQTL was used to determine correlation with gene expression, genotype and eQTL SNPs.

Studies revealed that specific risk alleles for calcific aortic valvular stenosis were specifically associated with decreased expression of PALM (palmdelphin) and severity of diseases increased as levels of PALMD mRNA decreased. Palmdelphin is a cytosolic protein that plays roles in membrane dynamics.

Thériault et al. initially carried out studies on cases from Quebec. They then carried out replication studies on the UK Biobank cohort and confirmed association of calcific aortic valvular stenosis with PALMD.

The Palmdelphin discovery potentially opens the way for development of new therapies. Bosse et al.[14] noted that treatments that increase expression of PLAMD must be evaluated.

Factors to take into account in connecting genotype and phenotype: Variable penetrance

A key premise in genetic testing is that if a particular genomic variant, frequently referred to as a mutation, leads to specific deleterious effects including a specific phenotype in a number of people, one can assume that it will lead to similar deleterious effects in other individuals. However, this is not always the case. The same genetic mutation in a specific gene, does not always lead to the same phenotypic effects in all individuals. Furthermore, the degree of severity of the phenotypic manifestations arising as a result of a specific mutation may vary in different individuals. This situation is referred to as reduced penetrance.

Cooper et al.[15] reviewed the molecular bases of reduced penetrance. They proposed that it resulted from interaction between genes and from environmental factors. Reduced penetrance is most often seen in autosomal dominant diseases though it may also occur in autosomal recessive diseases.

Cascade testing refers to follow-up genetic studies in directly related family members of an individual with a specific disorder due to a defined genomic alteration, particularly a known pathogenic mutation, to determine if family directly related family member carry the same mutation and are at risk for the same disorder.

Clinically studies sometimes reveal that closely related individuals with the same genetic mutation have different clinical features from each other. In addition, sometimes directly related individuals with the same mutation do not manifest features of the disease originally attributed to the specific gene mutation.

Reduced penetrance may be related to age differences in the individuals who carry a specific pathogenic mutation, some mutations determine genetic diseases that only manifest late in life.

Another important consideration is whether specific environmental factors offer protection against development of disease manifestation in the mutation carrier.

Other important considerations are whether modifier genes alter presentation of specific disease feature in the mutation carrier. Cooper et al. noted that modifier genes likely play key roles in determining the pathogenicity of specific gene variants. However, identification of modifier genes is difficult in human conditions.

Aspects of reduced penetrance are important to consider when genetic testing is been widely used in healthcare settings.

Sequencing studies and bioinformatic analyses in populations have led to the realization that each individual has multiple potential damaging genetic variants. However, many particularly damaging variants have been reported to be associated with pathological features.

Cooper et al. noted that a number of mutations that lead to dominant genetic disorders (disorders that manifest in heterozygotes) are associated with reduced penetrance. One example they presented involves a specific mutation, the Factor V Leiden mutation that predisposes to venous thrombosis. This mutation occurs with a frequency of 2–5% in European populations. Although this mutation has been shown to be associated with a six-fold increase in the risk of venous thrombosis, Cooper et al. noted that the majority of individuals with this mutation do not manifest symptoms, however, in females there is an important additional risk factor that influences penetrance. Thrombophilic genotypes such as factor V Leiden increase risks of venous thromboembolism in users of combined hormonal contraception.[16]

An important example of a variant with variable penetrance is the mutation Cys282Tyr in the HFE (homeostatic iron regulator) protein that influences predisposition to hemochromatosis. In homozygous form this mutation is strongly associated with hemochromatosis. However, in the case of heterozygotes the HFE pathogenic mutation, gender and age impact the frequency of disease manifestations. Prior to menopause women are relatively protected from disease manifestations.

Hemochromatosis can result from mutation in genes other than HFE and hemochromatosis will be discussed in more detail in a subsequent section.

Cooper et al.[15] discussed evidence that pathogenic variants in a specific gene may manifest different clinical phenotypes or different degrees of disease severity. One important example is familial Mediterranean fever where three different mutations in the MEFV protein have been associated with disease, but each mutation varies in its degrees of penetrance. There is also evidence that some individuals with pathogenic MEFV mutation may not manifest symptoms.

Allelic exclusion

Another important factor that can influence disease manifestations, particularly in autosomal dominant conditions due to damaging mutation, is whether or not in a specific cell gene expression occurs from both copies of that gene (i.e., the normal gene and the mutant gene), or whether the gene is only expressed from one chromosome. If in some cells the allele with the mutation is not expressed pathogenic effects may not necessarily result. Silencing of one allele of a gene may result from regulatory factors.

Massah et al.[17] reported that in mammals 6–10% of genes on non-sex determining chromosomes are selected for monoallelic expression. Allelic exclusion may be dependent on epigenetic mechanisms that silence a specific allele.

In their review Cooper et al.[15] noted several examples of dominant disorders where clinical penetrance was impacted by allelic silencing. Examples include retinitis pigmentosa due to pre-mRNA processing factor 31 (PRPF31) mutations where asymptomatic carriers of the mutation expressed high levels of the wildtype PRPF31 allele.[18]

Clinical manifestations of the autosomal dominant disorder erythropoietic protoporphyria were shown to be reduced by increased expression of the normal allele.

Segmental genomic copy number variants and disease manifestations

The pathological outcome of specific genomic copy number variants is now known to be influenced by the sensitivity of specific genes in the implicated segment to dosage alteration. Databases have been constructed that give information on whether or not dosage alteration, either reduction or extra copies, give rise to pathogenic manifestations, e.g., Clin gene dosage sensitivity map (https://www.ncbi.nlm.nih.gov/projects/dbvar/clingen/).

Modifier genes and networks

In a 2017 review on modifier genes Riordan and Nadeau[19] emphasized that most genetic variants function in the context of networks and that the ultimate phenotype that arose was not totally dependent on a single genetic variant.

They also noted that taking into account the fact that genes function in networks possibly provides additional therapeutic opportunities when function impairments result from defects in a particular gene.

Riordan and Nadeau emphasized that phenotypes generally result from target gene activity integrated with genetic background and environmental influences and that target gene activity might include a specific gene or multiple genes.

Modes of effects of modifier genes

Riordan and Nadeau defined the major modes of effects of modifier genes. They noted that modifiers could impact penetrance and or expressivity of a target gene mutations. Modifiers could also impact the pleiotropic phenotypic effects of mutation in a particular target gene.

Penetrance may be defined as the proportion of individuals in a population who have the specific mutations in a target gene and who also manifest the specific mutation related phenotypic abnormality.

Riordan and Nadeau noted that there is abundant evidence from mouse studies that the degree of penetrance of a specific mutation in a particular gene differs in different mouse strains. They noted that there are also examples of penetrance difference in specific human gene mutations.

Modifiers of complex traits

Riordan and Nadeau noted that modifier genes described in the literature were primarily modifiers of single gene disorders. They emphasized however that most phenotypes are complex traits resulting from the effects of multiple genes including modifier genes and effects of environmental factors. They also noted that modifying effects are sometimes due to variants in quantitative trait loci.

They noted that the identification of specific modifier genes and their sequences could provide insights into molecular interactions and networks. Important knowledge of modifier genes and of genetic networks could potentially provide new insights into therapies for disease due to specific damaging mutations.

Riordan and Nadeau noted in addition, that modifier genes and networks likely also serve to counteract the deleterious effects of certain environmental exposures.

Disorders that can result from environmental factors, single rare gene defects or common gene variants

Nephrotic syndrome

This condition is characterized by generalized edema, proteinuria, reduced serum protein levels and elevated serum cholesterol levels. This syndrome may develop followed acute infections in some cases.

Some forms of the disorder that occur in children particularly respond to steroid therapy and the steroid responsive forms have been associated with specific genetic variants in major histocompatibility loci particularly HLA-DQA1 and HLADQB1 and defects in adaptive immunity are possibly involved in the pathogenesis of the disorder.[20]

Steroid resistant form of nephrotic syndrome in children have been found to be associated with specific single gene defects. Tan et al.[21] reported results of studies in 78 children with this disorder and they identified pathogenic mutations that were likely disease causing in six genes described below.

NPHS1	Encodes nephrin cell adhesion molecules that functions as a glomerular filtration barrier
NPHS2	Encodes podocin that plays a role in the regulation of glomerular permeability
WT1	Encodes a transcription factor
MYO1E1	Encodes a myosin protein that functions as actin-based molecular motor
TRC6	Encodes a transient receptor potential cation channel subfamily C, a calcium channel
INF2	Encodes a protein that functions in actin filament polymerization and depolymerization

In a study in China, Wang et al.[22] established a genetic etiology in 28.3% of cases of steroid non-responsive nephrotic syndrome. Their study implicated NPHS1, NPHS2, WT1 and ADCK4 (also known as COQ8B), a protein is found associated with in the catalytic domain of certain protein kinases.

In a study in South Africa, Asharam et al.[23] reported finding homozygosity for a specific NPHS2 variant in 8 out of 30 Black children with steroid resistant nephrotic syndrome.

Earlier reports by Bouchireb et al.[24] reported results of on 101 children with steroid resistant nephrotic syndrome that led to identification of 101 previously identifies NPHS2 mutations and 25 new mutations in that gene.

The continuum of rare and common diseases

It is important to note that rare or common variants in a specific gene may lead to disease.

A specific protein that is highly abundant in urine has been known for >100 years. At different times this protein has been known by different names. Devuyst et al.[25] reported that in 1873 an Italian physician, Rovida, described the most highly abundant protein in urine as cilindrina. In 1950 the most highly abundant protein in urine was described a s a mucoprotein and was given the name of the investigators who analyzed it and it became known as the Tamm-Horsfall protein. In 1987 the protein was designated uromodulin by

Pennica et al.[26] The uromodulin gene was cloned and it was found to be exclusively expressed in the kidney.

Devuyst et al., noted that rare mutations in uromodulin (UMOD) have been reported to lead to autosomal dominant tubulointerstitial kidney disease. In addition, common variants in UMOD were reported to increase the risk of chronic kidney disease.

Devust et al. reported that uromodulin is particularly abundant in the thick ascending loop of the nephron. Several physiological roles have been proposed for uromodulin. There is evidence that it influences electrolyte handing and that it protects against crystal aggregation and stone formation. Through activation of a potassium cotransporter NKCC2 (solute carrier (SLC12A1)) and through activation of the potassium channel ROMK (potassium voltage-gated channel also known as KCNJ1). Uromodulin also plays a role in blood pressure regulation.

Uromodulin facilitates reabsorption of calcium through its impact on transient receptor potential cation channel TRPV5. Uromodulin also impacts the innate immune response through its impact on the Toll receptor TLR4.

Specific Mendelian disorders due to uromodulin mutations include medullary cystic kidney disease type 2 (MCKD2) and familial juvenile hyperuricemic nephropathy.

Devust et al., noted that a population-based study in Europe identified a common single nucleotide variant in the promoter regions of the UMOD encoding gene, rs12917707 where the minor allele T was associated with a reduction in risk of chronic kidney disease and with an increase in the glomerular creatinine filtration rate. A variant that is in linkage disequilibrium with rs12917707 has found to be associated with a decreased risk of kidney stones in the Icelandic population.[27]

Personalized precision medicine in individuals with Mendelian disorders

Definitive diagnoses of Mendelian disorders and inborn errors of metabolism

The specific laboratory tests to be undertaken in order to definitely diagnose Mendelian disorders, including inborn errors of metabolism, have long been based on the phenotypic manifestation and clinical findings in the patient and the use of knowledge bases, and publications with information on the specific phenotypic and in some cases biochemical information, on the characteristics of disorders that had genetic causation bases. Important knowledge bases included

Mendelian inheritance in Man, first published in 1966.[28] The metabolic Basis of Inherited Disease, first published by Stanbury et al.[29]

Quantitative and qualitative analyses of proteins and enzymes were carried out. For enzymes functional assays involved activity profiles relative to pH, stability of enzyme relative to temperature changes, substrate conversion parameters and activity in response to potential inhibitors. Analyses of urine and plasma and sometimes cellular analyses were carried out to determine levels of a limited number of organic molecules derived from metabolism

Initial DNA sequencing studies were designed to search for pathogenic variants in the coding sequence of potential target genes. The selection of genes for analysis was based on specific phenotypic features and results of biochemical tests.

Particularly In the second half of the twentieth century studies on genetic diseases focused in part on mapping genetic disease loci to specific human chromosomes through family studies, and the availability of relatively few chromosomal assigned genetic markers that showed detectable variants. In addition, identification of individuals with specific chromosome abnormalities and specific chromosome abnormalities also facilitates the mapping of disease genes to human chromosomes.

Following conclusion of the Human genome Project in 2003 and the extensive technological improvements in sequencing and development of microarray techniques to analyze DNA markers it became possible to identify genetic variants and disease associated nucleotide variants. Earlier genetic studies had failed to recognize the tremendous variation in the genome sequence in healthy humans. Knowledge of this diversity has evolved as a result of sequencing of thousands of individuals in different populations.

Another early caveat in use of DNA sequencing data for disease diagnoses was the initial almost exclusive focus on the protein coding genome. During the twenty-first century focus on the non-protein coding regulatory genome has increased.

The initial focus on clinical genetic studies was on relatively rare hereditary diseases. Intense studies on genetic factors that play roles in the pathophysiology of more common diseases came later. It is however important to note that early in the twentieth century Fisher[30] postulated that many genetic alterations, each of small effect, could lead to common diseases.

Discovery of genes that harbored variants that cause rare diseases

Progressive refining of variant classification has led to new discoveries related to the causation of rare disease. However careful analyses of variants and searches in databases with information on genetic variants are required to increase certainty that the variant is likely pathogenic and is related to the specific disease that the patient has. Follow up confirmatory studies include segregation analyses to determine if other affected members of a family carry the same variant. In addition, specific functional studies may be valuable. These will be discussed further below.

Problems in exome sequencing

These include lack of adequate capture of DNA from all genes, lack of capture of the less common exons of a specific gene. Key problems are related to interpretation of the pathological significance of a particular sequence variant.

Parameters used in the interpretation of likely pathogenicity include assessment of the variants on the encoded amino acid, determination of the likely impact of deletion of a specific amino acid or deletion of a series of amino acids on the function of the protein. Pathological significance interpretation is sometimes linked to determination of the evolutionary conservation of the particular amino acid in that protein in other vertebrates. Variants that lead to premature termination of transcription of a particular gene and variants the impacts appropriate splicing are considered particularly damaging.

Increasingly specific databases with cumulative information on genetic variants and on patient phenotypes associated with specific diseases are used in variant interpretation. In addition, databases with sequence information from individuals reported to be healthy are available, e.g., ExAC, are available and are useful in variant data interpretation. It is however important to note that healthy individuals in different population may differ with respect to the presence of absence of specific gene variants.

Sequence variants report standards have evolved over time to include classification of variants based on the strength of available evidence that takes into account not only the type of variant but also strength of documented evidence that variant was or was not associated with a specific disease in well documented cases. Variants may be classified as pathogenic, likely pathogenic, benign or as variants of unknown significance. In moving forward for potentially pathogenic variants to be considered to provide clinically useful information additional studies and addition information searches are necessary. Some of the important data bases are listed in the Table below.

Strande et al.[31] described an evaluation framework ClinGen to attribute relevance of specific gene variants in disease causation. The assessment strength of evidence was defined as, definitive, strong, moderate, limited, conflicting evidence. Strande et al. noted that in 2017 approximately 3000 of the approximately 20,000 genes in the human genome had been reported to have association with one or more Mendelian diseases (Fig. 1).

Assess clinical validity of genomic variants found and of their potential relevance to patient phenotype:

Genomic/Genetic Databases: OMIM, ClinVar. Disease specific databases

Search literature for information on condition, gene/genes implicated and known variants e.g. PUBMED

If clinically relevant defect in gene of major effect indentifield, search for information on variant penetrance, possible modifier genes

Search if impact of disease-causing variant can be modified by specific environmental factors, or dietary alterations

Search for evidence of specific pharmacological substances that can impact effects of major gene changes, also whether any relevant clinical trials are in progress

Determine of impact of gene defect can be altered by specific prophylactic measures

Consider value of cascade testing in family

Fig. 1 Follow-up of genotype studies and genotype phenotype correlations. Resources for analyses and procedures required to determine the potential clinical relevance of genomic study findings are listed.

Databases with information on human genes, mutations and disease relevance

Gene	https://www.ncbi.nlm.nih.gov/
Clin Var	https://www.ncbi.nlm.nih.gov/
PheGen1	https://www.ncbi.nlm.nih.gov/gap/phegeni
ExAC	http://exac.broadinstitute.org/
gnomAD	http://gnomad-old.broadinstitute.org/
ClinGen	https://www.clinicalgenome.org/
DisGeNet	http://www.disgenet.org/
HMG	http://www.hgmd.cf.ac.uk/ac/index.php
Gene Cards	https://www.genecards.org/
Phenovar	https://phenovar.med.usherbrooke.ca/

The ClinGen endeavor

The goal of this endeavor is to build "an authoritative central resource that defines the clinical relevance of genes and variants for use in precision medicine and research."

Strande et al.[31] described the ClinGen frame work to evaluate the validity of the gene disease attributions and to assign a semi-quantitative index to gene-disease association. The ClinGen framework was used primarily to classify autosomal dominant, autosomal recessive and X-linked disorders and it comprises four main evidence categories limited, moderate, strong and definitive. It is important to note that validation evaluates whether a specific pathogenic variant in a specific gene leads to disease.

Limited was the category used when there was at least one report that provided plausible evidence that a specific gene caused a disease.

The moderate category was assigned when multiple unrelated probands with a specific disease were found to have pathogenic variants in a particular gene.

The strong evidence category was assigned when there were several reports of pathogenic variants in a specific disease.

The definitive evidence category was assigned when both genetic and experimental data had been published and replicate and where after 3 years no contractor evidence had been published.

Genetic evidence for a gene variant disease association could be based on studies in cases with a specific disease or on the basis of case controls studies. For case control studies the variant needed to have evidence of disease relevance e.g. the variant impacted function of a gene product in affected individuals and the frequency of the variant was low in the general populations or reportedly healthy individuals.

A separate category of evaluation was attributed to data on statistical evaluation and segregations analyses. For population studies valid statistical analyses of differences in frequency of the specific variant in cases and controls needed to be provided.

Experimental evidence of the significance of a variant, included studies on biochemical function, protein interaction, level of expression altered functional parameters and studies of impact of the variant on model organisms or evidence of phenotype rescue.

Functional validation of genetic variants in cases with inborn errors of metabolism (IEM)

Rodenburg[32] reviewed validation of genetic variants discovered by their group in Nijmegen in cases of IEM. They initiated an initial analysis step to filter exome data to initially examine data on known genes associated with inborn errors of metabolism. Rodenburg noted that a caveat in this process was that detailed clinical information and sequence information is frequently not available in reports on patients with similar phenotypic manifestations.

They found it useful to carry out searches in data bases that included information on sequencing and phenotypic manifestations.

Mosaicism

Mosaicism, the presence in one individual of two or more populations of cells that differ in some aspect of their genomes, can lead to difficulties in molecular diagnosis if both cell types are not analyzed. Post-zygotic mosaicism may arise during the very early developmental stages leading to wide-spread occurrence of cells that are not identical in their genomes. Mosaicism may arise later in development leading to occurrence of genomic differences in cells of a specific lineage.

Mosaicism may involve differences in chromosome number, with loss or gain of a specific chromosome in a certain proportion of cells. It may also involve the presence in some cells of a structurally abnormal chromosome. In addition, a proportion of cells in an individual can have copy number variants of a specific genomic segment. DNA sequence alterations may also be present only in a proportion of cells in an individual.

Mosaicism that is arises in adult life is often restricted to blood cells.

In a review of mosaicism and human disease Biesecker and Spinner[33] noted that the best known form of somatic mosaicism is cancer, where somatic genomic alterations have led to malignancy.

It is important to note that in specific dominantly inherited genetic disorders mosaicism may occur. Mosaicism can arise because specific disease- causing mutation that was inherited is lost in a certain proportion of cells. Another form of mosaicism in dominantly inherited diseases. Involves the deletion of the normal version of the gene that was inherited from one parent. In a particular cell and its progeny there are then no normal functioning versions of a particular gene and this leads to tumor formation. This process is referred to as loss of heterozygosity and has been found to be present for example in specific tumors characteristic of Tuberous sclerosis.[34]

Monozygotic twins may also differ from each other in a certain proportion of cells as a result of post-zygotic processes.

RNA analyses to improve diagnosis

Byron et al.[35] reviewed RNA sequencing and its potential contributions to clinical diagnostics. They noted that RNA sequencing provided information on gene expression levels and could provide information on nucleotide variants in specific genes. They also noted that multigene mRNA signatures are were being incorporated into clinical assessments. Alterations in the multigene expression panel may provide insight into underlying biological processes or responses to therapy.

Analysis of RNA expression levels of a specific gene have been applied to provide information on the impact of a specific gene variant and also on to provide information on regulation of gene expression.

Cummings et al.,[36] Kremer et al.,[37] and Rodenburg[32] have drawn attention to the value of RNA transcript studies in the functional validation of DNA sequence variants.

A number of investigators, including Van Karnebeek et al.,[38] have reported on the value of combining exome studies and metabolomic studies, in the diagnosis of inborn errors of metabolism. The analytic methods applied in metabolomics include gas chromatography, liquid chromatography, mass spectroscopy, and nuclear magnetic resonance spectroscopy.

More technologically challenging methods to validate pathogenicity of a DNA sequence variant involve introducing the wild-type version of the gene, in vitro into cells of an individual with a particular variant. Other methods could involve introducing the specific variants into model organisms.

Multidisciplinary precision medicine

Wanders et al.[39] emphasized that that the interpretation of the biological significance of sequence variants detected on exome or genome sequencing was not trivial and that prediction programs on the impact of sequence variants (e.g., SIFT, Polyphen) were imperfect. They emphasized the importance of functional studies, including metabolomics, glycomics, lipidomics, enzymology, and studies in model organisms to confirm diagnoses.

The combined data that needed to be obtained to make appropriate diagnoses and to implements therapy include information of phenotype, metabolomic, proteomics, genomics enzymology and whole cell assays. They also emphasized the importance of transcription analysis.

Wanders et al., noted however that the application of sequence analyses had led to the discovery of previously unknown gene defects that lead to inborn errors of metabolism. In addition, new phenotypes had been discovered to be caused by genes with variants previously found to be involved metabolic defects.

Wanders et al. emphasized that manifestations of inborn errors of metabolism were due to the combined interactions of the primary gene defect, the genetic background of the patients, the environmental background.

Enzyme assays in inborn errors of metabolism

Wanders et al.[39] noted that improvements were need in enzymatic analyses since in the laboratory enzymes were frequently analyzed under artificial conditions that did not resemble condition in the organism. In addition, artificial substrates rather than the natural substrate of the enzyme were often used in the laboratory. They noted that whole cell assays were increasingly being used, for example in studies on mitochondrial function where oxygen consumption and glycolysis could be analyzed.

Gene panels are being designed to sequence the specific genes that have been found to harbor pathogenic mutations that lead to specific diseases.

Gonorozsky et al.[40] noted that applications of modern DNA sequencing technologies were reported to achieve diagnoses in up to 50% of cases of genetic disease. They proposed that frequent use of transcriptome studies could increase diagnostic yield. The caveat in transcriptome studies is however that they need to be carried out using disease relevant tissue.

Next-generation sequencing for diagnosis of genetic diseases

In a review of sequencing and disease diagnosis Adam and Eng[41] noted that panel sequencing methods target gene known to be associated with diseases that lead to a specific clinical phenotype, e.g., muscle diseases, skeletal dysplasia. The number of genes selected for inclusion varies for different phenotypes. The costs of panel sequencing are lower than the costs of exome sequencing and data analysis.

Adams and Eng reported examples of results obtained when panel sequencing was selected and noted that they furnished diagnoses in 30–51% of cases.

Adams and Eng also reviewed aspects of the clinical usefulness, next generation sequencing. On the basis of reports in clinical genetics publications they noted that in approximately 52% of cases received definitive diagnoses after sequencing and medical management was altered in 25% of cases.

They also considered the value of acquisition of genomic data in precision medicine. They note that sequencing data can potentially provide data on the risk for common diseases and data that indicate presence of variants that may predispose an individual to adverse responses to specific medications.

Additional factors to be taken into account following molecular diagnosis of Mendelian disease

It is important to note that not all pathogenic mutations in a specific disease- causing gene necessarily give rise to the same phenotypic manifestation or to the same degree of severity.

These include aspects of mutation penetrance and the existence of modifier genes. In a 2018 review Davidson et al.[42] emphasized that the classification of a Mendelian disease as monogenic (due to defect in a single gene), is clearly an oversimplification. In a monogenic disease a variety of different phenotypic manifestations can occur even in members of a family who all carry the same pathogenic mutation in a specific disease. These variations ae likely due to the modifying effects of other genes and also due to modifying effects of specific environmental factors.

Modifier genes are genes that interacts with a disease-causing gene and alters its effect, either through specific impacts on gene expression or through specific effects on the product of the gene.

The penetrance of a gene is determined by studying a number of individuals that each carry a specific pathogenic mutation in a specific gene and determining the ration of individuals who manifest the phenotypic manifestation characteristic of the disorder found in that disease.

In the case of high penetrance mutations, almost all individuals that carry that mutation have disease manifestations. For low penetrance mutations a significant proportion of individuals who carry the disease-causing mutation do not manifest disease manifestations. Of course, in some diseases age can influence penetrance.

Davidson et al. also noted that among individuals with a specific disease-causing mutation variable expressivity can occur, meaning that different individuals with the mutations may each have different phenotypic manifestations from the range of manifestations characteristic of the disease.

Modifier genes

Several different methods have been implemented to discover modifier genes. One method uses genome wide association studies of genetic variant in an attempt to identify variants that segregate with altered severity of the disease.

Genes that encode products that have functions related to the biological function of the primary disease gene may have variants that act as modifiers. Davidson et al. noted that a number of different genes that encode products related to the product of the Gaucher disease gene glucocerebrosidase 1, (glucosylceramidase beta, GBA1) have been reported as potential modifier of the Gaucher disease phenotype. These include genes that regulate lysosomal function and genes that influence interact in the glucocerebrosidase pathway.

Products of modifier genes in Gaucher disease.

TFEB	Transcription factor involved in lysosome biogenesis
PSAP	Located in the lysosome involved in glycosphingolipid breakdown
SCARB2	Transporter located in the lysosomal membrane
GBA2	Glucosylceramidase beta 2
GBA3	Glucosylceramidase beta 3

International collaborations on rare disease research

The definition of rare disease varies in different countries. In the USA any disease that affects fewer than 200,000, individuals in the country or 1 in 1500 individuals is defined as a rare disease. In Europe a disease that affects 1 in 2000 individuals is defined as a rare disease. In Japan a rare disease is defined as a disease that affects 1 in 2500. The IRDiRC considers a disease with a frequency of 5 in 10,000 to be a rare disease (see Dawkins et al.[43]).

Data gathered by the IRDiRC between 2010 and 2016 and reported by Dawkins et al. in 2018, revealed that between 6000 and 8000 rare diseases were known and that collectively rare diseases affected between 6% and 8% of the human population. Approximately 80% of rare diseases were reported to be of genetic origin.

The report also noted that for most rare disease therapies did not exist or if available, were very expensive. In addition, patients with rare disease required higher utilization of medical facilities. However regulatory and economic incentives, such as the Orphan Drug Act passes in the USA in 1983, were reported to be proving successful in encouraging the biopharmaceutical industry in the USA and in Europe, to invest in potential treatments for rare diseases. By 2010, 247 new therapeutic agents for rare diseases were developed.

Dawkins[43] reported that 49% of clinical trials carried out in collaboration with the IRDiRC were funded by companies 63% of these were phase 1 or 2 trials, 31% were phase 3 trials and 5% were phase 4 trials.

The IRDiRC 2018 report (see Dawkins[43]) noted that great progress had been made in development of diagnostic tests for rare disease.

The Centers for Mendelian Genomics (CMG) report[44] report emphasizes that rare disease research that leads to gene discovery can expand possibilities for molecular diagnoses, enhance patient management and provide information on disease recurrence risk. In addition, through gene discovery and analyses of gene product function, possibilities for disease therapies improve.

In documenting discoveries, the centers reported that diagnoses had been achieved in 47% of cases submitted; of these 1101 diagnoses were defined as high confidence discoveries of new disease associated genes. In 40% of cases the gene defect found in the patient corresponded to a previously reported disease associated genes. In 13% of cases a previously known disease associated gene was found to be associated with a new phenotype, a situation referred to as phenotype expansion.

In addition to contributions to accurate molecular diagnoses of genetic diseases, the CMG projects have contributed to development of bioinformatic resources. These include tools for data sharing, www.matchmakerexchange.org for comparing genetic variants and phenotypic information and resources for recording phenotype data using standardized terminology (ontology).

The CMG report also documented a number of challenges that require further work. One of these challenges included investigation of the molecular basis of variable penetrance of specific gene mutation. There is evidence that variants in genes other than the specific gene with a defined pathogenic allele. An example of a two- locus phenotype was reported by Timberlake et al.[45] They described results of a study on the occurrence of abnormal cranial suture fusion, craniosynostosis They studied 191 probands with non-syndromic craniosynostosis and in 132 cases they were able to study the proband and both parents in 59 cases they studied the proband and one parent trios. Seventeen of the affected probands were found to have rare damaging variants in the gene SMAD6 that encodes signal transducers and transcriptional modulators that influence multiple signaling pathways. Damaging variants in SMAD6 were reported to be only 60% penetrant.

Studies revealed that 14 of the 17 also inherited a risk allele in the BMP2 gene. The pathogenic SMAD2 allele was inherited from one parent and the BMP2 risk allele rs1884302 was inherited from the other parent. The BMP2 pathogenic variant is a rare variant. The risk allele is a common allele. However, the risk variant significantly increased the risk of craniosynostosis. Timberlake et al. reported that this was an example of epistatic interactions between SMAD2 and BMP2 loci. They concluded that common variants may impact the penetrance of rare alleles.

Genetic disorders in which causation cannot be easily resolved by genomic studies

Disorders that exhibit marked clinical heterogeneity, may be due to defectives in different genes in different patients or they may be due to combined effects of variants in multiple genes. In some cases, insight into underlying pathogenic mechanism may be obtained through transcriptome studies that evaluate gene expression, particularly in a cell or tissue type known to be impacted in the disorder.

One example of the value of transcription studies in providing information on the molecular pathogenesis of a disorder, is a report of studies on the transcriptome in skin fibroblasts obtained from patients with Ehlers Danlos hypermobility syndrome. This syndrome is associated with joint hypermobility, joint instability and fragile skin. Studies on skin fibroblasts by Chiarelli et al.[46] revealed altered transcription of genes involved in determining extra-cellular matrix architectures and homeostasis and altered transcription of gene involved in cell-cell adhesion. In addition, there was evidence of altered transcription of genes involved in the immune response. Products of genes that showed altered transcription in these studies are listed below.

TGM2	Transglutaminase, catalyzes cross linking of protein
MMP16	Matrix metalloproteinase 16, involved in breakdown of extra-cellular matrix
GPC4	Glypican 4, plays a role in cell division and growth
CDH2, CDH10	Cadherins, calcium-dependent cell-cell adhesion molecules
PCDH9	Protocadherin 9, mediates cell adhesion
CLDH11	Claudin 11, component of tight junction strands, involved in cell polarity
FLG	Fillagrin, aggregates keratin intermediate filaments in mammalian epidermis
DSP	Desmoplakin, component of desmosomes that promote cell- cell adhesions
COLEC12	Collectin 12, has functions associated with host defense

Gene environment interactions in causation of congenital malformation

Sparrow et al.[47] reported that congenital scoliosis due to verte-bral defects occurs in 1 in 1000 live births. They noted that haplo-insufficiency of genes in the Notch signaling pathway can cause these defects. In affected cases in two unrelated families Sparrow et al. iden-tified pathogenic mutations in the transcription factor encoding genes HES7 and MESP2. Pathogenic mutations in these genes have been de-scribed as dominant but weakly penetrant.

Sparrow et al., developed mouse models with defects in these genes. The mutations were found to be weakly penetrant. Further studies in the mouse model revealed that exposure to mild hypoxia during gestation dramatically increased the incidence of vertebral de-fects particularly in the Hes.7 +/− mice.

Sparrow et al. concluded that results of their study provided evi-dence for gene mutation/environmental interactions in the genera-tion of congenital abnormalities.

Importance of genome studies in providing insight into underlying gene mechanisms

Synofzik et al.[48] reviewed the genetic characterization of autosomal recessive cerebellar ataxia and steps that led to the elucidation of un-derlying disease pathways. They noted that through definition of un-derlying disease pathways approaches to therapy could be designed.

Ataxia are progressive neurodegenerative diseases. Genes mutated in number of these diseases had been characterized prior to development of next generation sequencing methods, however, development of these methods greatly expanded information on genetic defects that lead to these disorders. Specific ataxia disease causing mechanism have been identified. Synozik et al. noted that key mechanisms included impaired mitochondrial metabolism, altered DNA repair leading to genome instability. Specific defects in lipid metabolism also led to ataxia.

Eleven forms of autosomal recessive cerebellar ataxia (ARCA) have been found to be due to defects in genes involved in mitochondrial metabolism. In several of these disorders, defects in mitochondrial structural proteins or defects in components of mitochondrial metabolic functions occur. Defects in replication of mitochondrial DNA replication occur due to deficiency of POLG (Polymerase G). Friedreich ataxia in an autosomal recessive cerebellar ataxia due to defective biogenesis of mitochondrial iron sulfur clusters.

At least 10 different forms of ARCA have been found to be due to defects in repair of nuclear DNA damage, including double or single strand DNA repair or nucleotide excision repair.

Fourteen different forms of ARCA have been found to be due to defects in genes involved in lipid metabolism, Specific defect involve genes that function to degrade complex lipids including ceramides, cerebrosides, and gangliosides. In addition, abnormal processing of lipids in peroxisomes have been found to lead to ARCA.

Synofzik et al. presented strategies for therapeutic approaches to treatment of ARCA. These included supplementation compounds that facilitate toxic metabolite reduction, chaperone replacement therapies, modifiers of gene expression and gene transfer.

Rare diseases and development of therapies

In a 2018 review Kaufman et al.[49] emphasized that although progress has been made, fewer than 5% of the 7000 known rare disease had treatments available. They noted that the slow rate of treatment development is particularly disappointing given the advances in discovery of rare disease- causing mutation and advances in discovery of disease pathogenesis.

Kaufman et al., documented important strategies to improve therapeutic developments. These included, infrastructure building, patient engagement, advancing information technology, sharing of information. Different phases in the cycle of therapeutic developments included the discovery phase of potential therapy, preparation for clinical trials including FDA processes and the post FDA approval phase.

Predicting phenotype from genotype

A primary goal of precision medicine is to correlate genotype with phenotype, and to elucidate the pathogenesis of phenotypic defects. Through knowledge of gene functions and the impact of gene disruptions, insights can potentially be derived that promote therapies.

As knowledge grows in clinical and laboratory genetics, it has become increasingly clear that even in a single family the exact phenotypical and functional consequence of a specific gene mutation cannot accurately and precisely be predicted. In an important review in 2017 Riordan and Nadeau[19] noted that only rarely can genotype and phenotype be completely correlated. It has become clear that genetic background in an individual case, including other rare variants or even more common variants, and epigenetic and environmental factors influence phenotypic manifestations even in individuals who share the same genetic defect.

Riordan and Nadeau drew attention to a number of interesting modifier genes that impact phenotype in specific Mendelian disorders.

Autosomal dominant retinitis pigmentosa was reported in some families to be due to mutation in PRPF31 a component of the spliceosome complex. Evans et al.[50] reported that even within a specific family all individuals with the same pathogenic mutation in PRPF31 did not manifest blindness. A region of the genome that segregated with protection in carriers of the deleterious PRPF31 mutation was identified. Venturini et al.[51] reported that the gene that encoded the transcription factor CNOT3 mapped in this region and a specific variant in CNOT3 modified the impact of the deleterious PRPF31 mutation. The CNOT3 variant was found to increase expression of the wildtype PRPF31 allele.

In patients with specific single gene disorders, genome wide association studies have sometimes been carried out and have revealed variants in gene distance from disease gene that influence phenotype. Corvol et al.[52] carried out studies on 6365 patients with cystic fibrosis and discovered 2 distant loci that impacted severity of lung disease in the patients.

Riordan and Nadeau noted that variants in functionally related networks may interact to influence phenotype.

Hemochromatosis, rare and common gene variants and gene-environment interactions

Hemochromatosis has been reported to be one of the most common genetic diseases in populations of European ancestry. This condition was reviewed by Powell et al. in 2016. The most common cause of hemochromatosis in individuals of European ancestry is a variant in the homeostatic iron regulator gene product HFE, pCys282Tyr.

Powell et al.[53] reported that 1 in 10 individuals with European ancestry are heterozygous for this mutation. However, the disease manifestation association with this mutation, iron overload, is dependent on other factors, including environmental factors and other genetic factors.

More than 100 different mutations in HFE have been associated with altered function of HFE, leading impaired interaction with transferrin and increased iron storage and to increased iron absorption, through reducing hepcidin expression. The p. Cys282Tyr mutation is the most common. Another common HFE mutation is p. His63Asp.

The penetrance of the HFE p. Cys282Tyr mutation is reported to be incomplete. Penetrance can be indirectly measured by assessing the degree of iron saturation of transferrin or by measuring the levels of the iron storage protein ferritin.

Environmental factors that increase susceptibility to iron overload in individuals with this HFE mutation include excess iron intake (e.g., as dietary supplements) and excess alcohol intake.

Hepicidin is another protein that plays a key role in determining iron homeostasis within the body. Low levels of hepcidin leads to increased intestinal iron absorption and to increased parenchymal iron deposits elsewhere in the body. Hepcidin is encoded by the *HAMP* gene, the initial gene product is prohepcidin that is proteolytically cleaved to give rise to three isoforms with, 20, 22 or 24 amino-acids.

Gulec et al.[54] reported that a network of molecules that act as iron sensors in the body and they act by regulating transcription of the hepcidin encoding gene. Hepcidin is transported to the gut where is bind to the protein ferroportin on gut enterocytes and promotes internalization and degradation of ferroportin and this decreases iron absorption. Increased hepcidin levels are therefore correlated with increased iron absorption.

Under conditions of decreased iron levels in the body hepcidin levels are decreased so that iron absorption can be increased.

Transfer of iron to the liver cells occurs through the binding of transferrin to the transferrin receptor. HFE gene product can bind to transferrin and reduce binding to the transferrin receptor.

HFE is one member of the network of iron regulators, other members of this network include hemojuvelin. HFE normally acts to regulate the production of hepcidin. Polymorphic variants in other genes that encode iron homeostasis regulators can influence the impact of the HFE gene variants.

Organ damage induced by excess iron levels occurs primarily in the liver, cardiac damage may also occur. Powell et al. noted that reactive oxygen species can be generated from iron. Reactive oxygen species have various damaging effects that include lipid peroxidation.

Gene defects with low penetrance except under specific environmental conditions

Porphyrias

Porphyrias are disorders in heme biosynthesis. Three forms of porphyria are defined as autosomal dominant disorders with variable penetrance include acute intermittent porphyria (AIP). Variegate porphyria (VP) and hereditary coproporphyria (HCP); Yasuda et al.[55] classified these forms as hepatic porphyrias. They noted that these disorders can present with prodromal neurological symptoms described as "brain Fog," insomnia, fatigue, acute abdominal pain. Other manifestations follow including hypertension, motor and sensory abnormalities.

Singal and Anderson[56] noted that variegate porphyria is also characterized by blistering of skin exposed skin on the hands and face, scarring and thickening of skin and alter pigmentation. Variegate porphyria is due to defects in the enzyme protoporphyrinogen oxidase PPOX. Exacerbating external factors include intake of specific medications particularly barbiturates, sulfonamide, rifampin and anticonvulsants including phenytoin carbamazepine and alcohol. Endocrine factors also play roles and high dose birth control medications should be avoided.

Singal and Anderson noted that at risk family members should be offered genetic testing. They noted that finding of a pathogenic PPOX mutation cannot be utilized as an indicator of when VP manifestations will commence.

The protoporphinogen oxidase enzyme is located in mitochondria and is involved in conversion of protoporphyrinogen to porphyrinogen.

In the ClinVar databases 23 PPOX mutations are defined as pathogenic and 28 are defined as likely benign or of unknown significance. It is important to note that there are specific founder populations with a high incidence of variegate porphyria due to specific PPOX mutations. In the swiss population members of a founder group were reported to have variegate porphyria due to a specific mutation PPOX c.1082–1083-ins. C.[57] In founder populations in the Netherlands and South Africa the recurrent PPOX mutation p. R59W was reported.[58]

It is interesting to note that in a report of cases investigated at Mount Sinai hospital in New York, Yasuda et al.[55] identified 6 missense mutations, each different from the common founder mutation.

Acute intermittent porphyria (AIP)

In this disorder heme biosynthesis is affected due to mutation in the enzyme hydroxymethylbilane synthesis (HMBS) also known as porphobilinogen deaminase (PBGD). It is inherited as an autosomal dominant disorder with variable penetrance. Lenglet et al.[59] reported that worldwide 400 different mutation have been reported to lead to this condition and that many of these mutations are family specific. Founder mutations have been reported; pARG16TRP in the Netherlands; p. TRP28X in Switzerland in Sweden.

AIP heterozygotes can have acute attacks characterized by neuro-visceral symptoms, neuropsychiatric, motor and sensory symptoms, abdominal pain nausea, vomiting and tachycardia also occur. It is important to note that most heterozygotes with pathogenic HMBS mutations do not manifest symptoms. Symptoms may be induced by specific medications, dehydrogenation, hyponatremia, caloric restriction and excess alcohol intake.

Yatsuda et al.[55] noted that although penetrance of pathogenic HMBS mutations is generally low there are families where penetrance is higher, and it is not clear if this higher penetrance is due to specific environmental factors or if it is due to the impact of variant in other genes.

Barreda-Sanchez et al.[60] reported the presence of a possible modifier gene in AIP. They carried out a study to investigates the penetrance of HMBS mutation c. 669-698del30. In their study in the Spanish population they defined penetrance as 52%. They reported that specific alleles in the cytochrome oxidase CYPD6 likely constituted a penetrance modifier.

Coproporphyria

The third form of hepatic porphyria is coproporphyria due to heterozygous mutations in the gene that encode Coproporphyrinogen III oxidase (CPOX). The clinical manifestations in Coproporphyria resemble those found in the other hepatic porphyrias. However, there is some evidence that motor impairments may be more severe than in the other forms of hepatic porphyria. Wang and Bissell[61] reported that photosensitivity and skin fragility occurred in approximately 20% of individuals with coproporphyria. They reported that manifestation may be triggered by medications, dehydration, hyponatremia, or endocrine factors particularly progesterone.

Twenty-one different pathogenic mutations in CPOX are reported in the ClinVar database; 63 mutations are reported to be benign or of uncertain clinical significance.

Rare cases of homozygous or compound heterozygous pathogenic CPOC mutation have been reported to lead to a condition in neonates that is associated with photosensitivity, anemia and jaundice.[62]

References

1. Hood L, Flores M. A personal view on systems medicine and the emergence of proactive P4 medicine: predictive, preventive, personalized and participatory. *N Biotechnol* 2012;**29**(6):613–24. https://doi.org/10.1016/j.nbt.2012.03.004. 22450380.

2. Collins FS, Varmus H. A new initiative on precision medicine. *N Engl J Med* 2015;**372**(9):793–5. https://doi.org/10.1056/NEJMp1500523. 25635347.

3. Jameson JL, Longo DL. Precision medicine—personalized, problematic, and promising. *N Engl J Med* 2015;**372**(23):2229–34. https://doi.org/10.1056/NEJMsb1503104. 26014593.

4. Bayer R, Galea S. Public health in the precision-medicine era. *N Engl J Med* 2015;**373**(6):499–501. https://doi.org/10.1056/NEJMp1506241. 26244305.

5. Davis MM, Shanley TP. The Missing-Omes: proposing social and environmental nomenclature in precision medicine. *Clin Transl Sci* 2017;**10**(2):64–6. https://doi.org/10.1111/cts.12453. 28105795.

6. Morgan AA, Crawford DC, Denny JC, Mooney SD, Aronow BJ, Brenner SE. Precision medicine: data and discovery for improved health and therapy. *Pac Symp Biocomput* 2017;**22**:348–55. https://doi.org/10.1142/9789813207813_0033. 27896988.

7. Whitsel LP, Wilbanks J, Huffman MD, Hall JL. The role of government in precision medicine, precision public health and the intersection with healthy living. *Prog Cardiovasc Dis* 2019;**62**(1):50–4. https://doi.org/10.1016/j.pcad.2018.12.002. Review, 30529579.

8. Ginsburg GS, Phillips KA. Precision medicine: from science to value. *Health Aff (Millwood)* 2018;**37**(5):694–701. https://doi.org/10.1377/hlthaff.2017.1624. 29733705.

9. Bycroft C, Freeman C, Petkova D, Band G, et al. The UK Biobank resource with deep phenotyping and genomic data. *Nature* 2018;**562**(7726):203–9. https://doi.org/10.1038/s41586-018-0579-z. 30305743.

10. Warren HR, Evangelou E, Cabrera CP, Gao H, Ren M, et al. Genome-wide association analysis identifies novel blood pressure loci and offers biological insights into cardiovascular risk. *Nat Genet* 2017;**49**(3):403–15. https://doi.org/10.1038/ng.3768. 28135244.

11. Giri A, Hellwege JN, Keaton JM, Park J, Qiu C, et al. Trans-ethnic association study of blood pressure determinants in over 750,000 individuals. *Nat Genet* 2019;**51**(1):51–62. https://doi.org/10.1038/s41588-018-0303-9.

12. Ren K, Jiang T, Zheng XL, Zhao GJ. Proprotein convertase furin/PCSK3 and atherosclerosis: new insights and potential therapeutic targets. *Atherosclerosis* 2017;**262**:163–70. https://doi.org/10.1016/j.atherosclerosis.2017.04.005. 28400053.

13. Theriault S, Gaudreault N, Lamontagne M, Rosa M, Boulanger MC, et al. A transcriptome-wide association study identifies PALMD as a susceptibility gene for calcific aortic valve stenosis. *Nat Commun* 2018;**9**(1):988. https://doi.org/10.1038/s41467-018-03260-6. 29511167.

14. Bossé Y, Mathieu P, Thériault S. PALMD as a novel target for calcific aortic valve stenosis. *Curr Opin Cardiol* 2019. https://doi.org/10.1097/HCO.0000000000000605. 30608251.

15. Cooper DN, Krawczak M, Polychronakos C, Tyler-Smith C, Kehrer-Sawatzki H. Where genotype is not predictive of phenotype: towards an understanding of the molecular basis of reduced penetrance in human inherited disease. *Hum Genet* 2013;**132**(10):1077–130. https://doi.org/10.1007/s00439-013-1331-2. Review, 23820649.

16. Bergendal A, Persson I, Odeberg J, Sundström A, Holmström M, et al. Association of venous thromboembolism with hormonal contraception and thrombophilic genotypes. *Obstet Gynecol* 2014;**124**(3):600–9. https://doi.org/10.1097/AOG.0000000000000411. 25162263.

17. Massah S, Beischlag TV, Prefontaine GG, et al. *Crit Rev Biochem Mol Biol* 2015;**50**(4):337–58. https://doi.org/10.3109/10409238.2015.1064350.

18. Rivolta C, McGee TL, Rio Frio T, Jensen RV, et al. Variation in retinitis pigmentosa-11 (PRPF31 or RP11) gene expression between symptomatic and asymptomatic patients with dominant RP11 mutations. *Hum Mutat* 2006;**27**(7):644–53. 16708387.

19. Riordan JD, Nadeau JH. From peas to disease: modifier genes, network resilience, and the genetics of health. *Am J Hum Genet* 2017;**101**(2):177–91. https://doi.org/10.1016/j.ajhg.2017.06.004. Review, 28777930.

20. Lane BM, Cason R, Esezobor CI, Gbadegesin RA. Genetics of childhood steroid sensitive nephrotic syndrome: an update. *Front Pediatr* 2019;**7**:8. https://doi.org/10.3389/fped.2019.00008. eCollection 2019, 30761277.

21. Tan W, Lovric S, Ashraf S, Rao J, Schapiro D, et al. Analysis of 24 genes reveals a monogenic cause in 11.1% of cases with steroid-resistant nephrotic syndrome at a single center. *Pediatr Nephrol* 2018;**33**(2):305–14. https://doi.org/10.1007/s00467-017-3801-6. 28921387.

22. Wang Y, Dang X, He Q, Zhen Y, He X, Yi Z, Zhu K. Mutation spectrum of genes associated with steroid-resistant nephrotic syndrome in Chinese children. *Gene* 2017;**625**:15–20. https://doi.org/10.1016/j.gene.2017.04.050. 28476686.

23. Asharam K, Bhimma R, David VA, Coovadia HM, Qulu WP, et al. NPHS2 V260E is a frequent cause of steroid-resistant nephrotic syndrome in black South African children. *Kidney Int Rep* 2018;**3**(6):1354–62. https://doi.org/10.1016/j.ekir.2018.07.017. eCollection 2018, 30450462.

24. Bouchireb K, Boyer O, Gribouval O, Nevo F, Huynh-Cong E, et al. NPHS2 mutations in steroid-resistant nephrotic syndrome: a mutation update and the associated phenotypic spectrum. *Hum Mutat* 2014;**35**(2):178–86. https://doi.org/10.1002/humu.22485. 24227627.

25. Devuyst O, Olinger E, Rampoldi L. Uromodulin: from physiology to rare and complex kidney disorders. *Nat Rev Nephrol* 2017;**13**(9):525–44. https://doi.org/10.1038/nrneph.2017.101. Review, 28781372.

26. Pennica D, Kohr WJ, Kuang WJ, Glaister D, Aggarwal BB, et al. Identification of human uromodulin as the Tamm-Horsfall urinary glycoprotein. *Science* 1987;**236**(4797):83–8. 3453112.

27. Gudbjartsson DF, Holm H, Indridason OS, Thorleifsson G, Edvardsson V, et al. Association of variants at UMOD with chronic kidney disease and kidney stones-role of age and comorbid diseases. *PLoS Genet* 2010;**6**(7):e1001039. https://doi.org/10.1371/journal.pgen.1001039. 20686651.

28. McKusick VA. *Mendelian inheritance in man.* Johns Hopkins University Press; 1966.

29. Stanbury JB, Wyngaarden JB, Fredrickson DS. *The metabolic basis of inherited disease.* McGraw Hill; First published. 1960.

30. Fisher RA. The correlation between relatives on the supposition of Mendelian inheritance. *Trans Roy Soc Edinb* 1918;**52**:399–433.

31. Strande NT, Riggs ER, Buchanan AH, Ceyhan-Birsoy O, DiStefano M, et al. Evaluating the clinical validity of gene-disease associations: an evidence-based framework developed by the clinical genome resource. *Am J Hum Genet* 2017;**100**(6):895–906. https://doi.org/10.1016/j.ajhg.2017.04.015. 28552198.

32. Rodenburg RJ. The functional genomics laboratory: functional validation of genetic variants. *J Inherit Metab Dis* 2018;**41**(3):297–307. https://doi.org/10.1007/s10545-018-0146-7. 29445992.

33. Biesecker LG, Spinner NB. A genomic view of mosaicism and human disease. *Nat Rev Genet* 2013;**14**(5):307–20. https://doi.org/10.1038/nrg3424. 23594909.

34. Green AJ, Smith M, Yates JR. Loss of heterozygosity on chromosome 16p13.3 in hamartomas from tuberous sclerosis patients. *Nat Genet* 1994;**6**(2):193–6. https://doi.org/10.1038/ng0294-193. 8162074.

35. Byron SA, Van Keuren-Jensen KR, Engelthaler DM, Carpten JD, Craig DW. Translating RNA sequencing into clinical diagnostics: opportunities and challenges. *Nat Rev Genet* 2016;**17**(5):257–71. https://doi.org/10.1038/nrg.2016.10. Review, 26996076.

36. Cummings BB, Marshall JL, Tukiainen T, Lek M, Donkervoort S, et al. Improving genetic diagnosis in Mendelian disease with transcriptome sequencing. *Sci Transl Med* 2017;**9**(386):eaal5209. https://doi.org/10.1126/scitranslmed.aal5209. 28424332.

37. Kremer LS, Bader DM, Mertes C, Kopajtich R, Pichler G, et al. Genetic diagnosis of Mendelian disorders via RNA sequencing. *Nat Commun* 2017;**8**:15824. https://doi.org/10.1038/ncomms15824. 28604674.

38. van Karnebeek CDM, Wortmann SB, Tarailo-Graovac M, Langeveld M, Ferreira CR, et al. The role of the clinician in the multi-omics era: are you ready? *J Inherit Metab Dis* 2018;**41**(3):571–82. https://doi.org/10.1007/s10545-017-0128-1. 29362952.

39. Wanders RJA, Vaz FM, Ferdinandusse S, van Kuilenburg ABP, Kemp S, et al. Translational metabolism: a multidisciplinary approach towards precision diagnosis of inborn errors of metabolism in the omics era. *J Inherit Metab Dis* 2018. https://doi.org/10.1002/jimd.12008. 30723938.

40. Gonorazky HD, Naumenko S, Ramani AK, Nelakuditi V, Mashouri P. Expanding the boundaries of RNA sequencing as a diagnostic tool for rare Mendelian disease. *Am J Hum Genet* 2019;**104**(3):466–83. https://doi.org/10.1016/j.ajhg.2019.01.012. 30827497.

41. Adams DR, Eng CM. Next-generation sequencing to diagnose suspected genetic disorders. *N Engl J Med* 2019;**380**(2):201. https://doi.org/10.1056/NEJMc1814955. 30625069.

42. Davidson BA, Hassan S, Garcia EJ, Tayebi N, Sidransky E. Exploring genetic modifiers of Gaucher disease: the next horizon. *Hum Mutat* 2018;**39**(12):1739–51. https://doi.org/10.1002/humu.23611. 30098107.

43. Dawkins HJS, Draghia-Akli R, Lasko P, Lau LPL, Jonker AH, et al. Progress in rare diseases research 2010-2016: an IRDiRC perspective. *Clin Transl Sci* 2018;**11**(1):21–7. https://doi.org/10.1111/cts.12500. Epub 2017. Review, 28796445.

44. Posey JE, O'Donnell-Luria AH, Chong JX, Centers for Mendelian Genomics, et al. Insights into genetics, human biology and disease gleaned from family based genomic studies. *Genet Med* 2019. https://doi.org/10.1038/s41436-018-0408-7. 30655598.

45. Timberlake AT, Choi J, Zaidi S, Lu Q, Nelson-Williams C. Two locus inheritance of non-syndromic midline craniosynostosis via rare SMAD6 and common BMP2 alleles. *Elife* 2016;**5**:e20125. https://doi.org/10.7554/eLife.20125. 27606499.

46. Chiarelli N, Carini G, Zoppi N, Dordoni C, Ritelli M, et al. Transcriptome-wide expression profiling in skin fibroblasts of patients with joint hypermobility syndrome/Ehlers-Danlos syndrome hypermobility type. *PLoS One* 2016;**11**(8):e0161347. https://doi.org/10.1371/journal.pone.0161347. eCollection 2016, 27518164.

47. Sparrow DB, Chapman G, Smith AJ, Mattar MZ, Major JA, et al. A mechanism for gene-environment interaction in the etiology of congenital scoliosis. *Cell* 2012;**149**(2):295–306. https://doi.org/10.1016/j.cell.2012.02.054. 22484060.

48. Synofzik M, Puccio H, Mochel F. Schöls autosomal recessive cerebellar ataxias: paving the way toward targeted molecular therapies. *Neuron* 2019;**101**(4):560–83. https://doi.org/10.1016/j.neuron.2019.01.049. Review, 30790538.

49. Kaufmann P, Pariser AR, Austin C. From scientific discovery to treatments for rare diseases—the view from the National Center for Advancing Translational Sciences—Office of Rare Diseases Research. *Orphanet J Rare Dis* 2018;**13**(1):196. https://doi.org/10.1186/s13023-018-0936-x. 30400963.

50. Evans K, Al-Maghtheh M, Fitzke FW, Moore AT, Jay M, et al. Bimodal expressivity in dominant retinitis pigmentosa genetically linked to chromosome 19q. *Br J Ophthalmol* 1995;**79**(9):841–6. 7488604.

51. Venturini G, Rose AM, Shah AZ, Bhattacharya SS, Rivolta C. CNOT3 is a modifier of PRPF31 mutations in retinitis pigmentosa with incomplete penetrance. *PLoS Genet* 2012;**8**(11):e1003040. https://doi.org/10.1371/journal.pgen.1003040. 23144630.

52. Corvol H, Blackman SM, Boëlle PY, Gallins PJ, Pace RG, et al. Genome-wide association meta-analysis identifies five modifier loci of lung disease severity in cystic fibrosis. *Nat Commun* 2015;**6**:8382. https://doi.org/10.1038/ncomms9382. 26417704.

53. Powell LW, Seckington RC, Deugnier Y. Haemochromatosis. *Lancet* 2016;**388**(10045):706–16. https://doi.org/10.1016/S0140-6736(15)01315-X. 26975792.

54. Gulec S, Anderson GJ, Collins JF. Mechanistic and regulatory aspects of intestinal iron absorption. *Am J Physiol Gastrointest Liver Physiol* 2014;**307**(4):G397–409. https://doi.org/10.1152/ajpgi.00348.2013. Review, 24994858.

55. Yasuda M, Chen B, Desnick RJ. Recent advances on porphyria genetics: inheritance, penetrance & molecular heterogeneity, including new modifying/causative genes. *Mol Genet Metab* 2018. pii: S1096-7192(18)30645-0. https://doi.org/10.1016/j.ymgme.2018.11.012. Review, 30594473.

56. Singal AK, Anderson KE. *Variegate porphyria*. GeneReviews Internet; 2013. https://www.ncbi.nlm.nih.gov/books/NBK121283/. PMID: 23409300.

57. Van Tuyll Van Serooskerke AM, Schneider-Yin X, Schimmel RJ, Bladergroen RS, et al. Identification of a recurrent mutation in the protoporphyrinogen oxidase gene in Swiss patients with variegate porphyria: clinical and genetic implications. *Cell Mol Biol (Noisy-le-Grand)* 2009;**55**(2):96–101. 19656457.

58. van Tuyll van Serooskerken AM, Drögemöller BI, Te Velde K, Bladergroen RS, Steijlen PM, Poblete-Gutiérrez P, et al. Extended haplotype studies in South African and Dutch variegate porphyria families carrying the recurrent p.R59W mutation confirm a common ancestry. *Br J Dermatol* 2012;**166**(2):261–5. https://doi.org/10.1111/j.1365-2133.2011.10606.x. 21910705.

59. Lenglet H, Schmitt C, Grange T, Manceau H, Karboul N, et al. From a dominant to an oligogenic model of inheritance with environmental modifiers in acute intermittent porphyria. *Hum Mol Genet* 2018;**27**(7):1164–73. https://doi.org/10.1093/hmg/ddy030. 29360981.

60. Barreda-Sánchez M, Buendía-Martínez J, Glover-López G, Carazo-Díaz C, Ballesta-Martínez MJ, et al. High penetrance of acute intermittent porphyria in a Spanish founder mutation population and CYP2D6 genotype as a susceptibility factor. *Orphanet J Rare Dis* 2019;**14**(1):59. https://doi.org/10.1186/s13023-019-1031-7. 30808393.

61. Wang B, Bissell DM. Hereditary coproporphyria. In: *GeneReviews® [Internet]*. Seattle, WA: University of Washington, Seattle; 1993-2019; 2012 December 13 [updated 2018 November 8] 23236641.

62. Hasegawa K, Tanaka H, Yamashita M, Higuchi Y, et al. Neonatal-onset hereditary coproporphyria: a new variant of hereditary coproporphyria. *JIMD Rep* 2017;**37**:99–106. https://doi.org/10.1007/8904_2017_20. 28349448.

PERSONALIZED MEDICINE. PRECISION MEDICINE: PART B MULTIFACTORIAL DISEASES, GENES, ENVIRONMENTS, INTERACTIONS

Cardiac conditions

Long QT syndrome

This disorder is diagnosed on 12 lead electrocardiograms when prolonged interval (>440 ms) between the Q and T waves is detected. The normal QT interval is 400–440 ms. Prolonged Q-T intervals may change to a specific pattern of ventricular tachycardia known as torsade de pointes that can lead to syncope and to sudden cardiac death. Long QT intervals can also be associated with other forms of arrhythmia.

Long Q-T syndrome can be induced by certain medications. Hereditary forms of this disorder also occur.[1] They reported that the population frequency for this disorder was 1 in 2000. In 2019 fifteen different forms of long Q-T syndrome are known https://ghr.nlm.nih.gov/gene. All 15 forms are listed in OMIM as having autosomal dominant inheritance. Defects leading to these disorders occur predominantly in genes that encode ion channels. Important gene products found to be mutated in long QT syndrome are listed below.

Type	Protein	Function
LQT1	KCNQ1	Voltage-gated potassium channel repolarization of the cardiac action potential
LQT2	KCNH2	Potassium voltage-gated channel subfamily H member 2
LQT3	SNC5A	Sodium voltage-gated channel alpha subunit 5
LQT4	ANK2	Ankyrin 2, integral protein in spectrin-actin cytoskeleton

Continued

Gene Environment Interactions. https://doi.org/10.1016/B978-0-12-819613-7.00013-X

Type	Protein	Function
LQT5	KCNE1	Potassium voltage-gated channel subfamily E regulatory subunit 1
LQT6	KCNE2	Potassium voltage-gated channel subfamily E regulatory subunit 2
LQT7	KCNJ2	Potassium voltage-gated channel subfamily J member 2 Andersen syndrome
LQT8	CACNA1C	Calcium voltage-gated channel subunit alpha1 C, Timothy syndrome
LQT9	CAV3	Caveolin 3 Caveolin proteins are proposed to be scaffolding proteins
LQT10	SCN4B	Sodium voltage-gated channel beta subunit 4, changes sodium channel kinetics
LQT11	AKAP9	A kinase anchoring protein binds to the regulatory subunit of protein kinase A
LQT12	SNTA1	Syntrophin alpha 1, associates with sodium channels
LQT13	KCNJ5	Potassium voltage-gated channel subfamily J member 5, allows K flow into cell
LQT14	CALM1	Calmodulin 1 member of the EF-hand calcium-binding protein family
LQT15	CALM2	Calmodulin 2 calcium binding protein

Founder mutations and the study genetic and environmental modifiers

Brink and Schwartz[2] reported results of studies on 26 South African families with long Q-T syndrome and a genetic mutation in the KCNQ1 gene with evidence that this mutation had been inherited from a distant ancestor. They noted founder mutation families for long Q-T syndrome had also been found in Finland and in Quebec and that founder mutation families represented important resources for analyses of genetic and environmental factors that modify clinical manifestations.

The long Q-T predisposing mutation in South African families, KCNQ1 A341V was shown to trigger arrhythmia particularly under stress conditions, including emotional stress and physical stress (particularly swimming). In the most severe cases, individuals experience cardiac events even in childhood and sudden cardiac death sometimes occurred before or during their twenties.

Among 86 carriers of the KCNQ1 A341V mutation Brink and Schwartz noted that Q-T interval values ranged between 406 and 676 ms. Even in individuals with Q-T intervals in the lower ranges there was an incidence of sudden cardiac death before 40 years.

Brink and Schwartz determined that among the KCNQ1 A341V mutation carriers, individuals with a low basal heart rate were at decreased risk for adverse cardiac events. This led them to analyze other genes including those with variants that influenced heart rate. They determined that LQT1 mutations in subjects in whom heart rates underwent rapid changes were particularly at risk for adverse events. Specific mutations in the adrenergic receptors ADRA2B and ADRA2C influenced heart rate. Deletion mutation ADRA2C del 322–325 and ADRB1-R389, were found to be present in some members of the South African founder families and the mutations influenced the basal heart rate. Adrenergic responses were particularly increased in individuals who carried these mutations.

Another genetic mutation that impacts manifestation in Long Q-T patients occurs in the gene that encodes NOS1AP a nitric oxide synthase adaptor protein. Two specific variants in the NOS1AP gene, rs4657139 and rs16847548, were found to influence severity of symptoms and risk for sudden death.

In 2016 Harmer and Tinker[3] reported advances in delineating genetics and pathophysiological mechanisms in long Q-T syndrome. They reported that LQT1 due to KCNQ1 mutations accounts for 40–45% of cases of long Q-T syndrome. This mutation occurs in the Romano-Ward syndrome (RWS), long-QT syndrome where only the heart rhythm is affected) and in cases and with Jervell Lange-Nielsen syndrome (JLNS) where heart rhythm defects and deafness occurs. LQT2 due to KCNQ2 mutation occurs in 30–35% of cases that present with RWS. Mutations in the sodium channel gene SCN5A were reported in 5–10% of cases of long Q-T syndrome and these cases present with RWS. The remaining defects in 12 other genes occurred and some of these cases included rare gene mutations.

In summary, long Q-T syndrome is primarily due to ion channel defects particularly in potassium ion channels, but it may also arise in cases with mutation in the sodium ion channels SCN5A, SCN4B and in the calcium ion channel CACNA1C. Rare cases are due to ion channel auxiliary sub-units, in proteins that interact with ion channels or proteins that modulate ion channels.

Harmer and Tinker concluded that aberrant trafficking through ion channel complexes appear to play the major role in long Q-T syndrome. They noted that the common thread in triggering arrhythmias in this condition is activation of the adrenergic system.

Winbo et al.[4] reported that sequence variants in the NOS1AP gene had been shown to influence clinical penetrance in long QT syndrome. They studied individuals from two founder populations. The families differed with respect to the specific pathogenic KCNQ1 mutation they carried. The Y111 mutation occurred in 148 members and 79 individuals carried the R518[ter] mutation. Winbo carried out genotyping of NOS1AP variants. Their results indicated that the NOS1AP variants did not influence the long QT interval in females with KCNQ1 mutations However, the NOS1AP variant did influence the Q-T interval length in males.

In summary then there is evidence that environmental factors that induce stress can impact presentation of long Q-T syndrome. In addition, there is evidence that variation in another modifier gene in association with sexual differences can impact severity of long Q-T syndrome.

Dilated cardiomyopathy

Bondue et al.[5] reviewed genotype phenotype correlations in dilated cardiomyopathy. They emphasized that environmental factors and genetic factors interact in the causation of this disease. Dilated cardiomyopathy (DCM) is an important cause of cardiac myopathy and may occur in young active adults.

Of particular importance is the fact that specific genetic variants identified as risk variants vary in their penetrance and their penetrance is impacted by specific environmental factors and by epigenetics. Bondue et al. also reviewed the molecular mechanisms through which the DCM risk variants lead to pathogenic changes in the heart.

Clinical features of dilated cardiomyopathy include left ventricular or biventricular dilation and impaired cardiac output in the absence of cardiac overloading or ischemic conditions. DCM may also be associated with cardiac arrhythmias.

Bondue et al. reported that 30–50% of affected individuals have a family history of DCM. This condition is genetically heterogeneous however, in a particular family it is primarily inherited as a monogenic autosomal dominant trait. Recessive inheritance was reported in a few cases. Genetic variants in >40 different genes have been implicated in DCM. The implicated genes predominantly encode proteins involved in cardiac muscle contraction or relaxation. The DCM causing gene products include cytoskeletal and sarcomere proteins, proteins that function in cardiac electric potential conduction, and proteins with mitochondrial and metabolic related functions.

Bondue et al. particularly emphasized the importance of 10 genes where the frequency of specific variants in the patient population was significantly increased above the frequency of the variants in databases of control populations. The products of these ten genes are listed below.

TCAP	Titin Cap binds and is a substrate for linkage to other proteins
TTN	Titin, abundant in sarcomere, has bind sites and acts as a template for contractile machinery
MYH7	Myosin heavy chain 7, encodes a subunit of cardiac myosin
TNNC1	Troponin C1, a subunit of troponin
TNNI3	Troponin component, exclusively cardiac

Continued

TNNT2	Troponin Ts cardiac type, Troponin is a thin filament that participates in muscle contraction
TPM1	Tropomyosin actin binding
LMNA	Lamin A, protein located near the nuclear membrane
TTN	Truncated modified form of titin
BAG3	BCL2 associated, molecular chaperone regulator

It is important to note that some of these genes have variants linked to forms of cardiomyopathy other than dilated cardiomyopathy and in some cases, they have variants associated with other forms of muscle disease. The cardiac disease associations are listed below. The following abbreviations are used: DCM dilated cardiomyopathy, HCM hypertrophic cardiomyopathy, LVNC left ventricular non-compaction, RCM, restrictive cardiomyopathy,

TCAP	DCM, HCM)
TTN	DCM, HCM, LVNC,RCM
MYH7	DCM, HCM, LVNC
TNNC1	DCM, HCM
TNNC1	DCM, HCM
TNNI3	DCM, RCM, HCM
TNNT2	DCM, HCM, LVNC
TPM1	DCM, HCM
BAG3	DCM RCM

It is also important to note that specific variants associated with cardiomyopathy, also occur as rare variants in large databases that list variants in controls. Specific mutations vary in their degree of penetrance.

Bondue et al. emphasized that the pathogenicity of specific mutations may be influenced by other factors, including viral infection, hypertension, increases in reactive oxygen species and by other toxic factors such as alcohol. Specific variants in alcohol dehydrogenase ADH1B, aldehyde dehydrogenase ALDH and cytochrome oxidase CYPE1 have been reported to increase susceptibility to ethanol induced cardiomyopathy, these include particular variants that lead to rapid conversion of ethanol to acetaldehyde and variants that slow the degradation of acetaldehyde. Certain antineoplastic drugs, including anthracyclines have cardiac damaging effects.

Hormonal factors can also have cardiac impact. Bondue et al. drew attention to peripartum cardiomyopathy that is particularly common in certain regions of Africa. There is also evidence that hormonal factors may interact with genetic factors to lead to peripartum cardiomyopathy.

Lamin mutations

The lamin a gene *LMNA,* encodes two proteins Lamin A and Lamin C. The Lamin A and Lamin C differ only in their 3' terminal regions. Lamin A represents the product of 12 exons while Lamin C is encoded by 10 exons. Lamin A and Lamin C form essential components of the nuclear membrane. Defects in the Lamin gene have been reported to lead to a number of different muscle disorders and to cardiac disorders. Nishiuchi et al.[6] reviewed the Lamin A gene mutations that lead to cardiac defects. They gathered multicenter data on 45 individuals with 37 Lamin A mutations and data on their relatives, thirty of the Lamin A mutations were defined as pathogenic. Pathogenic mutations included 26 truncating mutations, these included premature termination mutations, frame-shift, nonsense and splice-site mutations, the remaining pathogenic mutations were missense mutations.

Cascade testing was carried out on family member of cases with pathogenic mutations and 92% of individuals with pathogenic mutations were phenotypically affected. The most frequent cardiac disturbance was cardiac conduction disturbance; this occurred in 81% of individuals and low ventricular ejection volumes occurred in 45% of individuals with pathogenic mutations. Onset of cardiac malfunction occurred earlier in cases with truncation mutations. Nishimura et al. reported that device therapy was initiated in patients with cardiac conduction defects; therapies include cardiac resynchronization therapy or insertion of pace-markers or defibrillators

Testing of family members of patients with cardiomyopathy inducing variants

In a 2017 review of dilated cardiomyopathy McNally and Mestroni[7] considered discovery of titin truncating variants in dilated cardiomyopathy to represent a major advance especially given reports that 20–25% of cases of non-ischemic dilated cardiomyopathy are reported to be due to these variants that are predominantly inherited as autosomal dominant traits.

They noted that between 5 and 8% of cases of non-ischemic dilated cardiomyopathy are attributed to missense or truncating mutations in LMNA inherited as autosomal dominant traits. SCN4A mutations were identified in 2–3% of cases and were noted to be particularly important in dilated cardiomyopathy cases with arrhythmia.

McNally and Mestroni emphasized the importance of genetic testing in at risk family members. They did emphasize that the interpretation of the pathogenic significance of particular variants was complex and that it was important to take into account variant information from large control cohorts, e.g. in the ExAC database.

McNally and Mestroni noted that mitochondrial functional abnormalities had been reported in some cases of dilated cardiomyopathy and that both nuclear and mitochondrial DNA defects had been implicated in these cases.

Cardio-vascular diseases and precision medicine

It is important to note that precision medicine approaches to cardio-vascular medicine include genomic analyses of both common and rare variants, and health metrics assessments including biochemical and imaging evaluations.

Dainis and Ashley[8] reviewed aspects of precision medicine as applied to common and rare cardiovascular diseases. They emphasized that genetic studies were of value in cardiovascular disease. In addition to facilitation of diagnosis and management in a specific patient, following accurate diagnosis in a specific patient, screening could be offered to first-degree relatives of an affected patient to promote preventive measures. Knowles et al.[9] described the value of such cascade testing in cases of familial hypercholesterolemia. Cascade testing to promote preventive measures is of particular value in the case of diseases where the consequences of a particular genetic variant develop over time. Cardiovascular diseases where molecular diagnosis and cascade testing are important also include cardiac arrhythmias.

Dainis and Ashley noted that the small but additive effects of genetic variants also influence the occurrence of coronary artery disease. They reported that a specific study designated MIGenes determined genomic risk scores for myocardial infarction in specific individuals. Scores were determined using particular SNP polymorphic allelic variants that had been shown to be associated with increased risk of cardiovascular disease. Individuals with high risk scores were offered statin therapy and were given lifestyle information. At the end of the multi-year project patients who were given genetic risk score information and advice on lifestyle management had a more significant decrease in low lipoprotein levels than patients offered therapy alone.

A multi-year study reported by Khera et al.[10] also reported that genetic risk information and specific lifestyle information led to a significant decrease in adverse cardio-vascular events.

Rare genetic variants in familial hypercholesterolemia

Dainis and Ashley noted the importance of specific rare variants in causation of hypercholesterolemia and increased risk of cardiovascular disease. These included rare variants in low density lipoprotein receptor (LDLR) apolipoprotein B (APOB) and proprotein convertase subtilisin/kexin type 9 (PCSK9). They noted that genotype information is also useful in determining appropriate dosing of the medication Warfarin that is used in patients with cardio-vascular disease. Information of genetic variants in VKORC1 responsible for activation of Vitamin K and CYP2C9 cytochrome oxidase that impact drug metabolism) have been shown to be of value in some studies while other studies claim that the genetic information is not useful if patients are carefully managed clinically.

Studies of genetic variants associated with common disease and polygenic risk scores

In a number of complex common diseases that are known to manifest high heritability rates, researchers have carried out genetic studies in an attempt to identify loci with alleles that contribute to the risk of developing the disorder. These studies involve searches for allelic nucleotide variants at specific sites throughout the genome and are commonly referred to as genome wide association studies (GWAS).

Even when disease associated allelic variants were found in specific loci, they contributed minimally to disease risk. It has emerged that the disease associated genomic loci identified in GWAS are most frequently found in regulatory regions of the genome, rather than in protein coding regions. Follow-up studies are then required to establish the genes that have altered expressions as a consequence of the effects of the risk allele at a specific associated locus.

Additional studies have been undertaken that combine information on the occurrence of common variant risk alleles at loci across the genome to derive a cumulative polygenic risk score for a specific disorder, e.g. diabetes or type of disorder. e.g. cardiovascular disease.[11]

Questions arise as to the utility of informing an individual of his or her polygenic risk score for a particular disorder. Some studies indicate that knowledge of the risk score leads some individuals to modify lifestyle in an attempt to mitigate the genetic risk effect through reducing environmental risk factors.[11] It is also possible that information about the genetic polygenic risk score may reveal necessity for closer clinical monitoring or imaging studies.

The precision medicine concept promotes the right prevention, and the right treatment for a specific person taking into account determined genetic and environmental risks factors relevant to that person.

Polygenic risk scores can also be presented in the context of the overall population risk. McCarthy and Mahajan[12] reported that GWAS studies had identified >400 different loci with risk alleles that contributed to development of Type 2 Diabetes mellitus. The majority of the disease associated alleles at these loci contributed only minimally to increased risk (risk increased from 1× to 1.05×). A few loci increased risk slightly more, e.g. a specific allelic variant near the locus TCF7L alter risk from 1× to 1.3× −1.5× in the European population. McCarthy and Mahajan noted that the immediate clinical utility of this data was questionable.

Age related macular degeneration: Gene variants and environmental interactions

Age related macular degeneration (ARMD) is characterized by progressive degeneration in the macula, the neurosensory region of the retina that is responsible for central high- resolution vision. In a review of ARMD Maugeri et al.[13] noted that in the early stages of the disorder aberrant pigmentation occurs in the retinal pigment epithelium leading to formation of pigment deposits referred to as drusen. These lipid rich deposits are reported to impair function of the retinal pigment epithelium and to impair metabolic transfer between the retinal pigment epithelium and the choroid.

In the later stage of ARMD neovascularization of the choroid occurs in some cases. This is referred to as wet macular degeneration. In some cases, atrophy of the macula occurs in the absence of neovascularization and this is referred to as the dry form of macular degeneration.

Environmental factors in macular degeneration

Smoking is known to be a significant risk factor for development of ARMD.[14] Factors reported to likely be protective against AMD include a diet rich in omega-3-fatty acids and anti-oxidants. A specific formulation composed of zinc, vitamins C and E, and carotenoids lutein and zeaxanthin, was reported to reduce progression of the dry form of MD in the AREDS2 study.[15]

There is some evidence that agents against vascular epithelial growth factor (VEGF) may be beneficial in treatment of the wet form of ARMD though the treatments are not without risk.

Maugeri et al.[16] and Cobos et al.[17] reported that variants in specific genes impact the risk of developing ARMD. Genes products with risk variants are listed below.

CFH	Complement factor H
ARMS2	ARMD susceptibility component of the choroid matrix
APOE	Apoliprotein E
VEGFA	Vascular endothelial growth factor
VEGFR1	Vascular endothelial growth factor receptor 1
SERPIN F1	This protein strongly inhibits angiogenesis
HTRA1	Member of the trypsin family of serine proteases

Specific complement factor H variants are considered to be the most significant risk factors. Complement system components have proteolytic activity. Maugeri et al.[15] emphasized that complement components play important roles in clearing immune complexes and components of apoptotic cells. Complement components are produced primarily in the liver, however there is evidence that complement synthesis occurs in the retina.[18] Local overactivation of complement likely plays a role in ARMD.[19] Complement regulators were isolated from drusen, these include vitronectin, clusterin and CD46 a negative regulator of complement. There is also evidence that amyloid beta accumulates in drusen. Accumulation of lipids and lipid peroxidation products also activate complement and lead to apoptosis.

Specific AMD risk variants

The complement factor variant rs1061170 has two allele C and T; the C allele is common and has a population frequency of 0.2666. The C allele leads to an amino acid change in complement factor H Y402H. The C allele was reported to increase ARMD 2.5× in heterozygous individuals and 6.0× in homozygotes. This variant increases the ARMD risk in Caucasian and Asian populations and it increases the risk for both wet and dry forms.[15] The homozygotes for the risk allele were reported to be less likely to respond to anti-VEGF treatment.

A protective variant in rs800292 has also been identified in the Complement H gene rs800292.

Age related neurodegenerative disorders

Alzheimer's disease

In 2018 Freudenberg-Hua et al.[20] reviewed Alzheimer's disease and considered the possibilities that precision could contribute to improve the diagnoses and therapeutics in Alzheimer's disease. They noted that worldwide 60–80% of dementia diagnoses were attributed to Alzheimer's disease. Furthermore. Alzheimer's disease is associated with significant financial costs primarily related to care since no definitive treatments have been developed.

Guidelines for diagnosis of Alzheimer were developed by the Alzheimer's disease association and by the National Institute of aging.[21] These guidelines included neurophysiological testing and analysis of specific biomarkers in cerebrospinal fluid.

In 2017 Frisoni et al.[22] reported that neuroimaging studies, including magnetic resonance imaging (MRI), and positron emission tomography (PET), were increasingly being used to assess neuronal loss and protein deposition in suspected Alzheimer cases. They developed diagnostic algorithms that included data from imaging studies and biomarker analyses. Parameters in these algorithms included MRI and PET scan findings. MRI findings included evidence of tissue loss and neurodegeneration as evidenced by decreased hippocampal and temporal lobe volumes. Pet scan findings of Alzheimer's disease included glucose hypometabolism and decreased uptake of radiolabeled fluorodeoxyglucose. Additional PET scan studies used a radiolabeled tracer that binds to amyloid and could reveal cortical deposition of amyloid.

Cerebrospinal fluid (CSF) biomarker analyses included assays of the ratio of amyloid beta forms Abeta 42 and Abeta 40 to determine if altered amyloid metabolites were present. Levels of CSF tau protein and hypophosphorylated tau were also measured; Alzheimer's disease is associated with accumulation of tau and of phosphorylated tau and levels of these rise in the CSF.

Different forms of Alzheimer's disease

The most definitive factor leading to Alzheimer's disease is aging. However, investigators have long distinguished between early onset Alzheimer's disease (EOAD) in which symptoms commence before 65 years of age and late onset Alzheimer's disease (LOAD). Freudenberg-Hua et al.[20] reported that worldwide data indicate that only 10–14% of all Alzheimer's disease cases are reported to be of the early onset type. Mutations in EOAD have been reported to involve Amyloid precursor protein (APP) or Presenilin 1 or Presenilin 2 protein, and these mutations can lead to autosomal dominant forms of the diseases. It is however important to note that evidence of inheritance of EOAD is not always present. Apparently sporadic cases of the disease occur. In sporadic cases it seems likely that new mutations occur, and furthermore sporadic Alzheimer's disease may be multigenic in origin. Lanoiselée et al.[23] reported results of genetic screening of Alzheimer cases in France and they reported finding sporadic cases of EOAD with de novo mutations in PSEN1, PSEN2 or APP.

Late onset Alzheimer's disease (LOAD) and APOE4

The association of the APOE4 of the protein apolipoprotein (APO) and Alzheimer's disease were first reported by the Roses group in 1995 (see Corder et al.[24]) >20 years later APOE4 remains the most significant risk factor in Alzheimer's disease. However, as will be demonstrated below, risk factors that impact at least 20 different genes and their products have been found to be associated with Alzheimer's disease. The mechanisms through which APOE4 increases Alzheimer's disease risk have not been clearly defined. Studies of these mechanisms are now becoming progressively important in light of recent lack of success in using strategies to decrease brain levels of amyloid beta protein to reduce impairments associated with Alzheimer's disease.

Huang and Mahley[25] analyzed APOE structure. They reported that APOE is produced in liver and also in brain. In brain it is apparently produced by astrocytes. APOE bind lipids, transports lipids and also interacts with receptors to promote uptake lipid into cells. The different forms of APOE protein differ at two key positions 112 and 158. The most common form APOE3 has cysteine at position 112 and arginine at position 158. APOE2 that may be protective against ALZ has cysteine at 112 and cysteine at 158, ApoE4 the form associated with Alzheimer's disease has arginine at position 112 and arginine at position 158, this substitution is thought to alter the three-dimensional protein structure.

It is important to note that individuals who have the APOE4 allele do not inevitably develop Alzheimer's disease. However, heterozygotes for APOE4 were reported to have two to three times increased risk of developing Alzheimer's disease and in APOE4 homozygotes the risk for Alzheimer's disease was five times greater than the population risk. In addition, the average age for development of the disease was 68 in APOE4 homozygotes and in individuals without APOE4 alleles the who developed Alzheimer diseases the average age of onset was reported to be 84 years.[26]

Brandon et al.[27] reported that the presence of APOE4 protein impacts brain metabolism, particularly brain glucose metabolism. Differences in brain glucose metabolism were demonstrated in APOE4 individuals even in the absence of cognitive impairment. APOE functions as a lipid carrier and Brandon noted evidence that APOE4 had increased lipid binding capacity. Lipids have been shown to bind to amyloid beta protein.

Brandon et al. also noted evidence that cerebral blood flow is reduced in APOE4 individuals even prior to development of cognitive impairment. APOE4 has also been shown to increase the incidence of cerebral amyloid angiopathy, a condition associated with the deposition of amyloid in the walls of small blood vessels in the brain.

Wadhwani et al.[28] developed pluripotent stem cells from APOE4 positive patients. From the pluripotent stem cells they developed excitatory forebrain neurons. They demonstrated that endogenous expression of APOE4 by these cells predisposed these cells to calcium dysregulation and death. This calcium dysregulation was also associated with increased release of phosphorylated tau protein.

Wadhwani et al. succeeded in carrying out gene editing in their pluripotent stem cells and they reported that conversion of the stem cells to the E3/E3 genotype corrected the observed functional abnormalities.

Analyses of gene variants in Alzheimer's disease

Results of a meta analyses of genetic data in Alzheimer's disease were published in 2019 by the researchers involved in the International Genomics of Alzheimer Project (IGAP), see Kunkel et al.[29] The final analyses included 94437 Alzheimer cases, including cases with clinical diagnoses and cases with autopsy diagnoses, and 59,163 controls. The study involved analyses of both common and rare risk variants. Common variants were variants with minor allele frequencies >0.01 in the population and rare variants with allele frequencies <0.01 in the population. Gene dosage changes were also analyzed.

The meta-analyses revealed variants that impacted following twenty-one gene loci

Gene	Significance	Function of gene product
APOE	1.2×10^{-881}	Apolipoprotein E, binds to a receptor essential for lipoprotein metabolism
BIN1	2.1×10^{-44}	Nucleocytoplasmic binding protein, involved in synaptic vesicle exocytosis
PALM	6×10^{-25}	Paralemmin, prenylated phosphoprotein associates with membranes
CLU	4.6×10^{-24}	Clusterin secreted chaperone protein
CRI	3.6×10^{-24}	Complement C3b/C4b receptor 1, binds to immune complexes
MS4A2	1.9×10^{-19}	Membrane spanning 4-domains A2, Immunoglobulin receptors
ABCA7	3×10^{-16}	ATP binding cassette subfamily A member 7 involved in membrane transport
TREM2	2.7×10^{-15}	Triggering receptor expressed on myeloid cells 2, involved in immune response
PTK2B	$6.3A10^{-14}$	protein tyrosine kinase 2 beta involved in activation of signaling pathway
SPI1	5.4×10^{-13}	Spi-1 proto-oncogene encodes transcription factor for T and B cell development

Continued

Gene	Significance	Function of gene product
SORL1	2.9×10^{-12}	Sortilin related receptor 1, plays roles in endocytosis and sorting
HLADRB1	1.4×10^{-11}	Major histocompatibility complex, class II, DR beta 1. Immune function
CD2AP	1.2×10^{-10}	CD2 associated protein, scaffolding molecule regulates actin cytoskeleton
NYAP1	9.3×10^{-10}	Neuronal tyrosine phosphorylated phosphoinositide-3-kinase adaptor 1
EPHA1	1.3×10^{-10}	EPH receptor A1, member of protein kinase family
INPP5D	3.4×10^{-9}	Inositol polyphosphate-5-phosphatase D, functions in signal pathways
FERMT2	1.4×10^{-9}	Fermitin family member 2, interacts with actin
SLC24A4	7.4×10^{-9}	Member of the potassium-dependent sodium/calcium exchanger protein family
ECHDC3	2.4×10^{-9}	Enoyl-CoA hydratase domain containing 3
ACE	7.5×10^{-9}	Angiotensin1 converting enzyme controls blood pressure fluid and electrolytes
CASS4	3.3×10^{-8}	Cas scaffold protein family member 4

Jansen et al.[30] carried out a study involving 71,880 Alzheimer cases and 383,378 controls. They identified 29 SNP loci, that potentially impacted 215 genes. They emphasized that significantly associated SNP loci were primarily located in intronic and intergenic gene regions and in regions active in chromatin. Their study led to identification of specific biological mechanism involved in Alzheimer disease. They highlighted lipid related processes and amyloid degradation pathways. The most significant loci identified, and their significance levels are listed below.

Locus	Significance level
APO	2.70×10^{-194}
BIN	3.58×10^{-29}
CLU	6.36×10^{-20}
PICALM	1.12×10^{-17}

Amyotrophic lateral sclerosis also known as Lou Gehrig disease, motor neuron disease

Amyotrophic lateral sclerosis (ALS) was reviewed by Mathis et al.[31] The incidence of this disorder in Western populations was reported to be 2 to 3 per 100,000 individuals and the incidence in Asia was lower.

The age of onset of ALS was reported to be 65 years though in some cases it presents earlier. ALS is reported to be fatal within 2–5 years of diagnosis.

ALS is characterized by degeneration of both upper and lower motor neurons leading to motor impairments, paralysis and eventually to impaired breathing and swallowing. Mathis et al. noted that non-motor symptoms sometimes occur, including in some cases cognitive impairments. In some individuals with ALS manifestations, fronto-temporal dementia occurs.

Both familial and sporadic forms of the disorder are now known. Sporadic forms are most common and 5–10% of cases are reported to be familial ALS. in addition to familial forms caused by single gene defects there may in some families be forms caused by the combined effects of deleterious variants in several genes. They emphasized that genetic counseling in ALS is complex.

Variants in >31 different genes have been associated with familial ALS and some of these variants may also occur in sporadic cases. Mathis et al. emphasized that the specific risk genes most frequently implicated differ in different populations. Both dominantly inherited and recessive variants can play roles in familial ALS. In addition, there is often evidence of variable penetrance of dominant ALS variants. It is also important to note that a number of specific genes have variants that can act as autosomal dominant risk variants and they also have variants that can acts as autosomal recessive traits.

In 2019, 31 different genes are known to have ALS causing variants are located in genes that map to 18 different human chromosomes and to the X chromosome. Mathis reported that there are 4 genes that are most commonly involved in ALS. The products of these genes and product functions are listed below.

SOD1	Superoxide dismutase 1, destroys superoxide radicals
FUS	RNA binding protein, involved in RNA splicing and RNA processing
TARDBP	Also known as TDP43, is a DNA binding protein acts as a transcriptional repressor
C9ORF72	C9orf72-SMCR8 complex subunit involved in regulation of endosomal trafficking

Mathis et al. noted that in ALS many of the histopathological features that are characteristic of other forms of neurodegenerative diseases are present. These include protein misfolding, protein aggregate deposition, accumulation of protein degradation products and presence of abnormally phosphorylated forms of proteins e g. phosphorylated TAR43. There is also evidence that abnormal proteins may disseminate along axonal pathways in a prion like fashion.

Evidence for RNA dysregulation in amyotrophic lateral sclerosis

Butti and Patten[32] noted that products of the key gene loci that are impacted by ALS causing variants have important functions related to RNA metabolism. TAR binding protein (TDP43) was reported to have two RNA recognition domains and to impact RNA splicing RNA stability and RNA transport. This protein impacted alternative splicing of RNAs transcribed from specific genes.

FUS protein was found to be present in RNA spliceosome complexes. Lagier-Tourenne et al.[33] reported that FUS variants altered splicing of 300 different brain mRNAs. There is also evidence that FUS variants alter transcription of specific genes.

Superoxide dismutase (SOD1) was reported to impact RNA metabolism and RNA stability. Altered function of SOD1 was found to promote aggregation of mRNA.

C9ORF72 harbors a hexanucleotide repeat expansion GGGGCC. This repeat was found to have undergone substantial expansion in specific cases of ALS and in cases with fronto-temporal dementia.[34] The precise mechanisms through which C9ORF72 expansions lead to neurodegeneration are still being elucidated. There is evidence that the repeat expansion sequesters RNA binding proteins and impacts nucleo-cytoplasmic transport.[35]

Possible gene specific approaches to therapy of ALS include use of antisense nucleotides to suppress specific mutations, e.g. SOD1 inducing mutations and to suppress C9ORF72 repeat expansions.

Gene-time-environment models

Yu and Pamphlet[36] considered the relevance of a gene-time environment model in ALS causation. They noted that no environmental factors have yet proven to be associated with ALS. Risk factors that have been investigated include infections, organophosphate exposures and heavy metal exposure. Yu and Pamphlet proposed that mitochondrial dysfunction could contribute to development of ALS. They emphasized that the motor neural system may be particularly susceptible to energy deficiency since it requires high metabolic rate and high ATP levels for adequate function. They propose that factors that impact mitochondrial function may serve as a trigger for initiation of ALS manifestations.

Mitochondrial damage due to environmental factors or age generated mitochondrial mutations may impair ability of the neural system to compensate for impaired function of products of genes that harbor risk variants.

Parkinson's disease (PD)

PD is an example of a disease that can result from impact of single rare gene defects, common gene variants and environmental factors, acting singly or in combination. The key pathological defects in PD include degradation of dopaminergic neurons in a specific brain region the Substantia nigra, and the accumulation of distinct protein aggregates referred to as Lewy bodies, Degradation of dopaminergic neurons leads to resting tremor, rigidity, postural instability, this disease is reported to occur in 3% of individuals older than 75 years.[37]

A key genetic defect in PD involves the gene that encodes the protein synuclein.[38] Extra copies of synuclein and specific synuclein mutations have been reported to lead to PD. Both Mendelian genetic and non-Mendelian genetic defects have been identified to contribute to PD causation in recent decades. Hernandez et al.[39] reviewed these genetic factors. Specific products of genes shown to play roles in Parkinson's disease are listed below.

Autosomal Dominant Parkinson's Disease

SNCA	Alpha synuclein, may serve to integrate presynaptic signaling and membrane trafficking
LRRK2	Leucine rich kinase, located in cytoplasm and on outer mitochondrial membrane
GIGY2	GRB10 interacting GYF protein, may be involved in tyrosine kinase receptor signaling
HTRA	Serine protease
VPS35	Involved in retrograde transport of proteins from endosomes to the trans-Golgi network

Autosomal Recessive Parkinson's disease

PRKN	Parkin RBR E3 ubiquitin protein ligase, involved in proteasomal degradation
PINK1	PTEN induced kinase 1, protects cells from mitochondrial stress induced dysfunction
DJ1	(PARK7) Parkinsonism associated deglycase, sensor of mitochondrial stress
ATP13A2	Transports inorganic cations as well as other substrates
PLA2G6	Phospholipase, catalyzes the release of fatty acids from phospholipids
DNAJC6	(HSP40 homolog) molecular chaperone prevents protein misfolding

Products of key Genes with Parkinson's disease risk variants and product functions	
SNCA	Alpha synuclein, may serve to integrate presynaptic signaling and membrane trafficking
LRRK2	Leucine rich kinase, located in cytoplasm and on outer mitochondrial membrane
GBA	Glucosylceramidase beta, cleaves the beta-glucosidic linkage of glycosylceramide

In addition, genetic linkage studies in families have led to chromosomal assignments for two forms of Parkinson's disease where the specific genes leading to the disease had not yet been identified.

Hernandez et al.[39] reported that five different SNCA mutations are particularly important in cases of synuclein related Parkinson's disease. In the LRRK2 protein a specific mutation p.G2019S was found to occur in many PD patients and in specific populations it was present in 42% of familial PD cases. These include Ashkenazi Jewish, Middle-Eastern and North-African populations. This same mutation was reported to occur in 2% of cases of sporadic PD in Northern European and USA populations. It is rare in Asian populations. Hernandez et al. noted that the penetrance of this mutation varied with age. At age 59 years the penetrance was reported to be 28%. At 79 years the mutation penetrance was 74%.

Other LRRK2 mutations also occur in autosomal dominant and sporadic cases of PD and they noted that LRRK2 mutations were the most common mutations found in both autosomal dominant and sporadic PD.

The vacuolar sorting protein 35 has been found to be a rare cause of PD.

In addition to the genes and gene products described above, results of genome wide association studies have revealed that specific single nucleotide variants at 28 loci contribute minimally to PD causation. Hernandez et al. concluded that data indicated that Parkinson's disease likely results from a combination of genetic, environmental and lifestyle factors.

Exome sequencing in Parkinson's disease

Robak et al.[40] reviewed results of exome sequencing in 1156 cases of Parkinson's disease and 1679 controls. Their study revealed a significant burden of damaging variants in genes associated with lysosomal storage diseases. Their study revealed 56% of cases had a damaging allele in one lysosomal disease associated genes, 21% of cases carried damaging alleles in multiple lysosomal storage disease related disease genes. In Parkinson's disease patients there was evidence for an excess

burden of heterozygous deleterious variants in genes that in homozygous or compound heterozygous states were associated with lysosomal storage diseases.

This study provided evidence that pleiotropic genes that cause
early onset monogenic disorders, may harbor heterozygotic variants
that act in combination to cause late onset complex genetic disorders.

GBA	lysosomal membrane protein that cleaves the beta-glucosidic linkage of glycosylceramide
SMPD1	lysosomal acid sphingomyelinase that converts sphingomyelin to ceramide.
CTSD	encodes a member of the A1 family of peptidases deficient in neuronal ceroid lipofuscinosis-10
SLC17A5	encodes neuronal ceroid lipofuscinosis-10 a
ASAH1	encodes N-acylsphingosine amidohydrolase 1, that is deficient in lipogranulomatosis,

There is evidence that glucosylceramide that accumulates in GBA
defects promotes aggregation of alpha synuclein, the overabundant
protein in Parkinson's disease.[41]

Environmental factors and Parkinson's disease

In 2003 Braak et al.[42] proposed a new hypothesis for PD causation.
They postulated that the olfactory system and the gut potentially transport pathogens that travel through retrograde axonal transport along
visceromotor projections, to the central nervous system. There the
pathogens lead to neuronal damage and they are subsequently transported to sub- cortical nuclei and ultimately to the cerebral cortex.
Ultimately diverse brain regions are involved in Parkinson's disease.

Over the years a number of studies have provided evidence that
individuals exposed to certain pesticides have increased incidence of
Parkinson's disease.[43]

In 2019 therapeutic approaches to PD remain designed to address
the motor defects and genetic discoveries have not yet been applied to
develop new therapies.

Traumatic brain injury and consequences

Wilson et al.[44] reported evidence that traumatic brain injury represents
a risk factor for a number of different disorders including Parkinson's disease, dementias, including Alzheimer's disease and dementia with Lewy
bodies, fronto-temporal dementia, amyotrophic lateral sclerosis, chronic
traumatic encephalopathy, epilepsy and psychiatric disorders.

They noted that older age at trauma was particularly associated with poorer outcomes. The best evidence for the association of repeated mild traumatic brain injury and the risk of neurodegeneration comes from studies on boxers and from studies on football and rugby players.

Psychiatric disorders

The Brainstorm consortium[45] carried out extensive genome wide studies on common allelic variants in 1,191,588 individuals to determine if there was any degree of overlap in risk variants associated with 25 different brain disorders.

Common risk variant overlap was established between psychiatric disorders attention deficit disorder (ADHD), bipolar disorder, major depressive disorder, and schizophrenia. They also demonstrated that the personality trait referred to as neuroticism was significantly associated with most psychiatric disorders and with migraine.

Investigators in The Brainstorm consortium concluded that the evidence for genetic risk correlations among the different psychiatric conditions suggested that the separated psychiatric diagnostic categories overlapped.

Post-Traumatic stress disorder risk did not show overlap with the other psychiatric disease risk factors. GWAS data on autism spectrum disorder risk variants and Tourette syndrome appeared more distinct from risk loci for the other psychiatric disorders, though more collection data on these disorders was recommended.

Personalized medicine and Cancer prevention

Germline mutations in cancer risk genes

The goal of testing individual for high risk cancer genes is to ensure that individuals who carry these mutations have appropriate and frequent clinical and radiologic screening to detect tumors at the early stages when they are most likely to be treated successfully. Patients with germline mutations in cancer risk genes often have a family history of cancer and particularly a history of cancer in their predecessors.

It is important to note that many of the genes that harbor mutations that increase cancer risk encode proteins that are involved in the repair of DNA damage. Barnes et al.[46] reported that humans are continually exposed to DNA damaging agents in the external environment. They noted however that damage from internally generated DNA damaging agents is likely more important.

Endogenous DNA damaging agents listed include reactive oxygen species, substances generated through lipid peroxidation, endogenous alkylating agents and metabolites derived from estrogen.

Barnes et al. emphasized that evidence for the roles of environmental or lifestyle factors in causation of DNA damage derive in part from evidence that populations who migrate from regions with low cancer risk to regions with high cancer risk develop cancer risk rates similar to those documented for the new environment.

Cancer gene panels

Increasingly sequencing panels have been designed to investigate the presence of pathogenic germline mutations that increase cancer risk. The specific genes included in the panels may differ slightly in panels offered by different companies. Slavin et al.[47] reported on multi-gene panels used in cancer screening and they also reported some of the problems encountered in interpreting the results obtained in panel analyses.

Classification of cancer risk genes

High risk genes are genes that have been shown most commonly to carry pathogenic mutation that increase risk for one or more types of cancer. The cancer types most common investigated for germ line mutations are, breast and ovarian cancer and gastro-intestinal cancers.

In addition to genes in the high risk category some cancer panels are designed to investigate genes defined as being in the moderate risk category. There are additional panels that investigate genes in the low risk category. It is important to note that only mutations that have been defined as pathogenic will be considered as being of potential clinical significance. In addition to data designed to search for evidence of germline mutation, clinicians and counselors gather data on family history of cancer.

Examples of genes designated as high risk genes for specific forms of cancer or listed below, meaning that germline pathogenic mutations have been found with relatively high frequency in patients with that form of cancer.[47] It is important to note that germline mutation in cancer predisposition genes have been found in patients with different pathological types of breast cancer including estrogen receptor positive cancer, epidermal growth factor receptor (HER2) positive cancer and triple negative breast cancer i.e. breast cancer that does not express estrogen receptor, progesterone receptor or HER2. High risk genes as reported by Slavin et al.[47] are listed below.

High risk Breast Cancer genes	
BRCA1	Breast Cancer 1
BRCA2	Breast Cancer 2
CDH1	Cadherin 1

Continued

High risk Breast Cancer genes

PTEN	Phosphatase and tensin homolog, tumor suppressor gene
STK11	Serine/threonine kinase 11, tumor suppressor gene
TP53	Tumor protein 53, contains transcriptional activation, DNA binding, and oligomerization domains

High risk Ovarian cancer genes

BRCA1 and *BRCA2*	
MLH1	mutL homolog 1, part of the DNA mismatch repair system
MSH2	Human homolog of the *E. coli* mismatch repair gene mutS,
STK11	Serine/threonine kinase 11, tumor suppressor gene

Moderate risk breast cancer genes, products and functions

ATM	ATM serine/threonine kinase, important cell-cycle checkpoint kinase
BRIP1	BRCA1 interacting protein C-terminal helicase 1, important in double-strand break repair
CHEK2	Check-point kinase 2, critical cell cycle regulator.
PALB2	Partner and localizer of BRCA2, permits the stable intranuclear localization of BRCA2

Moderate risk ovarian cancer genes, products and functions

RAD51C	Paralog of *RAD51*C, involved in the homologous recombination and repair of DNA
RAD51D	RAD51 paralog D, involved in the homologous recombination and repair of DNA
MSH6	Member of the DNA mismatch repair MutS family
PALB2	Partner and localizer of BRCA2, permits the stable intranuclear localization of BRCA2

Pathogenic BRCA mutations and incomplete penetrance

Incomplete penetrance refers to the fact that a specific genotype may not always give rise to a specific phenotype. Cancer specific pathogenic germ-line mutations that are known to give rise to cancer in some individuals do not give rise to cancers in other individuals who have the same mutations in their genomes.

In the case of breast cancer, specific BRCA1 mutations defined as pathogenic, were reported to give rise to cancer, by the age of 70 years in 57% of individuals who had those specific BRCA1 mutations in their germline.[48]

Downs et al.[49] carried out studies to investigate genetic factors that influence non-penetrance of BRCA1 mutations defined as pathogenic in databases. Identification of factors that influence penetrance are clearly important for genetic counseling and are also potentially important in in providing insight into avenues toward disease prevention and disease modification.

Downs et al. recruited from each family studies, pairs of individuals with and without breast cancer who both carried pathogenic BRCA1 mutations. In total 27 pairs of individuals were studied. Of the BRCA1 mutations studied 22 were defined as definitely pathogenic. The age range of breast cancer affected ranged from 23 to 59 (mean 34 years). The age of the unaffected breast cancer individuals in the pairs ranged from 51 to 82 (mean 65 years).

Downs et al. carried out exome sequencing on peripheral blood to identify potential germline sequence differences between the two groups, breast cancer affected and unaffected. Their study led to identification of 12 variants that potentially decreased the risk of breast cancer. The 12 variants that occurred in breast cancer unaffected individuals had a range of different functions.

Downs et al. studied the effect of a one variant found in unaffected individuals in ANLN, an actin binding protein that plays a role in cell growth and migration. They noted that the variant had decreased expression compared with the normal allele. The variant was shown to alter the localization of the protein and impacted the normal function, the variants in ANLN that occurred in the unaffected individuals has a population frequency of 0.113.

High risk Colorectal cancer genes products and functions

APC	APC, WNT signaling pathway regulator (adenomatous polyposis coli)
BMPR1A	Bone morphogenetic protein (BMP) receptor transmembrane serine/threonine kinases
EPCAM	Epithelial cell adhesion molecule
MLH1	mutL homolog 1, part of the DNA mismatch repair system
MSH2	Human homolog of the E. coli mismatch repair gene mutS,
MSH6	Member of the DNA mismatch repair MutS family.
MUTYH	mutY DNA glycosylase, glycosylase involved in repair of oxidative DNA damage repair
PMS2	mismatch repair system component, also corrects small insertions and deletions
SMAD4	Member of the Smad family of signal transduction proteins
STK11	Serine/threonine kinase 11, tumor suppressor gene

Moderate risk Colorectal cancer genes, products and functions

CHEK2	Check-point kinase 2, critical cell cycle regulator
PTEN	Phosphatase and tensin homolog, tumor suppressor gene
TP53	Tumor protein 53, contains transcriptional activation, DNA binding, and oligomerization domains

Cancer risk genes and genetic testing

In 2017 Okur and Chung[50] published a report on hereditary cancer genes. They noted that specific mutations in BRCA1 and BRCA2 genes are screened for not only in breast and gynecologic cancers panels but also in pancreas and prostate cancer screening panels.

The *CDH1* gene is included in breast and gynecologic cancer screening panels and also in colorectal and gastro-intestinal cancer screening panels. The *EPCAM, MLH1, MSH2, MSH6, PMS2, PALB2 PTEN, STK11 TP53, ATM* and *CHEK2* genes were also included in several different cancer screening panels.

A key challenge in cancer risk gene testing by sequencing analysis is that specific mutations identified on sequencing cannot always be definitively classified as pathogenic mutation and may be classified as variants of unknown significance.

In addition, Okur and Chung emphasized that specific cancer risk genes, including TP53 have a relatively high rate of de novo mutations and patients with de novo mutations will not necessarily give a history of cancer in predecessors.

Huang et al.[51] analyzed 10,389 cancer cases and they identified pathogenic cancer predisposition variants in 8% of cases.

The germline predisposition variant occurred in 33 different forms of cancer. They noted that these pathogenic predisposition variants included known cancer risk variants and also rare variants. The variant included single nucleotide variants, small deletions and insertions and larger copy number variants. Some tumors manifested bi-allelic changes so that both copies of a specific risk gene were damaged or in some case on copy of the gene has a pathogenic variants and deletion occurred in the second copy of that gene.

Huang et al. reported that most of the cancer predisposing gene changes occurred in tumor suppressor genes. However, some cancer predisposing mutations included activating changes in oncogenes, Oncogenes particularly impacted including *MET* (MET proto-oncogene, receptor tyrosine kinase). *RET* (Ret protooncogene trans-membrane receptor), and *PTPN11* protein tyrosine phosphatase, non-receptor type 11). Some *PTPN11* changes were shown to be tumor activating while other PTPN11 changes had tumor suppressing effects.

Huang et al. carried out expression studies and they established that low gene expression of tumor suppressor genes was found in at least half of the cases with tumor suppressor gene mutations. In cases with oncogenic mutations increased expression of oncogenes was found in 62% of cancers.

Huang et al. noted that that germline cancer predisposing genes often manifested mutations at sites the undergo post-transcriptional modifications.

Huang et al. concluded that germline cancer predisposing variants should be carefully evaluated at the level of downstream expression.

Lynch syndrome

The discovery of this syndrome began with careful documentation by a pathologist AS. Warthin in 1913,[52] of families with high incidences of cancers that impacted particularly the stomach, colorectum and uterus. In 1966 Lynch, Shaw and colleagues[53] published reports of families with similar manifestations to those described by Warthin and the syndrome became known as Lynch syndrome

Further analyses of the phenotypic spectrum of this cancer predisposition syndrome were carried out and these revealed that colorectal tumors occurred predominantly in the proximal colon and that one third of cancers were found in the cecum. The uterine cancer found in this syndrome were found to be predominantly endometrial cancers. In addition, cancers of the stomach, small bowel, hepato-biliary system, urological system and ovary have been found in families with Lynch syndrome. Other tumors that occur in these families include sebaceous skin tumors (Muir Torre syndrome). In addition, other rare tumors may occur including glioblastoma in the brain and breast, pancreas, prostatic and adrenocortical tumors have been reported in Lynch syndrome families.

Molecular pathogenesis of Lynch syndrome

Key discoveries of the molecular pathogenesis of Lynch syndrome began with the studies of tumor DNA and reports of microsatellite instability, i.e. wide variation in the number of dinucleotide repeat elements. This microsatellite instability was subsequently shown to be due to defects in the function of elements in the DNA mismatch repair system.[54]

It was known that mutations in specific yeast genes led to frameshift mutations in microsatellite repeats. Human homologs of the yeast genes were isolated initially MSH2 and MLH1, were identified by Strand in 1993.[55] Subsequently defects in these genes and related genes were identified in Lynch syndrome patients. Specific defects associated with Lynch syndrome include heterozygous mutations in genes that encode MLH1, MSH2, MSH6 and EPCAM.[56]

DNA mismatch repair and its role in cancer

In a 2018 review Baretti and Le[57] noted that the hypermutator phenotype resulted from alterations in short repetitive sequences and in some cases from single nucleotide substitutions. They noted further that germline mutations in mismatch repair genes were characteristic of Lynch syndrome. In sporadic cancer the mismatch repair genes were frequently silenced through epigenetic mechanisms.

The cancer risk in individuals with germline mutation in mismatch repair is very high and the cumulative life-time risk in males with these mutations was reported to be as high as 60–70. In females with these mutations the cumulative lifetime cancer risk was reported to be 40% to 50%. The population frequency of pathogenic mutations in mismatch repair genes was reported to be 1in 1000. DNA sequencing is currently most frequently used to screen for germline mutations.

A growing number of techniques have been developed to test for microsatellite instability in tumors. Screening for mismatch repair gene defects is important not only in Lynch syndrome cases, but also in other tumors since it has emerged that this information is of therapeutic relevance. The tumor DNA can be screened for microsatellite instability. In addition, immunohistochemistry is used to detect altered levels of mismatch repair proteins.

Mismatch DNA repair gene defects in tumors and immunotherapy

There is evidence that mismatch repair gene mutations in tumors lead to increased numbers of activated T lymphocytes. Cases with these tumors were found to respond well to treatment with checkpoint inhibitors.[57]

Identifying cancer risk mutations

Classifying genes as cancer risk mutations is based on the lifetime cancer risk conferred by germline mutations in that gene.

In a survey of 1058 individuals with colorectal Yurgelun et al.[58] reported that that of 1058 patients 105 individuals (9.9%) carried pathogenic germline mutation in high penetrance cancer susceptibility genes. 33 patients had mutations characteristic of Lynch syndrome, 74 had high risk mutations but were not classified as Lynch syndrome cases. In these patients, mutations occurred in APC in 5 patients, 3 patients had MUTYH mutations, 11 had BRCA2 mutations 2 had PALB mutation 1 had CDKN2A mutation and 1 had mutation in TP53.

Microbiome and colorectal cancer

Colorectal cancers arise from gene mutations in addition there was is evidence that incidence and progression of these cancers are impacted by gut microbiota.

Indications that microbiota influence colon cancer come in part from the predominant location of these cancers in the distal colon and in the rectum. In addition, there is evidence for the role of inflammation in colon cancer genesis since patients with inflammatory bowel syndrome have an increased incidence of colon cancer.[59]

The microbiota present in human colon include primarily bacteria, but the archaea, fungi, protozoa and viruses are also present. They emphasized the important role of intestinal epithelial cells as a barrier to inflammatory invasion and they noted that the intestinal epithelial cells undergo continuous proliferation.

Gao et al. reported that analyses have revealed difference in the microbiota between patients with colorectal cancers and controls.

Saxena and Yeretssian[60] noted specific aspects of immune functions in reducing the risk, and they noted that in this regard Toll receptors and NOD like receptors and inflammasomes were particularly important.

Multiple endocrine neoplasia syndromes

There are two syndromes due to germline mutations that give rise to multiple endocrine neoplasias, MEN1 and MEN2.[61]

The MEN1 syndrome due to defects in a gene designated MEN1 is characterized by the presence of tumors in different neuroendocrine tissues, including parathyroid and anterior pituitary foregut and pancreas. The tumors may be hormone secreting.

The MEN2 syndrome is due to germline mutations in the RET proto-oncogene. There are different clinical subtypes of MEN2 and there are different RET mutations associated with the different subtypes. Characteristic tumors occur in the thyroid gland, in the parathyroid gland, and also in the adrenal gland (pheochromocytomas). In addition, mucosal neuromas may occur on the lip and tongue and unusual ganglioneuromas may occur in the intestine occurs. One particular mutation in RET at position cys634arg occurs most frequently and has moderate penetrance.

Germline mutation in the succinate dehydrogenase subunit SDHB have been reported to lead to paragangliomas, pheochromocytomas and some forms of intestinal stromal tumors.[62]

Melanoma

Hayward et al.[63] reported results obtained from whole genome sequencing on different forms of melanoma. They noted that skin melanomas occur primarily in Europeans. However, melanomas in mucosal sites and melanomas on palms of hands or on soles and feet (acral melanomas), occur in individuals in different populations.

In skin melanomas Cytosine to Thymidine C to T transition mutations are most common and are a signature of ultra-violet radiation damage.[64]

Hayward et al.[62] noted that different mutation signatures are found in acral melanomas and in mucosal melanomas. Structural genomic changes also occur in these lesions. There is evidence that malignant melanomas can also arise from pre-existing pigmented lesions (naevi) in the skin.

The most frequently mutated genes in skin melanomas included BRAF (BRAF (B-Raf proto-oncogene, serine/threonine kinase), CDKN2A (cyclin dependent kinase inhibitor 2A), NRAS (NRAS proto-oncogene, GTPase), and TP53 (Tumor protein 53).

Hayward et al. reported that the genes most frequently found to be mutated in acral melanomas included BRAF, CDKN2A, NRAS and TP53. In acral mutations the most frequently mutated genes included NRAS, NF1 (Neurofibromatosis) and SF3B1 (splicing factor 3b subunit 1).Mutations in the promoter regions of TERT were also reported to be commonly found in melanomas.

Levy et al.[65] reported that mutations in the transcription factor MITF a master regulator of melanocyte homeostasis, have been found in some melanomas is associated with the development of melanoma.

Potrony et al.[66] reported that the germline mutation, leading to amino acid change MITF E318K is associated with high nevi count and that it acts as a risk gene mutation for melanomas.

References

1. Schwartz PJ, Stramba-Badiale M, Crotti L, Pedrazzini M, Besana A, et al. Prevalence of the congenital long-QT syndrome. *Circulation* 2009;**120**(18):1761–7. https://doi.org/10.1161/CIRCULATIONAHA.109.863209. 19841298.
2. Brink PA, Schwartz PJ. Of founder populations, long QT syndrome, and destiny. *Heart Rhythm* 2009;**6**(11 Suppl):S25–33. https://doi.org/10.1016/j.hrthm.2009.08.036. 19880070.
3. Harmer SC, Tinker A. The impact of recent advances in genetics in understanding disease mechanisms underlying the long QT syndromes. *Biol Chem* 2016;**397**(7):679–93. https://doi.org/10.1515/hsz-2015-0306. 26910742.
4. Winbo A, Stattin EL, Westin IM, Norberg A, Persson J, et al. Sex is a moderator of the association between NOS1AP sequence variants and QTc in two long QT syndrome founder populations: a pedigree-based measured genotype association analysis. *BMC Med Genet* 2017;**18**(1):74. https://doi.org/10.1186/s12881-017-0435-2. 28720088.
5. Bondue A, Arbustini E, Bianco A, Ciccarelli M, Dawson D, et al. Complex roads from genotype to phenotype in dilated cardiomyopathy: scientific update from the working Group of Myocardial Function of the European Society of Cardiology. *Cardiovasc Res* 2018;**114**(10):1287–303. https://doi.org/10.1093/cvr/cvy122. 29800419.
6. Nishiuchi S, Makiyama T, Aiba T, Nakajima K, Hirose S, et al. Gene-based risk stratification for cardiac disorders in LMNA mutation carriers. *Circ Cardiovasc Genet* 2017;**10**(6). pii: e001603. https://doi.org/10.1161/CIRCGENETICS.116.001603. PMID:29237675.
7. McNally EM, Mestroni L. Dilated cardiomyopathy: genetic determinants and mechanisms. *Circ Res* 2017;**121**(7):731–48. https://doi.org/10.1161/CIRCRESAHA.116.309396. 28912180.

8. Dainis AM, Ashley EA. Cardiovascular precision medicine in the genomics era. *JACC Basic Transl Sci* 2018;**3**(2):313–26. https://doi.org/10.1016/j.jacbts.2018.01.003. eCollection 2018. [Review].

9. Knowles JW, Rader DJ, Khoury MJ. Cascade screening for familial hypercholesterolemia and the use of genetic testing. *JAMA* 2017;**318**(4):381–2. https://doi.org/10.1001/jama.2017.8543. 28742895.

10. Khera AV, Emdin CA, Drake I, Natarajan P, Bick AG, et al. Genetic risk, adherence to a healthy lifestyle, and coronary disease. *N Engl J Med* 2016;**375**(24):2349–58. 27959714.

11. Lewis CM, Vassos E. Prospects for using risk scores in polygenic medicine. *Genome Med* 2017;**9**(1):96. https://doi.org/10.1186/s13073-017-0489-y.PMID:29132412.

12. McCarthy MI, Mahajan A. The value of genetic risk scores in precision medicine for diabetes. *Expert Rev Precis Med Drug Dev* 2018;**3**(5):279–81. https://doi.org/10.1080/23808993.2018.1510732.

13. Maugeri A, Barchitta M, Mazzone MG, Giuliano F, Agodi A. Complement system and age-related macular degeneration: implications of gene-environment interaction for preventive and personalized medicine. *Biomed Res Int* 2018;**2018**:7532507. https://doi.org/10.1155/2018/7532507. eCollection 2018. Review, 30225264.

14. Thornton J, Edwards R, Mitchell P, Harrison RA, Buchan I, Kelly SP. Smoking and age-related macular degeneration: a review of association. *Eye (Lond)* 2005;**19**(9):935–44. 16151432.

15. Bowes Rickman C, Farsiu S, Toth CA, Klingeborn M. Dry age-related macular degeneration: mechanisms, therapeutic targets, and imaging. *Invest Ophthalmol Vis Sci* 2013;**54**(14). ORSF68–80. https://doi.org/10.1167/iovs.13-12757. 24335072.

16. Maugeri A, Barchitta M, Agodi A. The association between complement factor H rs1061170 polymorphism and age-related macular degeneration: a comprehensive meta-analysis stratified by stage of disease and ethnicity. *Acta Ophthalmol* 2019;**97**(1):e8–21. https://doi.org/10.1111/aos.13849. 30280493.

17. Cobos E, Recalde S, Anter J, Hernandez-Sanchez M, Barreales C, et al. Association between CFH, CFB, ARMS2, SERPINF1, VEGFR1 and VEGF polymorphisms and anatomical and functional response to ranibizumab treatment in neovascular age-related macular degeneration. *Acta Ophthalmol* 2018;**96**(2):e201–12. https://doi.org/10.1111/aos.13519. 28926193.

18. Morgan BP, Gasque P. Extrahepatic complement biosynthesis: Where, when and why? *Clin Exp Immunol* 1997;**107**(1):1–7. Review, 9010248.

19. Kawa MP, Machalinska A, Roginska D, Machalinski B. Complement system in pathogenesis of AMD: Dual player in degeneration and protection of retinal tissue. *J Immunol Res* 2014;**2014**:483960. https://doi.org/10.1155/2014/483960. 25276841.

20. Freudenberg-Hua Y, Li W, Davies P. The role of genetics in advancing precision medicine for Alzheimer's disease—a narrative review. *Front Med (Lausanne)* 2018;**5**(108). https://doi.org/10.3389/fmed.2018.00108. eCollection 2018, 29740579.

21. McKhann GM, Knopman DS, Chertkow H, Hyman BT, Jack Jr CR, et al. The diagnosis of dementia due to Alzheimer's disease: recommendations from the National Institute on Aging-Alzheimer's Association workgroups on diagnostic guidelines for Alzheimer's disease. *Alzheimers Dement* 2011;**7**(3):263–9. https://doi.org/10.1016/j.jalz.2011.03.005. 21514250.

22. Frisoni GB, Boccardi M, Barkhof F, Blennow K, Cappa S, et al. Strategic roadmap for an early diagnosis of Alzheimer's disease based on biomarkers. *Lancet Neurol* 2017;**16**(8)661–76. https://doi.org/10.1016/S1474-4422(17)30159-X. 28721928.

23. Lanoiselée HM, Nicolas G, Wallon D, Rovelet-Lecrux A, Lacour M, et al. APP, PSEN1, and PSEN2 mutations in early-onset Alzheimer disease: A genetic screening study of familial and sporadic cases. *PLoS Med* 2017;**14**(3). e1002270. https://doi.org/10.1371/journal.pmed.1002270. eCollection 2017, 2835080.

24. Corder EH, Saunders AM, Pericak-Vance MA, Roses AD. There is a pathologic relationship between ApoE-epsilon 4 and Alzheimer's disease. *Arch Neurol* 1995;**52**(7):650–1. 7619017.

25. Huang Y, Mahley RW. Apolipoprotein E: structure and function in lipid metabolism, neurobiology, and Alzheimer's diseases. *Neurobiol Dis* 2014;**72**(Pt A):3–12. https://doi.org/10.1016/j.nbd.2014.08.025. 25173806.

26. Strittmatter WJ, Weisgraber KH, Huang DY, Dong LM. Salvesen GS binding of human apolipoprotein E to synthetic amyloid beta peptide: Isoform-specific effects and implications for late-onset Alzheimer disease. *Proc Natl Acad Sci U S A* 1993;**90**(17):8098–102. 8367470.

27. Brandon JA, Farmer BC, Williams HC, Johnson LA. APOE and Alzheimer's disease: Neuroimaging of metabolic and cerebrovascular dysfunction. *Front Aging Neurosci* 2018;**10**:180. https://doi.org/10.3389/fnagi.2018.00180. eCollection 2018, 29962946.

28. Wadhwani AR, Affaneh A, Van Gulden S, Kessler JA. Neuronal apolipoprotein E4 increases cell death and phosphorylated tau release in Alzheimer disease. *Ann Neurol* 2019. https://doi.org/10.1002/ana.25455. 30840313.

29. Kunkle BW, Grenier-Boley B, Sims R, Bis JC, Damotte V, et al. Genetic meta-analysis of diagnosed Alzheimer's disease identifies new risk loci and implicates Aβ, tau, immunity and lipid processing. *Nat Genet* 2019;**51**(3):414–30. https://doi.org/10.1038/s41588-019-0358-2. Epub 2019 Feb 28, 30820047.

30. Jansen IE, Savage JE, Watanabe K, Bryois J, Williams DM, et al. Genome-wide meta-analysis identifies new loci and functional pathways influencing Alzheimer's disease risk. *Nat Genet* 2019;**51**(3):404–13. https://doi.org/10.1038/s41588-018-0311-9. 30617256.

31. Mathis S, Goizet C, Soulages A, Vallat JM, Masson GL. Genetics of amyotrophic lateral sclerosis: a review. *J Neurol Sci* 2019;**399**:217–26. https://doi.org/10.1016/j.jns.2019.02.030. 30870681.

32. Butti Z. Patten SA RNA dysregulation in amyotrophic lateral sclerosis. *Front Genet* 2019;**9**:712. https://doi.org/10.3389/fgene.2018.00712. eCollection 2018, 30723494.

33. Lagier-Tourenne C, Polymenidou M, Hutt KR, Vu AQ, Baughn M, et al. Divergent roles of ALS-linked proteins FUS/TLS and TDP-43 intersect in processing long pre-mRNAs. *Nat Neurosci* 2012;**15**(11):1488–97. https://doi.org/10.1038/nn.3230. 23023293.

34. Renton AE, Majounie E, Waite A, Simón-Sánchez J, Rollinson S, et al. A hexanucleotide repeat expansion in C9ORF72 is the cause of chromosome 9p21-linked ALS-FTD. *Neuron* 2011;**72**(2):257–68. https://doi.org/10.1016/j.neuron.2011.09.010. 21944779.

35. Zhang K, Coyne AN, Lloyd TE. Drosophila models of amyotrophic lateral sclerosis with defects in RNA metabolism. *Brain Res* 2018;**1693**(Pt A):109–20. https://doi.org/10.1016/j.brainres.2018.04.043. 29752901.

36. Yu B, Pamphlett R. Environmental insults: critical triggers for amyotrophic lateral sclerosis. *Transl Neurodegener* 2017;**6**:15. https://doi.org/10.1186/s40035-017-0087-3. eCollection 2017, 28638596.

37. Strafella C, Caputo V, Galota MR, Zampatti S, Marella G, et al. Application of precision medicine in neurodegenerative diseases. *Front Neurol* 2018;**9**:701. https://doi.org/10.3389/fneur.2018.00701. eCollection 2018. Review, 30190701.

38. Nussbaum RL, Polymeropoulos MH. Genetics of Parkinson's disease. *Hum Mol Genet* 1997;**6**(10):1687–91. Review, 9300660.

39. Hernandez DG, Reed X, Singleton AB. Genetics in Parkinson disease: Mendelian versus non-Mendelian inheritance. *J Neurochem* 2016;**139**(Suppl 1):59–74. https://doi.org/10.1111/jnc.13593. Review, 27090875.

40. Robak LA, Jansen IE, van Rooij J, Uitterlinden AG, Kraaij R, et al. Excessive burden of lysosomal storage disorder gene variants in Parkinson's disease. *Brain* 2017;**140**(12):3191–203. https://doi.org/10.1093/brain/awx285. 29140481.

41. Moors T, Paciotti S, Chiasserini D, Calabresi P, Parnetti L, et al. Lysosomal dysfunction and α-Synuclein aggregation in Parkinson's disease: diagnostic links. *Mov Disord* 2016;**31**(6):791–801. https://doi.org/10.1002/mds.26562. 26923732.

42. Braak H, Rüb U, Gai WP, Del Tredici K. Idiopathic Parkinson's disease: possible routes by which vulnerable neuronal types may be subject to neuroinvasion by an unknown pathogen. *J Neural Transm (Vienna)* 2003;**110**(5):517–36. Review, 12721813.

43. Chen H, Ritz B. The search for environmental causes of Parkinson's disease: moving forward. *J Parkinsons Dis* 2018;**8**(s1):S9–17. https://doi.org/10.3233/JPD-181493. 30584168.

44. L, Stewart W, Dams-O'Connor K, Diaz-Arrastia R, Horton L, et al.. The chronic and evolving neurological consequences of traumatic brain injury. *Lancet Neurol* 2017;**16**(10):813–25. https://doi.org/10.1016/S1474-4422(17)30279-X. Review, 28920887.

45. Brainstorm Consortium, Anttila V, Bulik-Sullivan B, Finucane HK, Walters RK, et al. Analysis of shared heritability in common disorders of the brain. *Science* 2018;**360**(6395). pii: eaap8757. https://doi.org/10.1126/science.aap8757.PMID:29930110.

46. Barnes JL, Zubair M, John K, Poirier MC, Martin FL. Carcinogens and DNA damage. *Biochem Soc Trans* 2018;**46**(5):1213–24. https://doi.org/10.1042/BST20180519. Review, 30287511.

47. Slavin TP, Niell-Swiller M, Solomon I, Nehoray B, Rybak C. Clinical application of multigene panels: Challenges of next-generation counseling and cancer risk management. *Front Oncol* 2015;**5**:208. https://doi.org/10.3389/fonc.2015.00208. eCollection 2015, 26484312.

48. Chen S, Parmigiani G. Meta-analysis of BRCA1 and BRCA2 penetrance. *J Clin Oncol* 2007;**25**(11):1329–33. 17416853.

49. Downs B, Sherman S, Cui J, Kim YC, Snyder C, et al. Common genetic variants contribute to incomplete penetrance: evidence from cancer-free BRCA1 mutation carriers. *Eur J Cancer* 2019;**107**:68–78. https://doi.org/10.1016/j.ejca.2018.10.022. 30551077.

50. Okur V, Chung WK. The impact of hereditary cancer gene panels on clinical care and lessons learned. *Cold Spring Harb Mol Case Stud* 2017;**3**(6). pii: a002154. https://doi.org/10.1101/mcs.a002154. Print 2017, 29162654.

51. Huang KL, Mashl RJ, Wu Y, Ritter DI, Wang J, Oh C, et al. Pathogenic germline variants in 10,389 adult cancers. *Cell* 2018;**173**(2). 355–370.e14. https://doi.org/10.1016/j.cell.2018.03.039. 29625052.

52. Warthin AS. Heredity with reference to carcinoma. *Arch Intern Med* 1913;**12**:546–55.

53. Lynch HT, Shaw MW, Magnuson CW, Larsen AL, Krush AJ. Hereditary factors in cancer. Study of two large midwestern kindreds. *Arch Intern Med* 1966;**117**(2):206–12. 5901552.

54. Aaltonen LA, Peltomäki P, Leach FS, Sistonen P, Pylkkänen L, et al. Clues to the pathogenesis of familial colorectal cancer. *Science* 1993;**260**(5109):812–6. 8484121.

55. Strand M, Prolla TA, Liskay RM, Petes TD. Destabilization of tracts of simple repetitive DNA in yeast by mutations affecting DNA mismatch repair. *Nature* 1993;**365**(6443):274–6. https://doi.org/10.1038/365274a0. 8371783.

56. Lynch HT, Snyder CL, Shaw TG, Heinen CD, Hitchins MP. Milestones of Lynch syndrome: 1895–2015. *Nat Rev Cancer* 2015;**15**(3):181–94. https://doi.org/10.1038/nrc3878. Review, 25673086.

57. Baretti M. Le DT DNA mismatch repair in cancer. *Pharmacol Ther* 2018;**189**: 45–62. https://doi.org/10.1016/j.pharmthera.2018.04.004. 29669262.

58. Yurgelun MB, Kulke MH, Fuchs CS, Allen BA, Uno H, et al. Cancer susceptibility gene mutations in individuals with colorectal cancer. *J Clin Oncol* 2017;**35**(10):1086–95. https://doi.org/10.1200/JCO.2016.71.0012. 28135145.

59. Gao R, Kong C, Huang L, Li H, Qu X, et al. Mucosa-associated microbiota signature in colorectal cancer. *Eur J Clin Microbiol Infect Dis* 2017;**36**(11):2073–83. https://doi.org/10.1007/s10096-017-3026-4. 28600626.

60. Saxena M. Yeretssian G NOD-like receptors: Master regulators of inflammation and Cancer. *Front Immunol* 2014;**5**:327. https://doi.org/10.3389/fimmu.2014.00327. 25071785.

61. Hyde SM, Cote GJ, Grubbs EG. Genetics of multiple endocrine neoplasia type 1/multiple endocrine neoplasia type 2 syndromes. *Endocrinol Metab Clin North Am* 2017;**46**(2):491–502. https://doi.org/10.1016/j.ecl.2017.01.011. Review, 28476233.

62. Niemeijer ND, Rijken JA, Eijkelenkamp K, van der Horst-Schrivers ANA, et al. The phenotype of SDHB germline mutation carriers: a nationwide study. *Eur J Endocrinol* 2017;**177**(2):115–25. https://doi.org/10.1530/EJE-17-0074. 28490599.

63. Hayward NK, Wilmott JS, Waddell N, Johansson PA, Field MA, et al. Whole-genome landscapes of major melanoma subtypes. *Nature* 2017;**545**(7653): 175–80. https://doi.org/10.1038/nature22071. 28467829.

64. Alexandrov LB, Nik-Zainal S, Wedge DC, Aparicio SA, Behjati S, et al. Signatures of mutational processes in human cancer. *Nature* 2013;**500**(7463):415–21. https://doi.org/10.1038/nature12477. 23945592.

65. Levy C, Khaled M, Fisher DE. MITF: master regulator of melanocyte development and melanoma oncogene. *Trends Mol Med* 2006;**12**(9):406–14. 16899407.

66. Potrony M, Puig-Butille JA, Aguilera P, Badenas C, Tell-Marti G, et al. Prevalence of MITF p.E318K in patients with melanoma independent of the presence of CDKN2A causative mutations. *JAMA Dermatol* 2016;**152**(4):405–12. https://doi.org/10.1001/jamadermatol.2015.4356. 26650189.

10

INTEGRATING GENETIC, EPIGENETIC AND ENVIRONMENTAL INFORMATION TO IMPROVE HEALTH AND WELL-BEING

Integrating information on genetic variants and environmental factors in analyses of disease risk factors

In a 2019 review Ogino et al.[1] emphasized the importance of taking genetic variant into consideration in epidemiological studies designed to identify disease risk factors. Key studies that illustrated this concept included studies on whether or not aspirin altered patient mortality in colon cancer. Molecular analyses revealed that growth of colon cancer tumors that expressed a specific enzyme prostaglandin-endoperoxide synthase 2 (PTGS2) was suppressed by aspirin. However, aspirin had no effects on the growth of tumors that did not express PTGS2. Further studies revealed that PTGS2 interacted with the phosphatidylinositol-4, 5-biphosphate kinase pathway. A specific gene in this pathway *PIK3CA* is often mutated in colon cancer. *PIK3CA* encodes phosphatidylinositol-4,5-bisphosphate 3-kinase catalytic subunit alpha. Studies revealed that regular aspirin use increased survival and possibly prevented *PIK3CA* mutation positive colon tumors.

Ogino et al. noted that immunotherapies are proving useful in the treatment of several forms of cancer. They suggested that studies of immunomodulators, including dietary, microbial factors and pharmacological factors should be undertaken in oncology. In a gene by environment interaction study Nan et al.[2] reported that a specific nucleotide variant, rs16973225 near the interleukin encoding gene *IL16* interacted with intake of non-steroidal anti-inflammatory agents including aspirin (NSAIDs) to influence colon cancer risk. In individuals with AC or CC genotypes at rs16973225 the risk of colon cancer was not altered by NSAID use.

Gene Environment Interactions. https://doi.org/10.1016/B978-0-12-819613-7.00010-4

Song et al.[3] reported that intake of the immune-modulator omega 3-polyunsaturated fatty acid can decrease colon cancer risk. Vitamin D has also been reported to have immunomodulatory effects that lowered cancer risk.[4]

Ogino et al.[1] also noted that links existed between the presence of specific intestinal microbiota and colon cancer. Castellarin et al.[5] reported that *Fusobacterium nucleatum* was often the prevalent in individuals with colorectal cancer. In 2013 Rubinstein et al.[6] reported that this organism expedited signaling in the E-cadherin/beta catenin pathway that promotes colorectal carcinogenesis.

Expression quantitative trait loci

These are genomic loci that can undergo variation and can impact the level of expression of specific genes.[7] These loci are frequently but not always, located in proximity to the genes they influence. The eQTLs that are located on the same chromosome as the gene they impact are referred to as cis eQTLs. The eQTLs that are on a different chromosome than the locus they influence are referred to as trans eQTLs.

Specific nucleotide changes in a specific eQTL can influence expression of a specific associated gene. Epigenetic mechanisms can also impact the eQTLs and their expression effects. The eQTLs that are altered by methylation are sometime referred to as methylation quantitative trait loci mQTLs.

Zhang et al.[8] identified a methylation quantitative trait locus. This mQTL significantly impacted expression of these genes within the genomic region of the alcohol dehydrogenase genes ADH1B and ADH1C. Altered expression of ADH1B and ADH1C was in part related to alcohol dependence.

Aspects of hypertension

The key parameters measured to assess blood pressure are systolic blood pressure, diastolic blood pressure and pulse pressure which is related to the difference between systolic and diastolic pressure measurements.

Dart and Kingwell[9] reviewed pulse pressure and its clinical significance. They noted that increased pulse pressure is attributed to increases arterial stiffness of the aorta and large vessels. Pulse pressure increases particularly during aging an I related to arterial elasticity. A number of studies revealed alteration in the collagen to elastin ration in vessel walls as an important factor in the pathogenesis of arterial stiffness.

Interesting observations reported by Dart and Kingwell include the fact that in patients with hypercholesterolemia there was a less steep increase in arterial stiffness with age.

However, the Framingham heart study[10] and the Boston Veterans' heart study[11] revealed that height of pulse pressure is a more accurate risk of subsequent cardiovascular events than both systolic and diastolic pressure.

Systolic hypertension

Precise optimal blood pressure levels in older individuals vary in different studies. The European Society of Cardiology[12] defined systolic hypertension as systolic blood pressure > 140 and diastolic blood pressure, 90 mm HG. A linear increase in systolic and diastolic blood pressure was reported to often occur before the 5th or 6th decade of life while after that the diastolic blood pressure was reported to decrease and the systolic blood pressure increased.

Bavishi et al. noted that there was evidence that with increasing age there was increased deposition of arterial calcium and collagen associated with fraying of arterial elastin.

Carotid artery intima and media thickness and carotid plaque

In atherosclerosis lipid rich material accumulates and elicits an inflammatory reaction in the sub-intimal space of medium and large arteries. Genome wide association studies (GWAS) have been carried out to explore genetic risk factors that impact this process. Franceshini et al.[13] carried out a study to identify genetic variants associated with abnormal carotid artery intima thickening and plaque formation (cImT). They also analyzed loci that impacted level of expression of risk loci (expression quantitative trait loci, eQTLs). It is interesting to note that their study revealed that specific eQTLs altered expression of two genes, one that encoded LOXL4 and another that encoded ADAMTS9. LOXL4 is a lysyl oxidase that is involved in cross-linking of elastin and collagen in the extracellular matrix and this cross lining influences the mechanical properties of arteries. ADAMTS9 is a metallo-proteinase that plays roles in angiogenesis and thrombosis.

Monogenic factors in hypertension

In a 2017 review of arterial hypertension Rossi et al.[14] noted that the two main determinants of blood pressure are cardiac output and peripheral vascular resistance. Cardiac output is primarily a function of cardiac contractility and peripheral resistance is a function of vascular tone. Blood pressure can also be impacted by renal, neural and endocrine factors.

In earlier decades evidence for the role of genetic factors in blood pressure determination was derived through twin studies that demonstrated that monozygotic twin pairs had more similar levels of systolic and diastolic blood pressure than did dizygotic twin pairs.[15]

There are examples of deleterious monogenic variants that significantly impact blood pressure. However, Rossi et al. emphasized that genetic predisposition to altered blood pressure is primarily due to polygenic risk factors, gene-gene interactions and environmental factors. Nevertheless, determination of specific genes impacted in monogenic forms of hypertension provides insights into the pathophysiology of hypertension.

Monogenic forms of endocrine related hypertension: Adrenocortical hormones

Hypertension can arise due to impaired adrenocortical hormone synthesis. Specific gene defects have been found to lead to Cushing syndrome a disorder that arises due to excess levels of cortisol that is produced by the adrenal cortex. Cushing syndrome is characterized by accelerated growth in children, excessive weight gain, hypertension, muscle weakness, osteoporosis and hyperglycemia.

Cushing syndrome can result from excess production of the pituitary hormone ACTH or impaired feedback control between adrenal cortex and pituitary. In a review of adrenocortical lesions leading to Cushing syndrome Lodish and Stratakis[16] noted that 75–90% of cases of non-ACTH related Cushing syndrome were associated with adenomas of the adrenal cortex. They emphasized that the cyclic AMP-pyruvate kinase A pathway (cAMP-PKA pathway) plays key roles in adrenocortical cell proliferation and function. Steps in this pathway include:

ACTH binding to the ACTH receptor a 5-member G-protein coupled receptor encoded by the *MC2R* gene.

Stimulation of adenyl cyclase and production of Cyclic AMP.

Cyclic AMP (cAMP) release.

Activation of cAMP depended protein kinase (PKA) that has 2 catalytic and 4 regulatory subunits.

PKA then activates cortisol synthesis.

A small percentage cases of excess ACTH synthesis are due to hereditary gene mutations in:

CDKN1B	Cyclin dependent kinase inhibitor 1B effective in multiple endocrine neoplasia MEN4
PRKAR1A	Protein kinase cAMP-dependent type I regulatory subunit alpha (Carney syndrome)
DICER1	Ribonuclease represses gene expression, normal product acts as tumor suppressor
USP8	Ubiquitin specific protease enhances EGFR expression transcription of ACTH

Marques and Karbonits[17] reported that the majority of pituitary adenomas appear sporadically and often have mutations in USP8 ubiquitin protease and in GNAS guanine nucleotide bind protein. They emphasized the importance of germline mutation that led to pituitary adenomas noted that these adenomas may sometimes produce pituitary hormones other than ACTH. In addition, in some cases the germline mutations were present in mosaic forms. In addition to the gene listed above Marques and Karbonits noted that the following genes may be mutated in cases of pituitary adenomas.

MEN1	Encodes menin, a nuclear scaffold protein that regulates gene transcription
AIP	Arylhydrocarbon receptor interaction protein

Cushing syndrome arising from excess adrenocortical function

Lodish and Strathakis[16] reviewed adrenocortical causes of Cushing syndrome. They emphasized that cyclic AMP protein kinase signaling play key roles in cortisol secretion by the adrenal gland. Dysregulated and aberrantly increased levels of cyclic Amp protein kinase play major roles in increased cortisol generation and secretion. They distinguished three pathologically distinct forms of adrenocortical adenomas and the genetic defects associated with each. The three forms included primary bilateral micronodular adrenal hyperplasia (PBMAH), adrenal adenomas and primary pigmented nodular adrenocortical disease (PPNAD). The gene defects associated with these three forms include defects in proteins kinase subunits PRKACA, PRKAR1, phosphodiesterases PDE11A, PDE8A, PDE8B, GNAS2 and ARMC5 armadillo repeat containing protein.

The ARMC5 proteins has been found to be have germline mutations that can lead to adrenocortical adenomas and somatic mutation occur in this protein in sporadic adrenocortical adenomas.

The phosphodiesterases PDE11A and PDE8B normally decrease levels of cyclic AMP. Inactivating mutation in these proteins lead to accumulation of cyclic AMP and to adrenocortical hyperplasia.

GNAS gain of function mutations occur in McCune Albright syndrome. These mutations lead to constitutive activation of adenyl cyclase.

Carney complex is an autosomal dominantly inherited disorder associated with endocrine tumors including adrenal lesions that lead to Cushing syndrome.

Lodish and Strathakis reported that mutation in PRKAR1A the regulatory subunit or protein kinase A and PRKARC the catalytic subunit of protein kinase A can lead to constitutive activation of the PKA pathway.

Other enzyme defects that may lead to Cushing syndrome occur in HSD11B1 hydroxysteroid 11-beta dehydrogenase 1.

Other key enzymes in the steroid pathway include CYP17A1 Mutations in this gene can lead to adrenal hyperplasia. Mutations in CYP11B1 this gene cause congenital adrenal hyperplasia due to 11-beta-hydroxylase deficiency.

Hypertension can also arise due to defects in enzymes involved in the synthesis of mineralocorticoids. One example is Gordon hypertension hyperkalemia syndrome.

Blood pressure abnormalities due to mutations in ion channels

Liddle syndrome represents salt sensitive hypertension and arises due to defects in epithelial sodium channels. At least 3 different forms of Liddle syndrome are known each due to defect in a sodium channel encoding gene SCNN1B, SCNN1G or SCNN1A.

Sympathetic and central nervous systems in hypertension

Fisher and Paton[18] reported that the most common form of hypertension was neurogenic high blood pressure with sympathetic nervous system over drive and excessive angiotensin activity. They noted that chronic sympathetic nervous system activity had diverse effect including influences on the mechanical propertied of arteries.

They emphasized that non-pharmacological intervention including physical exercise and lifestyle interventions can reduce sympathetic nervous system activity and blood pressure.

Johnson and Xue in a 2018 review[19] noted that within the brain hypertensive responses can be sensitized by specific stimuli. They noted that hypertensive response sensitization involves the sympathetic nervous system and they reviewed physiological and psychosocial stressors that that elicited the hypertensive response sensitization (HTRS).

Johnson and Xue noted that particular stimuli can induce long-term modification in neural networks. The sympathetic nervous system plays key roles in initiating corrections to alteration in physiological homeostasis. The classically analyzed sympathetic nervous activated by threat includes the flight or flee response. Central nervous system region involved in flight or flee responses were found to involve the hypothalamus, amygdala, dorsal tegmentum and the central gray matter.

Johnson and Xue noted that psychosocial stressors that do not necessarily disrupt homeostasis may nevertheless be perceived as imposing threat to psychological or physical integrity. Psychological stressors were found to particularly involve the limbic system, hippocampus and the prefrontal cortex. A number of investigators have hypothesized that sustained or repetitive activation of the sympathetic nervous systems and central nervous system mechanisms pay important roles in essential hypertension.

Johnson and Xue cited evidence that only 10% of the total hypertensive population in the world likely have underlying medical causes for hypertension and that overactivity of the sympathetic nervous system is considered by many investigators to be a major cause of hypertension.

Specific studies have revealed elevated sympathetic nervous system activity and elevated noradrenaline levels in individuals with essential hypertension. There is evidence that in some patients the extent of sympathetic overdrive increases progressively. In some patients, sympatholytic medications lowered blood pressure.

Physiological and psychosocial stressors neural and humoral impacts increased vascular resistance, increased cardiac output and increased renal filtration that increases blood volume. Component of the neural network that controls blood pressure includes lamina terminalis in the sub-fornical organ that was determined to be the target organ for angiotensin. The lamina terminalis occurs in the wall of the forebrain between the foramen of Monro and the optic nerve stalk. This region includes the sub-fornical organ SFO and the medial preoptic nucleus and the organum vasculosum. Structures in the lamina terminalis have connection to the paraventricular nucleus of the hypothalamus. The medial preoptic nucleus is likely involved in integrating information regarding fluid and blood pressure. The paraventricular nucleus is thought function in integrating information on cardiovascular fluid levels with the paraventricular nucleus that has extensions to the sympathetic neurons including those in the intermediary column in the spinal cord.

Johnson and Xue[19] noted that the neural network and sympathetic nervous system components that responds to psychosocial stress are less well defined. However, the amygdala is the key central structure that responds to stress and it has connection to the hippocampus. There are connections from the dorsomedial hypothalamus and the sympathetic neurons in the spinal cord.

Johnson and Xue concluded that a range of physiological and psychosocial challenge that can interact with hypertension response sensitization mechanism likely play roles in essential hypertension.

Renal factors involved in blood pressure control

The Renal Angiotensin Aldosterone system (RAAS) has been determined to constitute a major system controlling blood pressure.[20] Renin is a product of the juxta glomerular cells in the kidney and is released into the circulation Renin then acts on angiotensinogen that is produced primarily in the liver, but is also produced in other tissues including the kidney. Renin cleaves angiotensinogen at its N terminal end to form angiotensin I.

Subsequent cleavage of angiotensin 1 leads to the generation of angiotensin II. Cleavage of angiotensin 1 is carried out by the angiotensin converting enzyme ACE that is produced by vascular endothelial cells. Nishiyama and Kabori noted that other enzymes have also been reported to play roles in angiotensin cleavage.

The key function of angiotensin II (ANGII) is that binds to specific receptors in the adrenal gland AT 1 or AT2, AT1 receptor in humans is encoded by the *AGTR1* gene and the AT2 receptor is encoded by the *AGTR2* gene. Different responses occur depending on whether ANGII binding to AT1 or AT2. Binding of ANGII to AT1 receptor leads to the production of aldosterone. Collectively the different components of this system constitute the Renin angiotensin aldosterone system RAAS.

Angiotensin II acts as a vasoconstrictive protein. The aldosterone impacts the circulation volume since it increases re-absorption of sodium and water in the kidney. Aldosterone promotes urinary excretion of potassium.

A homologue of the angiotensin converting enzyme known as ACE2 leads to a different cleavage product of angiotensin. This cleave generates a heptapeptide that binds to a different receptor the MAS related G protein coupled receptor that plays roles in multiple processes including smooth muscle relaxation.

Nishiyama and Kobari[20] noted that angiotensin 11 reaches higher levels in the renal interstitial fluid than in the circulation. Inappropriate activation of angiotensin productions and of the RAAS system leads to hypertension and kidney damage.

Kidney disease and hypertension

Chronic kidney disease can lead to hypertension and primary hypertension can lead to kidney disease. Hamrahian and Falkner[21] reported that pathogenic mechanisms involved included sodium dysregulation, altered renin angiotensin aldosterone pathway activity and sympathetic nervous system activity.

Epigenetics and hypertension

Friso et al.[22] reviewed evidence for the role of epigenetic factors in altering the levels of expression of specific genes that encode products involved in determination of arterial blood pressure. They emphasized the dynamic interplay among genetic factors and environmental factors that impact biological pathways involved in blood pressure determination.

Friso et al. briefly reviewed molecular processes involved in epigenetics including DNA methylation that occurs primarily at cytosine-guanine (CpG) dinucleotides in DNA and involves addition of methyl group to cytosine. S-adenosyl methionine serves as the donor of methyl groups to DNA. The folate one carbon pathway is essential for S-adenosyl methionine generation.

In reviewing reports of epigenetic modifications related to hypertension, Friso noted evidence of altered methylation of specific genes based on studies of peripheral blood cell of hypertensive patients. Increased levels of methylation of 3 genes were reported:

HSD11B2 hydroxysteroid 11-beta dehydrogenase 2 involved in the conversion of cortisol to cortisone.

sACE catalyzes the conversion of angiotensin I into a physiologically active peptide angiotensin II.

Unusually low levels of methylation in hypertension were reported in ADD1 adducin 1 acts as a substrate for protein kinases A and C.

Friso et al. reviewed potential roles of non-protein coding RNAs in hypertension. They noted that specific microRNAs miR-143, miR 145 and miR 133 mir1 have been implicated in expression of genes in vascular muscle cells. High levels of expression of these microRNAs were found to be correlated with increased diastolic blood pressure.

High blood pressure predisposition, polygenic and environmental effects

In 2018 Pazoki et al.[23] studied 277,005 individuals aged 40–69 years and individuals were followed for between 6 and 11 years. Their study involved analyses of 314 loci with variants previously shown to be associated with hypertension risk. These loci included variants that had been found to influence pulse pressure, systolic or diastolic blood pressure.

In addition, Pazoki et al. collected data on lifestyle factors known to impact the risk of hypertension. These include body mass index, diet, sedentary life style. Smoking and salt intake as measured by urinary salt excretion.

Results of their study demonstrates that healthy life style most significantly influences blood pressure independently of underlying genetic risk significance was $p \times 10^{-320}$.

This study also revealed that healthy lifestyle significantly lowered the risk of cardiovascular disease.

Stroke

Stroke can be caused by lack of blood flow to a specific area of the brain or by intra-cerebral hemorrhage. Lack of flow to a specific area can arise due to embolism in an artery. It may be due to small blood vessel disease. In 2018 results of a genome-wide association study with multi-ancestry history of stroke were studied by the Mega Stroke Consortium. The study included analyses of single nucleotide polymorphisms (SNPs) and small insertion deletion polymorphisms (indels). This study led to identification of 32 genome wide significantly stroke associate loci, including 10 loci that had been identified in other studies. Significant differences in the specifically implicate loci were demonstrated between Asian and European population groups.

In addition to identification of loci significantly associated with stroke, the investigators carried out pathway and biosystem analyses to identify functional pathways of significant loci. The most significant pathway identified was in the coagulation system. Other significant pathways included the cardiac pathway, muscle fate pathway and nitric oxide metabolic pathway.

Studies were also carried out to determine if specific stroke associate loci in intergenic locations influenced expression of nearby genes, either as expression quantitative loci (eQTLs), as protein quantitative trait loci (pQTLs) or as methylation quantitative trait loci (mQTLs).

Meta-analyses revealed that significant genetic association was found with blood pressure. Eleven of the associated genes had functions not previously implicated in stroke pathology.

Malik et al.[24] reported results of a genome wide association study that involved study of individuals with European ancestry and data available from the UK biobank and MEGASTROKE consortium. The studies 72,147 stroke patients and 823,869 controls.

Key loci found to be significant in their study included an exon polymorphism in the nitric oxide synthase gene NOS3. This polymorphism designated rs1799983 leads to a substitution p. Glu298Asp Other significantly associated variants occurred in the COL4A1 gene and near in the DYRK1A gene.

Malik et al.[25] noted that nitric oxide signaling constitutes and important regulator of vascular tone. The COL4A1 protein is an important structural component of basement membrane in blood vessels. The variant they identified potentially interferes with the assembly or biological function of COL4A1. The DYRK1A gene encodes a dual specificity tyrosine phosphorylation regulated kinase and was reported to be involved in angiogenic responses in vascular endothelial cells. The DYRK1A gene is located on chromosome 21. DYRK1A is over-expressed in Down syndrome and the risk of stroke has been reported to be increased in Down syndrome.[26]

Polygenic factors life-style factors and stroke

Rutten-Jacobs et al.[27] reported results of a follow-up study 306,473 individuals aged 40–73 years with no previous history of stroke. The follow-up study included physical, medical and life-style information.

In addition, they derived a polygenic risk score for stroke based on previously established information on 90 polymorphic markers.

By the end of 7 years strokes had occurred in 2077 cases. The stroke risk was 35% higher individuals who had risk score in the upper-third of the polygenic risk score range Unfavorable life style led to a 66% increased risk of stroke.

Single gene defects, small vessel disease and stroke

Small vessel brain disease is an important cause of stroke. Verdura et al.[28] reported that the majority of cases of small vessel brain disease was related to hypertension. There are however families where small vessel brain disease occurs as an apparently autosomal dominant disease trait. Verdura identified a heterozygous variant HTRA1 p.R166L in affected members of a family with small vessel brain disease HTRA1, a serine proteinase that is a regulator of cell growth. They subsequently found the same mutation in 201 unrelated probands with small vessel brain disease.

Di Donato et al.[29] undertook *HTRA1* gene sequence analysis in 142 patients with cerebral small vessel disease who were negative for *NOTCH3* gene mutation. They identified 5 different HTRA1 mutations in nine patients from five different families. All variants were reported to alter highly conserved amino-acids and were absent from 320 chromosomes in matched controls. The mutations identified were reported to be pathogenic in in silico analyses with 4 and 5 of the different tolls used.

Di Donato et al. noted that their patients had clinical manifestations similar to those described in other patients with HTRA1 mutations. Clinical features included transient ischemic attacks, and cognitive decline beginning in the 6th decade of life. Magnetic resonance imaging of the brain revealed areas of high signal intensity in the periventricular area and in deep white matter with multiple lacunar infarcts in the external and internal capsule.

Cadasil

Cadasil is an acronym, for cerebral autosomal dominant arteriopathy with sub-cortical infracts and leukencephaly.[30] It was discovered to be due to pathogenic mutations in the *NOTCH3* gene. Ferrante et al.[31] reported that the predominant clinical manifestations include migraine headaches with aura, recurrent stokes of an ischemic nature, progressive white matter loss a memory loss, debilitating dementia and multiple psychiatric symptoms.

Ferrante et al. noted that histological studies in Cadasil patients led to identification of granular osmiophilic deposits near blood vessels.

Molecular genetic studies revealed that NOTCH3 mutation leading to Cadasil impacted one of the 34 extracellular domains of the NOTCH3 protein. Mutations usually added or removed a cysteine residue. Pathogenic NOTCH3 mutation lead to degeneration of smooth muscle cells in blood vessels. Ferrante et al., reported that >230 NOTCH3 mutations leading to Cadasil have been reported.

Analysis of normal NOTCH3 function reveals that is a receptor for growth factors. NOTCH3 signaling was reported in early life to be involves in blood vessel formation. In adult life NOTCH3 plays roles in vascular homeostasis.

Cerebral amyloid angiopathy

Carpenter et al.[32] carried out studies on genetic variants in APOE and their relationship to cerebral amyloid angiopathy. The *APOE* gene that maps to human chromosome 19q13.32 encodes lipoprotein E and there are 3 common allelic forms of APOE, A2, A3 and A4. The APOEA4 allele is a major risk factor for cerebral amyloid angiopathy, a condition where amyloid accumulates in the walls of medium sized cerebral arteries. This accumulation is known to weaken the blood vessels and to increase their liability to rupture.

Carpenter et al. also noted that that individuals with the APOE4 allele had higher levels of blood cholesterol than individuals with the APOE3 genotype. The APOE2 apparently also increased the risk of vascular changes. It is important to note that the presence of the APOE4 risk allele increased risk with an odds ratio of 2.2 and risk alleles in other genes increased with odds ratio <2.

Carpenter et al. also noted that variants that increase the risk for spontaneous intra-cerebral hemorrhage occurred in the genes that encoded the following proteins:

ACE	Angiotensin converting enzyme
COL4A	Collagen 4A
MTHFR	Methylenetetrahydrofolate reductase
PMF1	Polyamine modulated factor 1
SLC25A4	Solute carrier family 25 member 4, transports ADP from cytoplasm into mitochondria
TRHDE	Thyrotropin releasing hormone degrading enzyme

Coagulation and atherothrombosis

Diseases in which atherothrombosis occurs, include coronary artery disease and stroke. Olie et al.[33] reviewed the coagulation system, including platelets and coagulation proteins, that play roles in atherothrombosis.

They noted that although therapeutic methods in thrombotic conditions have primarily focused on platelets, anti-coagulant therapy continues to gain attention.

Ollie et al. traced the progression of atherothrombotic occlusion processes noting that in atherosclerosis there is disruption of endothelial cells, chronic inflammatory responses. These include binding of monocytes to endothelial lesion, transformation of monocytes to macrophages that then ingest lipoproteins and become foam cells. Processes induced by foam cells include proliferation of vascular smooth muscle cells and fibrin deposits.

Atherothrombotic events are initiated by disruption of atherosclerotic plaques, adherence of platelets to damaged sites and activation of platelets. Activation of platelets leads to release of factors including Von Willebrand factor and thromboxane 2. In addition, a specific transmembrane receptor binds to blood coagulation factor VII a and thrombin formation occurs. These leads to generation of fibrin from fibrinogen.

Ollie et al. noted that exposure of the transmembrane receptor TF also known as tissues factor is thought to be the key factor that triggers the coagulation cascade.

In subclinical atherosclerotic lesion increased intima media occurs (IMT) Studies have established a link between IMT and TF levels.

Coagulation tissue factor (TF) is encoded by the F3 genes and is sometime referred to as coagulation factor III. It is a cell surface glycoprotein. The binding of coagulation factor 11 to coagulation factor VIII a generates a process that initiates the coagulation protease cascade.

Ollie et al. noted that there is a duel pathway of interaction involving both coagulation factors and activation of the thrombin receptor referred to as PAR1. PAR1 is encoded by the *F2R* gene. The PARs are G-protein coupled receptors that play key roles in coagulation.

Thrombin induced changes in platelets including release of serotonin, thromboxane, chemokines and growth factors. Other pro-aggregation factors are also released from platelets. Thrombin was also noted to regulate blood vessel diameter.

Interaction between coagulation factors and platelet specific receptors and platelet factors therefore play key roles in thrombotic processes.

Morikawa et al.[34] reported that single nucleotide variant rs773902 in the PAR4 encoding gene F2RL3 leading to a missense mutation Ala120Thr was found to influence platelet aggregation in specific individuals in the Japanese population.

Morikawa et al.[34] carried out analyses of the rs773902 PAR alleles in stroke patients in the Stroke Genetic Networks Study. They reported a significant association of modest effect between the rs773902 allele and stroke.

Rannikmäe et al.[35] reported modest evidence for association of specific variants in COL4A1 and HTRA1 with lacunar ischemic stroke.

Environmental factors and stroke

Graber et al.[36] obtained evidence from 21 different publications that air pollution, and particularly toxic small particulate matter in polluted air contributed to the incidence of stroke.

Qian et al.[37] carried out analyses of fatal cases of intracerebral hemorrhagic stroke that occurred in Shanghai China during the period between June 2012 and May 2014. They also investigated air pollution levels during that period.

Their study revealed that the incidence of fatal intra-cerebral hemorrhage was significantly associated with the levels of particulate matter pM2.5 in polluted air.

Coronary heart disease (CAD)

Coronary heart disease represents a leading cause of death throughout the world. Comprehensive studies have been carried out over several decades to identify the key genetic and environmental factors involved in the pathogenesis of this disease. Specific clinical features of the disease include acute heart attacks, myocardial infarction but also atrial fibrillation and heart failure.

Musunuru and Kathiresan[38] reviewed progress in understanding the genetics of this disease and efforts to translate this information to the clinic. Genetic studies on CAD include studies on rare genetic variants and studies on common variants. They emphasized that the common predisposing variants have small effect sizes. Therefore, genome wide association studies to identify common variants requires large population sizes. Studies on common variants have been carried out on hundreds of thousands of people and have led to identification of >150 nucleotide variants associated with CAD that exceed statistical significance of $P = 4 \times 10^{-8}$.

Key follow up studies of these variants aim to determine the biological significance of the genetic variants and to determine how specific genetic variant impacts disease phenotype. Studies on the variants associated with heart disease indicate that they impact the following proteins or functions; low density lipoproteins, cholesterol, triglyceride rich lipoproteins, insulin resistance, thrombosis, inflammatory reactions, cell adhesion processes, cellular proliferation, vascular tone vascular remodeling, nitric oxide signaling (Fig. 1).

History, phenotype	Clinical studies	Laboratory studies	Follow-up, therapy
Patient/Family concerns	Study selection based on clinical manifestations may include:	Routine serum chemistry proteins, electrolytes	Family studies based on findings in patient, e.g., DNA sequencing, metabolite analysis. Functional analysis on defective enzyme or proteins, may involve studies on stem cells, model organisms.
Life History, to include: Pregancy, birth history post-natal history	Radiological studies	Special studies based on history and physical exam may include:	
	Cardiac studies		
Developmental history: Growth, developmental milestones	Neurological follow-up: EEG, muscle evaluation Hearing assessment	Genomic microarrays chromosome studies metabolomics. DNA sequencing, exome whole genome RNA analysis methylations studies	Design therapy to compensate for metabolic defect, to replace deficiency Consider specific gene therapy, e.g., antisense
Illnesses, infections exposures Morphological examination	Neuro-behavioral studies		

Fig. 1 Phenotype and genotype correlations adaptations and therapies. Important steps in clinical and laboratory evaluations of patients with diseases that are potentially genetically determined or genetically influenced are listed. In addition, follow-up guidance is presented including adaptations and possible therapies.

Musunuru and Kathiresan noted that coding gene variants that significantly impact CAD risk have been detected in genes that encode the following proteins:

LDLR	Low density lipoprotein receptor
APOA5	Apolipoprotein a5 plays an important role in regulating the plasma triglyceride levels
APOC3	Apolipoprotein C3, protein component of triglyceride (TG)-rich lipoproteins
LPA	Lipoprotein A, attaches to atherosclerotic lesions and promote thrombogenesis
PCSK9	Proprotein convertase subtilisin/kexin type 9, involved in cholesterol and fatty acid metabolism
ANGPTL4	Angiopoietin like 4, regulates glucose homeostasis, lipid metabolism, and insulin sensitivity
LPL	Lipoprotein lipase triglyceride hydrolase and in receptor-mediated lipoprotein uptake
ASGR1	Asialoglycoprotein receptor involved in degradation of glycoproteins

Heritability of coronary artery disease

This has been estimated to be between 40% and 50% and 38% of the heritability is determined to be due to common variants.

Musunuru and Kathiresan noted that the majority of large studies in coronary heart disease had been carried out in European populations. They stressed the importance of studies in other population groups. By 2016 81% of reported studies were carried out in European populations. Frequencies of specific pathogenic variants were found to be significantly higher in other populations. Examples presented by Musunuru and Kathiresan included specific PCSK9 variants Y142X and C679X that were more common in African populations than in European populations. In an Inuit population a specific variant CPT1A p479L was present in 70% of the population. This variant occurs with a frequency of 0.001% in Europeans. This variant impacts fatty acid metabolism.

They also note that the disease predictive power of polygenic risk scores also varied in different populations.

With respect to determining the relationship of risk variants to function, Musunuru and Kathiresan noted that the highest-ranking risk locus for coronary heart disease is on chromosome 9p21. However, it has not been possible to determine the functional mechanism through which this locus impacted coronary heart disease.

Treatment of broadly relevant causes and specific drivers of coronary heart disease

Musunuru and Kathiresan noted that elevated levels of low-density lipoprotein are a key contributor to coronary heart disease and that statin therapy that decreases LDL cholesterol decreases coronary artery disease risk. They emphasized that treating broadly relevant risk factors may be more productive than hyper-personalized treatment of specific drivers.

Polygenic risk scores

The numbers of SNP variants analyzed to determine polygenic risk scores for coronary heart disease has ranged from hundreds of variants to 6.6 million variants. Musunuru and Kathiresan concluded that polygenic risk score determination may offer complementary information in risk prediction, particularly when combined with other information used to predict risk.

Core genes and peripheral genes in complex common diseases

Liu and Pritchard[39] presented a model of the genetics of complex common diseases that included core genes and peripheral genes. They noted that core genes may be trans regulated by a number of different peripheral genes.

They noted that key observations in genome wide association studies include evidence that even loci that are calculated to be significantly associated with a particular disease have minimal effects on disease risk. Examples they included were studies on diabetes mellitus where 18 different loci yield highly significant GWAS scores but only explained 6% of the heritability of the disorder.

Another key observation noted was that for a particular condition the associated loci, were widely distributed in the genome.

Liu and Pritchard proposed the core genes encoded products that had direct cellular or organismal effect while peripheral genes had regulatory effects.

Modifiable risk factors in coronary heart disease

Pencina et al.[40] carried out an assessment to quantify the effects of modifiable risk factors in coronary heart disease. The risk factors they assessed included, systolic blood pressure, lipid levels. Diabetes mellitus and smoking.

Results of their analyses revealed that lowering systolic blood pressure to below 130 decreased risk by 28%. Lowering non-HDL cholesterol levels to below 130 mg/deciliter, (3.37 milllimoles per mL) decreased risk by 17%.

Hypercholesterolemia

Three genes have been implicated in autosomal dominant familial hypercholesterolemia, *LDLR, APOB* and *PCSK9;* Sharifi et al.[41] reported that 40% of patients with raised levels of LDL cholesterol had mutations in these genes. Polygenic inheritance is now reported to be responsible for abnormal LDL cholesterol levels in 60% or more of patient with abnormally high LDL cholesterol levels.

Polygenic risk studies and complex common diseases

Khera et al.[42] stressed the importance of identifying individuals who are at increased risk for serious common disorders so that preventive measures and treatment can be initiated. They noted that a number of common disorders have now been shown to be due to polygenic risk factors.

Khera et al. reported a system that analyzed alleles at >6 million single nucleotide polymorphic sites and results from these studies these were then entered into a polygenic predictor program.

Khera et al. emphasized that polygenic risk score determination differs from monogenic risk score determination. In the case of monogenic conditions sequence analyses of individual specific disease related genes are carried out. In polygenic risk score analyses relevant genotypes are assessed through microarray analyses and a number of different conditions can be simultaneously assessed.

Gene interactions

Even in the early days of Mendelian genetics, researchers including, William Bateson[43] and R.A. Fisher[44] noted that mutations in one gene may mask the effect of mutations in another gene. Such interactions were some referred to as epistatic interactions.

In 2019 Gifford et al.[45] reported and interesting example of rare gene interactions leading to cardiomyopathy in 3 children in a specific family. DNA sequencing analyses revealed that each affected child had inherited from the unaffected mother rare pathogenic mutations in two genes *MYH7* and *MKL2*. They had also inherited from the father pathogenic mutation in the gene *NKX2-5*. These genes encode the following products: MYH7 myosin heavy chain 7, MKL2 also known as MRTFB myocardin related transcription factor B, NKX2-5 a homeobox transcription factor involved in regulation of heart development.

Using gene editing techniques Gifford et al. generated these missense mutations in mice. Their studies revealed that if in the mouse all three of these gene mutations were present each in heterozygous form cardiomyopathy resulted. If each gene mutation was separately induced in a mouse cardiomyopathy did not result.

The specific histological features of the cardiomyopathy in the mice with all 3 mutations resembled the histological defect identified in heart tissue of a child from the family who died in consequence of the cardiomyopathy.

Gifford et al. concluded that NKX2-5 gene product modifies functional input of the MYH7 and MRTFB gene products.

Regulation of gene expression

Key aspects of biological functions include the ability of specific genomic elements or loci to influence the expression of specific target genes. Predominant genomic elements involved in regulation of gene expression include gene promoters, enhancers, inhibitors transcription factor binding sites and chromatin regulators.

Enhancer elements may be located close to or at some distance from the genes they regulate. They may occur in intergenic regions or they may be located in introns of genes. When active enhancers display specific modifications, these include specific histone modifications, acetylation and methylation marks, H3K27ac and H3K4me. Active enhancers and active regions of gene expression tend to be located within regions of open chromatin, where nucleosomes are not tightly packed.

In recent decades chromatin looping processes that bring enhancer genomic regions into close proximity to the promoters of the genes they regulate has been shown to be an important regulatory mechanism.[46] Specific proteins also facilitate the connections between enhancers and gene promoters. Among these are mediator, Cohesin and CTCF a protein that binds to a specific nucleotide sequence (CCCTC).

Analyses of chromatin looping and specific nucleotide sequences involved have been facilitated by development of new techniques designated as chromatin capture techniques.

Regulatory element variations

Vockley et al.[47] reviewed genomic regulatory elements that impact gene expression. These variations can involve single nucleotide variants. Some of these impact transcription factor binding sites. Variants can also include epigenetic modifications, variants that impact binding of RNA polymerase to promoter regions. In addition, structural genomic variations, deletion, insertions, translocations can alter the relationship between regulatory elements and the genes they control.

Vockley noted that efforts are underway to identify regulatory element in the genome and to establish the specific target genes for specific regulatory elements. It is important to note that particular regulatory elements may impact more than one target gene. A specific target gene may be impacted by a number of different regulatory elements.

Vockley noted that genome wide association studies contributed to identification of potential variants in regulatory elements in the non-protein coding genome. In vitro techniques, including reporter assays permit analyses of the effect of specific genomic elements in the regulation of target genes. In addition, gene editing techniques facilitate analyses of the effect of deletions of regulatory elements on gene expression.

Examples of studies that led to the identification of a specific regulatory element and its target gene include identification of a variant in a transcription factor binding site that impacts expression of the SORTL1 gene that impacts cholesterol metabolism.

Transcription analyses

Analyses of mRNA transcripts are increasingly being undertaken to identify the consequences of genomic defects that potentially lead to altered gene expression. Transcript analysis does requires access to the specific cell type in which the gene to be analyzed is expressed.

Detailed information on cell and tissue specific RMA expression is being documented through the ENCODE project www.encodeproject.org.

Transcriptome wide association studies

Gallagher and Chen-Plotkin[48] reviewed data on the correlation of gene expression and genome wide association study findings. They noted that most disease associated signals observed in GWAS mapped to the non-protein coding genome. Many of these signals were hypothesized to lie with regulatory loci and signals were sometimes designated as expression quantitative trait loci. They noted that there were examples of functional studies that elucidated the specific functional effects of specific GWAS loci. One example was that of Musunuru et al.[49] who discovered that a GWAS locus rs12740374 impacted the level of expression of the SORT1 gene and that levels of expression of SORT1 impacted levels of plasma lipids and lipoprotein and lipoprotein metabolism.

A specific variant located in an intron of the FTO locus was reported to be associated with increased tendency to obesity. Smemo et al.[50] reported that this variant impacted expression of a distant gene IRX3 that encodes a homeobox.

Gallagher and Chen-Plotkin noted that interaction of a variant in a regulatory locus with its target gene may involve long range interaction, e.g., chromatin looping between the two genomic regions. Such long-range interactions have been demonstrated in some cases.

There is evidence that expression of the fetal hemoglobin gene on human chromosome 11 is impacted by three different loci.[51] One of these locus occurs within the BCL11A gene locus on chromosome 2p16.1; the second locus occurs in an intergenic region between HBS11 and MYB on chromosome 6q23.3 and the third locus is within the beta globin gene locus.

Special investigations have also been carried out in some cases to determine if a specific GWAS associated allele alters binding of transcription factors. Wainberg et al.[52] reviewed aspects of transcriptome wide studies TWAS designed to integrate genome wide association studies with transcriptome analyses.

Genome wide studies in common metabolic diseases

Barroso and McCarthy[53] reviewed progress made in identifying common genomic variants that play roles in the causation of late onset metabolic diseases including type 2 diabetes, obesity and lipid metabolism as related to cardiovascular diseases.

They emphasized that accumulation of information from large datasets have revealed that heritability of these disorders is determined by allelic variants at large numbers of genomic loci and that many associated loci have very small effect sizes.

Mahajan et al.[54] reported that common allelic variants are responsible for approximately 20% of the heritability of diabetes mellitus. Extensive data from families indicated that the genetic contribution accounted for approximately 40% of the risk for diabetes mellitus.

Population differences occur in the specific alleles associated with diabetes mellitus. These differences involve rare alleles and common alleles. Some risk alleles that are more common in one population than in others. A specific risk allele rs7903146 in the transcription factor TCF7L2 encoding locus was found to be present in 30% of Europeans and in 2% of Asians.

Barroso and McCarthy noted that marked population differences were generally not found in diabetes mellitus associated loci.

They noted that specific clinical features of disease may display specific associations. Chasman et al.[55] identified 6 loci with variants that impacted kidney function.

Barroso and McCarthy noted evidence that only 5–10% of the risk variants that influence common metabolic disease map to coding sequence, the 90–95% of metabolic associated risk loci in the non-coding genome likely impact the transcription of coding gene. Furthermore the closest gene to the risk variant may not be the effector gene. This fact has been illustrated by the obesity risk locus FTO. The FTO risk locus impacts expression of IRX3 and IRX5 that are involved in adipocyte differentiation.

They also drew attention to the fact that the same gene may be dysregulated in more than one disease. It is also to bear in mind that regulation of expression of a gene may be controlled by different regulatory elements depending on the cell and tissue type.

Barroso and McCarthy noted that GWAS signals have some level of physiological specificity in that association signals for hyperlipidemia are connected with lipid metabolic pathways. Loci associated with obesity and body mass index are connected to genes involved in appetite control and genes involved in energy balance adipogenesis and insulin signaling.

Interactions between allelic variants and the environment

Some alleles may be advantageous under certain environmental conditions but be disadvantageous in other environments. Barroso and McCarthy noted that intense cold and high fat intake in Artic population likely account for the high frequency there of a specific allele at the carnitine palmitoyl transferase CPT1 locus that encodes a protein that functions in the importation of long chain fatty acids from cytoplasm to mitochondria where fatty acid oxidation takes place. The specific allele CPT1 p. Pro479Leu rs80356779 reaches frequencies of 40% in certain Eskimo, Aleutian and Siberian populations. The same allele occurs at very low frequencies in populations in temperate zones. This same allele was reported to be damaging in children living in temperate zone leading to hypo-ketotic hyperglycinemia.

Another example of the role of environments and specifically diet as a selective force occurs in the gene CREBF CREB3 regulatory factor where a specific allele rs373803828 p. Arg475Gln occurs in 50% of Samoans and is very rare in other populations.[56] This allele was reported to increase fat storage. An interesting interpretation for selection of this allele was put forward namely that this allele promoted fat storage that was of survival value in ancient ocean-going populations.

Barrosa and McCarthy also considered aspects of the internal environment including association between the gut microbiome and metabolic phenotypes. They noted evidence that the microbiome composition may be influenced by genetic variants in the host. There is evidence that lactase variants impact the gut microbiome composition.

They emphasized the importance of identification of potentially modifiable non-genetic factors that impact disease predisposition.

Barroso and McCarthy concluded that studies of genetic variants served to highlight the complexity of human biology

Rare genetic defects that increase the impact of specific environmental factors

Infectious agents

Parvovirus B19

Human Parvovirus B19 was first discovered in 1975 by Cossart[57] during the course of electron microscopic studies. In most individuals, infections with Parvo B19 leads to mild disease. In children it often leads to erythema reddening of checks and a rash on trunk and limbs. However, in individuals who have hematological conditions with a compensatory active bone marrow response Parvovirus B19 can lead to severe aplastic anemia.

Landry[58] reviewed Parvovirus infections and noted that during the acute infection stage this virus may occur in the nasopharynx, skin, synovia of joints and in the myocardium, Parvovirus B19 has also been found in erythroblasts, granulocytes, macrophages, dendritic cells, and in B and T lymphocytes. There is also evidence that the virus persists in tissue after infections and is then dormant if the host has mounted an active antibody response.

Parvovirus B19 was reported to be common in the population and antibodies to the virus were reported to be present in 40–60% of the population.

Landy noted that in individuals who have chronic hemolysis and compensatory reticulocytosis parvovirus infections can lead to dramatic fall in hematocrit and to aplastic anemia. Genetic conditions that lead to chronic hemolysis include mutations in globin genes, e.g., sickle cell disease and other hemoglobinopathies and a condition known as hereditary spherocytosis. Five different genetic defects have been found to give rise to spherocytosis 4 forms are due to mutations in specific red cell membrane proteins. Mutations occur in Spectrin A, Spectrin B, Ankyrin1, PA; Erythrocyte membrane protein band 4.2 an ATP-binding protein. One form of hereditary spherocytosis is due to defects in a specific solute carrier protein SLC4A1 that transports ions across the red cell membrane.

Cystic fibrosis

This is an autosomal recessive disorder due to pathogenic gene mutations in the cystic fibrosis transmembrane conductance regulator (CFTR). Many different pathogenic variants have been identified in this gene. There is evidence for phenotypic variability. In a 2016 review Brennan and Schrijver[59] noted that the phenotypic variability cannot totally be accounted for by different pathogenic CFTR variants. They emphasized that mutations in modifier genes, epigenetic factors and environmental factors impact the disease manifestation. There are marked population difference in cystic fibrosis disease frequency. The frequency in non-Hispanic whites in the USA is 1 in 2300; in Hispanic whites the frequency is 1 in 13,500; in African American the frequency is 1 in 15,000; in Asian American the frequency is 1 in 35,100.

Brennan and Schrijver noted that the CFTR protein has two membrane spanning domains that form a transporter pore for transport of chloride and bicarbonate and this also impacts sodium transport. Defects in chloride transport lead to high levels of chloride in secretion in sweat and other body fluids. The secretions and mucus in patients with cystic fibrosis tend to be viscous.

Clinical manifestations of CF can include impaired respiratory functions, bronchopulmonary disease, rhinosinusitis. Patients may also have abnormal metabolic features due to impaired pancreatic function.

The thickened accumulated bronchial secretions often lead to infections with different microorganisms. Granchelli et al.[60] reported that approximately 30,000 individuals in the USA have cystic fibrosis (CF) and that the life expectancy in CF is reduced by approximately 50%. Reduced life expectancy is due to declining lung function and infections. They noted that a number of different types of microorganisms occur in lung infections in CF patients.

Chief among these are *Staphylococcus aureus, Pseudomonas aeruginosa*, often these organisms manifest antibiotic resistance. In addition, a number of unusual organisms also lead to lung infections in CF patients. These include *Burkholderia cepacia, Stenotrophomonas malrophilia*, non tuberculosis type Mycobacteria and fungal organisms including *Aspergillus* and *Candida, Achromobacter xylosoxidans*.

Granchelli noted that infection with *Burkholderia cepacia* precipitates great decline in health. They noted further that the different infecting microorganisms apparently interact with each other.

Lack of diversity of organisms and predominance of a particular organism was often associated with severely compromised of lung function.

Adjusting the environment particularly nutrition to compensate for inborn errors of metabolism

Accurate diagnosis is key to management and treatment of inborn errors of metabolism. Earlier diagnostic testing for these disorders involved primarily chemical analyses of the levels of specific metabolites, protein studies enzyme studies and in some cases histopathological studies of affected cells and tissues.

Later, diagnostic methods involved the study of specific genes likely to be mutated in a particular disorder, through implementation of nucleotide sequence analysis through Sanger and/or implementation of polymerase chain reaction techniques to analyze nucleotide sequence in specific gene segments.

Wanders et al.[61] reviewed multidisciplinary approaches to diagnosis of inborn errors of metabolism that have taken place since the 1990s. They emphasized analyses of metabolites including acylcarnitine analyses that facilitate diagnosis of disorders of beta oxidation. Levels of specific forms of carnitine and detection of abnormal forms of acyl carnitine were facilitate by mass spectroscopy. Development of mass spectroscopy techniques also facilitated analyses of aminoacids and organic acids and facilitated analyses of metabolites in dried blood spots and the establishment of newborn screening programs.

Platforms were also developed for analysis of lipids, including neutral lipids, ceramides, sphingolipids and gangliosides that reach abnormal levels or have abnormal composition in several inborn errors of metabolism. Liquid chromatography and mass spectroscopy techniques facilitate lipid studies.[62]

Glycomics panels were designed to analyzed complex carbohydrates include O-glycans and N-glycans that are attached to proteins and lipids. Congenital abnormalities of glycosylation were identified by Jaeken.[63]

In 2018 Abu Bakar et al.[64] reported methods of glycan analysis. For N-glycan analysis glycans are released from substrates by enzyme treatment. For O-glycans chemical methods are used to release glycan from substrates to which they are bound. Chromatographic and electrophoretic techniques and ionization techniques are then used to analyze glycan composition.

Jaeken and Peanne[65] reported that combinations of glycomics analyses and gene analyses led to identification of 105 distinct glycosylation disorders. These disorders include primary and secondary glycosylation defects.

Metabolic studies

Studies of metabolites can include comprehensive metabolome analyses or studies on targeted sets of metabolites. Currently these studies are carried out on extra-cellular fluids, plasma, urine or cell culture medium. Wevers and Blau[66] reported that the intra-cellular metabolome has been largely unstudied but could potentially provide important information. They emphasized the importance of metabolomic, glycomics and lipidomics in analysis of unidentified diseases that likely represent inborn errors of metabolism.

Wevers and Blau noted that proteomic analyses continue to be important and that analysis of protein complexes is an important new development.

Sequence analysis

In the past two decades next generation sequencing techniques, including exome sequencing and whole genome sequencing have become more widely used for diagnosis and for confirmation of diagnoses made on the basis of other studies.

Rodenburg[67] noted that exome sequencing was particularly useful for accurate diagnosis in disorders where a defect in any one of a large number of different genes could potentially lead to the same clinical manifestations. Exome sequencing data can lead to identification of pathogenic nucleotide variants that likely represent the disease-causing mutation in a specific case. However, exome sequencing also reveals nucleotide variants that cannot be definitely classified as being pathogenic or benign and a referred to as variants of unknown significance (VUS).

Rodenburg emphasized the importance of establishing the functional consequences of variants that based on sequence information alone cannot be definitively clarified as pathogenic. Criteria used for establishing that a mutation as pathogenic depend on the specific nucleotide changes and on reported information of occurrence of the same nucleotide mutations in other patients with the same disease. Nucleotide changes including deletions or insertion that alter the reading frame of the gene transcripts, premature stop codon mutations and some missense variants that alter the amino acid sequence can potentially alter the function of the gene product.

Other important sequence features to consider when determining the pathogenic significance of a variants include determination as to whether that particular sequence change is rare in sequences derived from healthy human populations and if the variant occurs in sequence that is highly conserved evolutionarily in other vertebrate species.

Segregation of a particular gene sequence variant in all members of a family affected with a particular disorder and absence of that variant in unaffected family members is useful information but not definitive evidence of pathological significance of the variant.

It is important to note that exome sequencing analysis does not give information on all splicing defects or information on elements that regulate gene expression. Such information may be obtained from whole genome sequence information that includes information on introns of genes and information on sequence in intergenic regions. However, analysis of whole genome sequence requires intense application and information on altered nucleotide sequence can often not be readily assessed.

Rodenburg emphasized the importance of carrying out function studies to determine the likely connections between abnormal results obtained on sequencing studies. Methods to integrate results of DNA sequencing with results obtained for example in targeted metabolic studies or analyses of levels of functions of the particular protein encoded by the gene that bears the mutation, are particularly important.

Functional studies may in some cases involve the use of model organisms. Techniques exist to induce the same mutation in DNA of a model organism, e.g., yeast, *C. elegans*, mouse or in cultured human cells, to determine if the specific mutation actually alters gene product function and possible leads to definable phenotypic changes.

Model organisms and cultured cells potentially can be used to investigate the efficacy of specific therapeutic applications in modifying the deleterious functional or histopathological changes induced by the mutation.

Transcriptome studies

A number of investigators have emphasized that in a significant number of individuals with suspected genetically determined disorders exome sequencing studies have not led to successful accurate diagnoses. Cummings et al.[68] noted that this was the case in 50–75% of cases. They demonstrated the utility of transcriptome analyses including RNA sequencing.

Cummings et al. demonstrate the utility of transcriptome analyses in diagnoses of patients with rare muscle diseases. Their RNA studies revealed the underlying molecular defects that led to disease. The molecular defects included exon skipping, introduction of variants leading to novel splice sites in exons, in introns or in intergenic regions and nucleotide changes that eliminated splice sites.

Protein complex profiling

This involves separation of specific protein complexes e.g. through use of antibodies, chromatography, ultra-centrifugation, followed by tryptic digest of proteins to derive peptides that can then be separated to yield information on the composition of the protein complexes. Such studies have, for example yielded information on the aberrant sub-unit composition of mitochondrial complex 1.[69] Specific bioinformatic programs have been developed to investigate the molecular origin of peptides derived from digestions of protein complexes.

The components of the mitochondrial electron transfer systems are examples of multicomponent protein complexes. Guerrero-Castillo et al.[70] analyzed the assembly of the components of mitochondrial complex 1.

This complex is composed of 44 different sub-units each encoded by a different gene; 37 sub-units are encoded by nuclear genes and 7 components are encoded by mitochondrial genes. In addition, there are genes that encode proteins that act as assembly units. A number of additional proteins are also required to add prosthetic groups to the sub-units.

Inborn errors of metabolism due to defects in mitochondrial function

Maldonado et al.[71] defined primary mitochondrial disease as multigenic and multi-phenotypic disorders. They noted that even in patients with the same genetic defect that impacts mitochondrial function the clinical manifestations of the disease can differ.

Mitochondria carry out a variety of different functions including synthetic and metabolic functions, and maintenance of calcium balance. Metabolic functions include activities related to the tricarboxylic acid cycle, fatty acid beta oxidation. Their key function is oxidative phosphorylation and energy production through activity of the electron transport system that drives ATP synthesis.

Patients with mitochondrial disorders have functional impairments in multiple different organs, particularly organs with high energy demands, brain heart, muscle.

Simon et al.[72] noted that mitochondrial Complex 1 deficiency is the most common mitochondrial respiratory chain defect. They noted that complex 1 deficiency can be caused by defects in the structural units of complex 1 and it can be caused by defects in the complex 1 assembly factors.

Simon et al. described patients with mutation in the nuclear gene on chromosome 7q31.32 that encodes the mitochondrial assembly factor NDUFA5. They described three different disease-causing mutation in NDUFA5 and noted that patients in different families were compound heterozygotes for damaging mutations. They noted considerable heterogeneity in the clinical manifestations of patients with NDUFA5 mutations.

Emergence of the phenotype in a specific metabolic disease

Argmann et al.[73] noted that within the inborn errors of metabolism field, as in complex diseases, it is important to consider how the phenotype in an individual emerges, taking into account not only the primary mutation that is causative, but also the background genome in which that mutation occurs.

Argmann noted that although inborn errors of metabolism are primarily due to single gene mutations that follow Mendelian inheritance patterns there is evidence that in many cases there is not a clear correlation between the severity of the mutation and the disease phenotype.

Scriver and Waters[74] proposed that inborn errors of metabolism be viewed as complex traits. They noted further that "the whole organism phenotype is more than the sum of its parts."

A number of investigators have drawn attention to the importance of considering modifying factors in disorders classified as single gene disorders. The modifying factors can include variants in other genes, epigenetic factors the microbiome and the environment. There are however considerable difficulties inherent in identifying modifying genome changes.

Aspects of gene interactions and the existence of modifying variants in other genes also has relevance to the phenomenon of variable penetrance of a specific gene mutation. Variable penetrance is the term applied to the phenomenon when in a particular individual the specific pathogenic mutation does not have the same pathophysiological manifestations expected based on findings in other individuals with the same gene mutation.

Compensatory mechanisms that can reduce the impact of specific gene mutations

El-Brolosy and Stainier[75] noted evidence gained particularly from studies in model organisms. If one gene is rendered non-functional through mutations other genes in the same pathway or other genes with related function may be upregulated to compensate. One example presented was that mice in which the ribosomal gene Rp122 was non-functional the mice did not have defects because a paralogous gene was upregulated.

Important information on compensatory effects has emerged from very large population studies where individuals have been identified who have disease causing functional mutations often in one or both copies of a specific gene known to be responsible for certain genetic diseases. Chen et al.[76] described such findings in 13 individuals in studies of 589,306 genomes. The specific mutations identified would have been suspected to lead to the following conditions: cystic fibrosis, Smith Lemli-Opitz syndrome, Epidermolysis bullosa campomelic dysplasia, Pfeiffer syndrome.

Narasimhan et al.[77] also found cases of homozygous deleterious mutations in unaffected individuals in a large sequencing study in the UK.

Mitochondrial related disorders and compensatory mechanisms

Leber's Hereditary optic neuropathy (LHON) occurs as a result of homoplasmic point mutations in a specific segment of the mitochondrial genome that encodes a sub-unit of mitochondrial respiratory complex 1. Giordano et al.[78] reported that variable penetrance occurs so that individuals with the same mutation do not manifest LHON.

They investigated mitochondrial DNA copy number and mitochondrial mass in individuals from 3 pedigrees and in individual LHON cases. They determined that mutation bearing individuals who did not manifest disease had higher mitochondrial copy number and high mitochondrial DNA content indicating that they had a higher capacity for activating mitochondrial biogenesis.

Treatment of inborn errors of metabolism

A number of general principles apply to treatment of different inborn errors of metabolism.

Attention to environmental factors are key to managements, general recommendations are avoidance of fasting, aggressive treatment of illnesses and fever, and appropriate nutrient supplies with nutritional supplementation. Specific dietary recommendations differ in some of the inborn errors of metabolism.

The treatment of mitochondrial diseases remains primarily supportive with supplementation with vitamins and coenzyme Q. Important supplements include Thiamine (B1), Riboflavin B2 Nicotinamide B3, Folic acid B9, Carnitine.

Kuszak et al.[79] reported that Resveratrol was reported in model organisms to promote mitochondrial biogenesis. It is a natural phenol related substance that is produced by plants. Several pharmacological agents are in preclinical and early clinical trials to treat mitochondrial disorders.

Dietary recommendation for other forms of inborn errors are tailored to the specific disorder. Close collaboration of families with physicians and nutritionists, dieticians are important.

Newborn screening and inborn errors of metabolism

Newborn screening was designed to detect conditions which if left untreated in early life would greatly compromise health and development. Specific congenital conditions that impair metabolism of proteins, particular amino-acid, carbohydrates or fat metabolism were screened for, since alterations in dietary intake could improve chances for normal development and health. By 2010 24 such conditions were screened for in newborns in States in the USA.[80] Many newborn screening programs also screened for congenital hypothyroidism.

The follow-up care for infants who screen positive for any of the inborn errors of metabolism requires close interactions between parents, physicians and dieticians.

Analyzing utility of nutritional interventions

The National Institutes of Health USA set in place a specific initiative "Nutrition and Dietary Supplement Intervention for Inborn Errors of Metabolism." The goal of this initiative is to identify knowledge gaps and to analyze utility of nutritional intervention for specific Inborn Errors of Metabolism. Studies are needed to measure the outcomes of specific nutritional modifications.[81]

Interactions

Integrating gene expression and environmental nutrient conditions

Haro et al.[82] reviewed mechanisms that detect levels of specific nutrients and integrate responses through control of gene transcription and/or mRNA translation. Haro et al. noted that glucose homeostasis was impacted by hormone signaling. A key factor in glucose and fructose response included CHREBP carbohydrate response binding protein. In the presence of high levels of glucose or fructose CHREBP binds to a transcription factor encoded by *MLX* and this complex activates genes with carbohydrate response elements.

PPARs peroxisome proliferator response elements act as lipid sensors. They act as transcription factors that bind to specific sequence elements, nuclear receptor response elements. Binding of PPARs to nuclear response elements lead to activation of genes involved in lipid oxidation and in body fat synthesis.

Amino acid may be derived from nutrients or from breakdown of body proteins. Haro et al. emphasized importance of the GCN2 (EIF2AK4, eukaryotic translation factor) ATF 4 (activating transcription factor 4) pathway and FGF21 growth factor in sensing low levels of amino acids and slowing down mRNA translation and growth.

High levels of amino acids are sensed by and can activate MTOR pathway to stimulate protein and lipid synthesis while decreasing protein breakdown.

Gene product modifications

Product of different genes interact and products of a specific gene are modified through activities of proteins encoded by other genes. Mutations in genes that encode protein modifying factors may lead to defective function of protein product that they modify.

Interesting examples of this situation have been discovered during the course of investigations on the disease Osteogenesis Imperfecta (Brittle Bone Disease). Osteogenesis imperfecta most commonly arises due to mutations that lead to defects genes *COL1A1* or *COL1A2* that encode type I collagen. The *COL1A1* or *COL1A2* mutations lead to osteogenesis imperfecta with autosomal dominant inheritance patterns. However, there are cases of this disorder that are not found to have mutations in COL1A1 or in COL1A2 and cases that do not have evidence of dominant inheritance patterns.[83]

Detailed studies of collagen biosynthesis, its intra-cellular biology and secondary modification of the collagen type 1 protein have revealed that collagen proprotein and collagen protein undergoes a series of important protein modifications.

Ishikawa et al.[84] identified a specific protein complex located in the endoplasmic reticulum that is involved in hydroxylation of a specific amino acid proline in the collagen protein; this modification is essential to collagen type 1 functions. Analyses of the protein complex revealed that it is composed of 3 separate proteins each encoded by a separate gene. The proteins include CRTAP (cartilage related proteins) encoded by the CRTAP gene; P3H1 protein prolyl 3-hydroxylase 1 encoded by the *LEPRE*1 gene and peptidylprolyl isomerase B also known as cyclophilin encoded by the *PPIB* gene. The cyclophilin protein was reported to participate in a specific hydroxylation modification of type 1 collagen protein. This hydroxylation was shown to be essential for cross linking of collagen fibers.

It turns out the mutations in any of these 3 genes can lead to osteogenesis imperfecta with autosomal recessive inheritance. Patients with Osteogenesis imperfecta due to homozygous or compound heterozygous mutations in any one of these genes fail to show mutation in the *COL1A1* or *COL1A2* gene.

There are additional gene products that can have mutations that lead to autosomal recessive Osteogenesis imperfecta. Nineteen different types of Osteogenesis imperfecta are currently known. X-linked recessive form of OI have also been reported; it is due to defects in MBTPS2 is a membrane-embedded zinc metalloprotease.[85]

Mitochondrial functions and adaptations

Herst et al.[86] reviewed the key roles that mitochondria play in bioenergetic and biosynthetic pathways to adjust to different metabolic requirements and different conditions. They emphasized the key role of mitochondria-nuclear crosstalk in these processes. This cross talk includes stress signals generated by mitochondria and nuclear responses to these signals.

Key functions of mitochondria include energy production via ATP generation. However, mitochondria are also involved in biosynthesis of certain compounds and can play roles in facilitating apoptosis, tissue degradation, under certain conditions. Key to apoptotic processes involve activation of the mitochondrial permeability transition pore and release of pro-apoptotic molecules under certain conditions.

Under conditions when there is increased energy demand, mitochondrial biogenesis can be increased. Alternately under condition where energy use is low mitochondrial numbers can be reduced through mitophagy.

Herst et al. noted that recent studies have established that specific segments of the mitochondrial 16S and 12S RNA sequences encode small peptides that play roles in energy homeostasis. One such peptide humanin can increase mitochondrial oxygen consumption rate and ATP production and mitochondrial biogenesis. Humanin is sometimes referred to as a mitohormone.

Mitosignals that play roles in mitochondrial nuclear crosstalk include increased levels of reactive oxygen species, decreased ATP generation, altered NADH to NAD ratios and altered calcium levels.

Homeostasis cofactors and their relationship to vitamins

Specific vitamins contribute to the formation of cofactors that play important roles in metabolism and also act as cofactors for activity of enzymes involved in the generation of epigenetic changes that impact gene expression.[87] There are serious consequences of inadequate vitamin intake.

Vitamins essential for cofactor generation include Niacin (vitamin B3), Folate (vitamin B9), Pyridoxine (vitamin B6) Riboflavin B2, Cobalamin (vitamin B12).

Niacin that gives rise to nicotinamide adenine dinucleotide (NAD) and nicotinamide adenine dinucleotide phosphate (NADP) and flavin adenine dinucleotide (FAD) derived from the riboflavin form cofactors or coenzymes are essential for oxidation reduction reactions including those essential for metabolism and energy generations.

With respect to epigenetics and control of gene expression, NAD acts as a cofactor for histone deacetylase enzymes include HADACs and Sirtuins.

S-adenosylmethionine is essential in carrying out transfer of methyl groups to chromatin in processes of gene expression regulation. Key to generation of S-adenosyl-methionine is the one carbon pathways and tetrahydrofolate. Cofactors essential in this pathway include riboflavin (B2) folic acid (B9) pyridoxine (B6) and cobalamin (B12) and aminoacids methionine and homocysteine.

Vitamin C (ascorbic acid) provided by various plants plays roles as antioxidant and has been shown to be involved in appropriate function of tissue repair systems, immune response and for function of certain neurotransmitters.

Sunlight

On exposure to sunlight and ultra-violet rays UVB 7dehydrocholesterol in skin can be converted to provitamin D. This provitamin is subsequently converted to 25 hydroxy vitamin D and the biologically active form of vitamin D 1,25 dihydroxy vitamin D in tissues including liver and kidney.[88] Vitamin D can also be obtained through dietary supplementation It is however important to avoid excessively high serum levels of 25-hydroxyvitamin D that can occur with over-supplementation.

Holick[89] noted that information related to sunlight exposure and cancer has led to human avoiding sunlight. He noted however that in regions with very low sunlight or individuals exposed to very little sunlight the frequency of chronic diseases was high.

Vitamin D deficiency leads to bone defects.

Genetic disorders where individuals have increased sensitivity to sunlight and UVB

Two categories of disorders are particularly important in this context. One category includes disorders with increased frequency of DNA damage particularly due to defects in systems of DNA repairs. A second category includes disorders with defects in porphyrin metabolism and particularly the cutaneous porphyrias.

Disorders with increased sensitivity to sunlight due to defects in repair of DNA damage repair include Xeroderma pigmentosum, Bloom syndrome, Rothmund Thomsen syndrome, Cockayne syndrome and Trichothiodystrophy.

Defects in enzymes involved in excision repair, ERCC2, ERCC4 and ERCCC5 enzymes are particularly important in Xeroderma pigmentosum type conditions. ERRC6 and ERCC8 are defective in Cockayne ERCC2 defects occur. Defects in DNA helicase REQL4, RECQL3 are seen in Bloom syndrome and Rothmund Thomsen syndrome.

Giordano et al.[90] emphasized that strict photoprotection remained necessary in these conditions while other forms of therapy were being investigated.

Porphyrias

In cutaneous porphyrias sun exposure can lead to blistering Mutations and deficiency in uroporphyrinogen decarboxylase enzyme are known to cause familial porphyria cutanea tarda and hepato-erythropoetic porphyria. Variegate porphyria (VP) is caused by heterozygous mutation in the gene encoding protoporphyrinogen oxidase (PPOX). This condition is associated with sun sensitivity, blistering, fragile skin and scarring.

Other forms of porphyria can also be associated with increased sunlight sensitivity, though it is perhaps less marked than in the cutaneous porphyrias. These include porphyrias due to Ferrochelatase deficiency.[91]

Genes, networks, society and patients

Systems perspectives

Ahn et al.[92] noted that reductionist approach to disease is to identify the single factor that is most likely causative of disease. Identification of the single elements then becomes the target of treatment.

Clinical medicine is also designed to search for aspects of impaired homeostasis that are associated with disease and then to initiate treatment to restore homeostasis. Altered levels of specific hormones or electrolytes represent aspects of altered homeostasis.

Ahn et al. noted that with respect to treatments of disease it is incumbent on physicians to also aim to identify risk factors for disease and to attempt to mitigate effects of these before disease manifestations present.

Since multiple factors may contribute to disease causation designing treatment to address a single factor may be of limited value.

They note that in the systems approach many factors that contribute to disease dynamics can be approached. These include several entities, such as genes and their function and gene interactions, and interactions of gene products. Other aspects include how specific modifications are made internally in response to certain conditions. These modifications can include altered regulation, altered transcription altered mRNA splicing altered RNA translation.

Ahn et al. emphasized that systems biology approaches facilitate understanding of how properties arise from non-linear interaction of multiple components. They noted that in considering network interactions it is also important to consider changes in interactions over time.

Loscalzo and Barabasi[93] emphasized that the key goals of Medicine include knowledge of pathobiology, identification of mechanisms of pathobiology and translation of this knowledge to design of preventive and therapeutic strategies.

William Osler[94] emphasized the value of correlating clinical presentation and underlying pathological findings. Loscalzo and Barabasi note that though valuable this approach does not take into account preclinical factors, susceptibility and modifying factors.

In addition, they noted that understanding of key single underlying pathological features does not explain variability in disease manifestation in one patient at different times or variability in different patients with the same underlying disease predisposing defect. Development of therapeutic targets requires knowledge of different aspects of disease -causing homeostasis disruptions.

Loscalzo and Barabasi emphasized the importance of defining networks and related components. These include molecular networks and phenotypic networks. Relevant molecular networks include regulatory networks, protein interaction and metabolic networks and transcription factor networks. They defined phenotypic networks as conditions where two different genes lead to diseases that manifest similar phenotypic manifestations.

Furthermore, different diseases may involve defects in the same gene, different diseases may impact the same metabolic pathway, different diseases may share pathological features. In addition, environmental factors may impact gene expression and disease manifestations.

Genes networks, society and patients

Greene and Loscalzo[95] noted that personalized precision medicine continues to reduce a patient to molecular sequences. They emphasized that monogenic causes explain only a minority of disease and that it remains important to focus on interactions, between genes, between proteins and influence of internal and external environments.

They noted the discoveries of causes of infectious diseases promoted concepts of reductionism and the treatments of specific diseases based on identification of the causative agent. Nevertheless, even in the case of several infectious diseases the importance of focusing some attention on factors other than the disease-causing organism has been clearly demonstrated. Green and Loscalzo drew to studies that revealed the improvements in nutrition and living conditions played significant roles in disease reduction, e.g. in the incidence of Tuberculosis. There is also clear evidence that environmental conditions impact the disease manifestation and severity even in monogenic disease.

Greene and Loscalzo noted that Utility Network Science is limited by what is included and what is excluded from datasets. In addition to characteristics of genes, gene expression, understanding of social and political contexts remain important and that it is important to include social medicine concepts in network medicine. Social medicine concepts include, anthropology, history, sociology, philosophy and epidemiology.

Health disease and society

Sturmberg et al.[96] emphasized that health is dependent on physiological mechanisms, societal domain and adaptive measures. They proposed that biology provides the anatomical and physiological elements, the bottom-up blueprint while top down constraints may be imposed by environmental and socio-cultural conditions.

They emphasized the importance of considering health as emerging from internal networks and external physical and social environments and also from the capacity to adapt to changes. They also noted the importance of resources to respond to stress and to cope with life demands.

In considering networks and interactions, Sturmberg et al. emphasized the importance of the autonomic nervous system and the hypothalamic pituitary axis and bioenergetics that involve mitochondrial and metabolic pathways.

They also noted that concepts of health include subjective health and subjective illness and the presence or absence of objective disease. Associated with these concepts are values, for example whether life is considered worth living and whether life constitutes a constant struggle.

Sturmberg emphasized the important role of the hypothalamic pituitary axis in initiating responses to stress through release of neuropeptides, neurohormones and neurotransmitters that in turn can modulate gene expression and also impact metabolic and proteomic networks.

They opined that a reductionist healthcare system is poorly able to compete with complex diseases and emphasized the lack of attention to biological networks in the health care system. In addition, they noted that healthcare systems failed to address social, political and economic inequities.

In reflecting on the opinions above on healthcare it is perhaps also necessary to take note of the fact that practicing clinicians tend to treat those aspects of disease for which they have treatment. Correcting of social, political and economic inequities may not be within their realms of possibility. However, in considering integrated systems perhaps there may be ways to engage broadly with other professionals who have access to appropriate services and resources.

References

1. Ogino S, Nowak JA, Hamada T, Milner Jr DA, Nishihara R. Insights into pathogenic interactions among environment, host, and tumor at the crossroads of molecular pathology and epidemiology. *Annu Rev Pathol* 2019;**14**:83–103. https://doi.org/10.1146/annurev-pathmechdis-012418-012818. 30125150.
2. Nan H, Hutter CM, Lin Y, Jacobs EJ, Ulrich CM, et al. Association of aspirin and NSAID use with risk of colorectal cancer according to genetic variants. *JAMA* 2015;**313**(11):1133–42. https://doi.org/10.1001/jama.2015.1815. 25781442.

3. Song M, Nishihara R, Cao Y, Chun E, Qian ZR, et al. Marine ω-3 polyunsaturated fatty acid intake and risk of colorectal cancer characterized by tumor-infiltrating T cells. *JAMA Oncol* 2016;**2**(9):1197–206. https://doi.org/10.1001/jamaoncol.2016.0605. 27148825.

4. Dou R, Ng K, Giovannucci EL, Manson JE, Qian ZR, Ogino S. Vitamin D and colorectal cancer: molecular, epidemiological and clinical evidence. *Br J Nutr* 2016;**115**(9):1643–60. https://doi.org/10.1017/S0007114516000696. 27245104.

5. Castellarin M, Warren RL, Freeman JD, Dreolini L, Krzywinski M, et al. *Fusobacterium nucleatum* infection is prevalent in human colorectal carcinoma. *Genome Res* 2012;**22**(2):299–306. https://doi.org/10.1101/gr.126516.111. 22009989.

6. Rubinstein MR, Wang X, Liu W, Hao Y, Cai G, Han YW. *Fusobacterium nucleatum* promotes colorectal carcinogenesis by modulating E-cadherin/β-catenin signaling via its FadA adhesin. *Cell Host Microbe* 2013;**14**(2):195–206. https://doi.org/10.1016/j.chom.2013.07.012. 23954158.

7. Nica AC, Dermitzakis ET. Expression quantitative trait loci: present and future. *Philos Trans R Soc Lond B Biol Sci* 2013;**368**(1620):20120362. https://doi.org/10.1098/rstb.2012.0362. Print 2013. Review, 23650636.

8. Zhang H, Wang F, Kranzler HR, Yang C, Xu H, et al. Identification of methylation quantitative trait loci (mQTLs) influencing promoter DNA methylation of alcohol dependence risk genes. *Hum Genet* 2014;**133**(9):1093–104. https://doi.org/10.1007/s00439-014-1452-2. 24889829.

9. Dart AM, Kingwell BA. Pulse pressure—a review of mechanisms and clinical relevance. *J Am Coll Cardiol* 2001;**37**(4):975–84. Review, 11263624.

10. Torjesen A, Cooper LL, Rong J, Larson MG, Hamburg NM, et al. Relations of arterial stiffness with postural change in mean arterial pressure in middle-aged adults: the Framingham heart study. *Hypertension* 2017;**69**(4):685–90. https://doi.org/10.1161/HYPERTENSIONAHA.116.08116. 28264924.

11. Laskey WK, Wu J, Schulte PJ, Hernandez AF, Yancy CW, et al. Association of arterial pulse pressure with long-term clinical outcomes in patients with heart failure. *JACC Heart Fail* 2016;**4**(1):42–9. https://doi.org/10.1016/j.jchf.2015.09.012. 26656142.

12. Bavishi C, Goel S, Messerli FH. Isolated systolic hypertension: an update after SPRINT. *Am J Med* 2016;**129**(12):1251–8. https://doi.org/10.1016/j.amjmed.2016.08.032. Review. Erratum in: Am J Med. 2017;130(9):1128, 2763987.

13. Franceschini N, Giambartolomei C, de Vries PS, Finan C, Bis JC, et al. GWAS and colocalization analyses implicate carotid intima-media thickness and carotid plaque loci in cardiovascular outcomes. *Nat Commun* 2018;**9**(1):5141. https://doi.org/10.1038/s41467-018-07340-5. 30510157.

14. Rossi GP, Ceolotto G, Caroccia B, Lenzini L. Genetic screening in arterial hypertension. *Nat Rev Endocrinol* 2017;**13**(5):289–98. https://doi.org/10.1038/nrendo.2016.196. Review, 28059156.

15. Havlik RJ, Garrison RJ, Katz SH, Ellison RC, Feinleib M, Myrianthopoulos NC. Detection of genetic variance in blood pressure of seven-year-old twins. *Am J Epidemiol* 1979;**109**(5):512–6. https://doi.org/10.1093/oxfordjournals.aje.a112708. 572137.

16. Lodish M, Stratakis CA. A genetic and molecular update on adrenocortical causes of Cushing syndrome. *Nat Rev Endocrinol* 2016;**12**(5):255–62. https://doi.org/10.1038/nrendo.2016.24. 26965378.

17. Marques P, Korbonits M. Genetic aspects of pituitary adenomas. *Endocrinol Metab Clin North Am* 2017;**46**(2):335–74. https://doi.org/10.1016/j.ecl.2017.01.004. 28476226.

18. Fisher JP, Paton JF. The sympathetic nervous system and blood pressure in humans: implications for hypertension. *J Hum Hypertens* 2012;**26**(8):463–75. https://doi.org/10.1038/jhh.2011.66. 21734720.

19. Johnson AK, Xue B. Central nervous system neuroplasticity and the sensitization of hypertension. *Nat Rev Nephrol* 2018;**14**(12):750–66. https://doi.org/10.1038/s41581-018-0068-5. Review, 30337707.

20. Nishiyama A, Kobori H. Independent regulation of renin-angiotensin-aldosterone system in the kidney. *Clin Exp Nephrol* 2018;**22**(6):1231–9. https://doi.org/10.1007/s10157-018-1567-1. 29600408.

21. Hamrahian SM, Falkner B. Hypertension in chronic kidney disease. *Adv Exp Med Biol* 2017;**956**:307–25. https://doi.org/10.1007/5584_2016_84. 27873228.

22. Friso S, Carvajal CA, Fardella CE, Olivieri O. Epigenetics and arterial hypertension: the challenge of emerging evidence. *Transl Res* 2015;**165**(1):154–65. https://doi.org/10.1016/j.trsl.2014.06.007. 25035152.

23. Pazoki R, Dehghan A, Evangelou E, Warren H, Gao H, et al. Genetic predisposition to high blood pressure and lifestyle factors: associations with midlife blood pressure levels and cardiovascular events. *Circulation* 2018;**137**(7):653–61. https://doi.org/10.1161/CIRCULATIONAHA.117.030898. Erratum in: Circulation. 20198;139(2):e2, 29254930.

24. Malik R, Chauhan G, Traylor M, Sargurupremraj M, Okada Y. Multiancestry genome-wide association study of 520,000 subjects identifies 32 loci associated with stroke and stroke subtypes. *Nat Genet* 2018;**50**(4):524–37. https://doi.org/10.1038/s41588-018-0058-3. 29531354.

25. Malik R, Rannikmäe K, Traylor M, Georgakis MK, Sargurupremraj M, et al. MEGASTROKE consortium and the International Stroke Genetics Consortium. Genome-wide meta-analysis identifies 3 novel loci associated with stroke. *Ann Neurol* 2018;**84**(6):934–9. https://doi.org/10.1002/ana.25369. 30383316.

26. Sobey CG, Judkins CP, Sundararajan V, Phan TG, Drummond GR. Srikanth VK risk of major cardiovascular events in people with down syndrome. *PLoS ONE* 2015;**10**(9):e0137093. https://doi.org/10.1371/journal.pone.0137093. 26421620.

27. Rutten-Jacobs LC, Larsson SC, Malik R, Rannikmäe K, MEGASTROKE Consortium, International Stroke Genetics Consortium, et al. Genetic risk, incident stroke, and the benefits of adhering to a healthy lifestyle: cohort study of 306 473 UK Biobank participants. *BMJ* 2018;**363**:k4168. https://doi.org/10.1136/bmj.k4168. 30355576.

28. Verdura E, Hervé D, Scharrer E, Amador Mdel M, Guyant-Maréchal L, et al. Heterozygous HTRA1 mutations are associated with autosomal dominant cerebral small vessel disease. *Brain* 2015;**138**(Pt 8):2347–58. https://doi.org/10.1093/brain/awv155. 26063658.

29. Di Donato I, Bianchi S, Gallus GN, Cerase A, Taglia I, et al. Heterozygous mutations of HTRA1 gene in patients with familial cerebral small vessel disease. *CNS Neurosci Ther* 2017;**23**(9):759–65. https://doi.org/10.1111/cns.12722. 28782182.

30. Dichgans M, Mayer M, Uttner I, Brüning R, Müller-Höcker J, et al. The phenotypic spectrum of CADASIL: clinical findings in 102 cases. *Ann Neurol* 1998;**44**(5):731–9. 9818928.

31. Ferrante EA, Cudrici CD, Boehm M. CADASIL: new advances in basic science and clinical perspectives. *Curr Opin Hematol* 2019;**26**(3):193–8. https://doi.org/10.1097/MOH.0000000000000497. 30855338.

32. Carpenter AM, Singh IP, Gandhi CD, Prestigiacomo CJ. Genetic risk factors for spontaneous intracerebral haemorrhage. *Nat Rev Neurol* 2016;**12**(1):40–9. https://doi.org/10.1038/nrneurol.2015.226. 26670299.

33. Olie RH, van der Meijden PEJ, Ten Cate H. The coagulation system in atherothrombosis: implications for new therapeutic strategies. *Res Pract Thromb Haemost* 2018;**2**(2):188–98. https://doi.org/10.1002/rth2.12080. 30046721.

34. Morikawa Y, Kato H, Kashiwagi H, Nishiura N, Akuta K, et al. Protease-activated receptor-4 (PAR4) variant influences on platelet reactivity induced by PAR4-activating peptide through altered Ca^{2+} mobilization and ERK phosphorylation in healthy Japanese subjects. *Thromb Res* 2018;**162**:44–52. https://doi.org/10.1016/j.thromres.2017.12.014. 29289806.

35. Rannikmäe K, Sivakumaran V, Millar H, Malik R, Anderson CD, et al. COL4A2 is associated with lacunar ischemic stroke and deep ICH: meta-analyses among 21,500 cases and 40,600 controls. *Neurology* 2017;**89**(17):1829–39. https://doi.org/10.1212/WNL.0000000000004560. 25653287.

36. Graber M, Mohr S, Baptiste L, Duloquin G, Blanc-Labarre C, et al. Air pollution and stroke. A new modifiable risk factor is in the air. *Rev Neurol (Paris)* 2019. https://doi.org/10.1016/j.neurol.2019.03.003. 31153597.

37. Qian Y, Yu H, Cai B, Fang B, Wang C. Association between incidence of fatal intracerebral hemorrhagic stroke and fine particulate air pollution. *Environ Health Prev Med* 2019;**24**(1):38. https://doi.org/10.1186/s12199-019-0793-9. 31153356.

38. Musunuru K, Kathiresan S. Genetics of common, complex coronary artery disease. *Cell* 2019;**177**(1):132–45. https://doi.org/10.1016/j.cell.2019.02.015. 30901535.

39. Liu X, Li YI, Pritchard JK. Trans effects on gene expression can drive omnigenic inheritance. *Cell* 2019;**177**(4):1022–34. e6, https://doi.org/10.1016/j.cell.2019.04.01431051098.

40. Pencina MJ, Navar AM, Wojdyla D, Sanchez RJ, Khan I, et al. Quantifying importance of major risk factors for coronary heart disease. *Circulation* 2019;**139**(13):1603–11. https://doi.org/10.1161/CIRCULATIONAHA.117.031855. 30586759.

41. Sharifi M, Futema M, Nair D, Humphries SE. Genetic architecture of familial hypercholesterolaemia. *Curr Cardiol Rep* 2017;**19**(5):44. https://doi.org/10.1007/s11886-017-0848-8. 28405938.

42. Khera AV, Chaffin M, Aragam KG, Haas ME, Roselli C, et al. Genome-wide polygenic scores for common diseases identify individuals with risk equivalent to monogenic mutations. *Nat Genet* 2018;**50**(9):1219–24. https://doi.org/10.1038/s41588-018-0183-z. 30104762.

43. Bateson W. *Mendel's principles of heredity.* Cambridge: Cambridge University Press; 1909.

44. Fisher RA. The correlation between relatives on the supposition of Mendelian inheritance. *Trans R Soc Edinburgh* 1918;**52**(2):399–433. https://doi.org/10.1017/s0080456800012163.

45. Gifford CA, Ranade SS, Samarakoon R, Salunga HT, de Soysa TY, et al. Oligogenic inheritance of a human heart disease involving a genetic modifier. *Science* 2019;**364**(6443):865–70. https://doi.org/10.1126/science.aat5056. 31147515.

46. Pennacchio LA, Bickmore W, Dean A, Nobrega MA, Bejerano G. Enhancers: five essential questions. *Nat Rev Genet* 2013;**14**(4):288–95. https://doi.org/10.1038/nrg3458. 23503198.

47. Vockley CM, Barrera A, Reddy TE. Decoding the role of regulatory element polymorphisms in complex disease. *Curr Opin Genet Dev* 2017;**43**:38–45. https://doi.org/10.1016/j.gde.2016.10.007. 27984826.

48. Gallagher MD, Chen-Plotkin AS. The post-GWAS era: from association to function. *Am J Hum Genet* 2018;**102**(5):717–30. https://doi.org/10.1016/j.ajhg.2018.04.002. Review, 29727686.

49. Musunuru K, Strong A, Frank-Kamenetsky M, Lee NE, Ahfeldt T, et al. From noncoding variant to phenotype via SORT1 at the 1p13 cholesterol locus. *Nature* 2010;**466**(7307):714–9. https://doi.org/10.1038/nature09266. 20686566.

50. Smemo S, Tena JJ, Kim KH, Gamazon ER, Sakabe NJ, et al. Obesity-associated variants within FTO form long-range functional connections with IRX3. *Nature* 2014;**507**(7492):371–5. https://doi.org/10.1038/nature13138. 24646999.

51. Galarneau G, Palmer CD, Sankaran VG, Orkin SH, Hirschhorn JN, Lettre G. Fine-mapping at three loci known to affect fetal hemoglobin levels explains additional genetic variation. *Nat Genet* 2010;**42**(12):1049–51. https://doi.org/10.1038/ng.707. 21057501.

52. Wainberg M, Sinnott-Armstrong N, Mancuso N, Barbeira AN, Knowles DA, et al. Opportunities and challenges for transcriptome-wide association studies. *Nat Genet* 2019;**51**(4):592–9. https://doi.org/10.1038/s41588-019-0385-z. 30926968.

53. Barroso I, McCarthy MI. The genetic basis of metabolic disease. *Cell* 2019;**177**(1):146–61. https://doi.org/10.1016/j.cell.2019.02.024. 30901536.

54. Mahajan A, Taliun D, Thurner M, Robertson NR, Torres JM, et al. Fine-mapping type 2 diabetes loci to single-variant resolution using high-density imputation and islet-specific epigenome maps. *Heart* 2019;**105**(2):144–51. https://doi.org/10.1136/heartjnl-2017-312932. 30242141.

55. Chasman DI, Fuchsberger C, Pattaro C, Teumer A, Böger CA, et al. Integration of genome-wide association studies with biological knowledge identifies six novel genes related to kidney function. *Hum Mol Genet* 2012;**21**(24):5329–43. https://doi.org/10.1093/hmg/dds369. 22962313.

56. Loos RJ. CREBRF variant increases obesity risk and protects against diabetes in Samoans. *Nat Genet* 2016;**48**(9):976–8. https://doi.org/10.1038/ng.3653. 27573685.

57. Cossart YE, Field AM, Cant B, Widdows D. Parvovirus-like particles in human sera. *Lancet* 1975;**1**(7898):72–3. https://doi.org/10.1016/s0140-6736(75)91074-0. 46024.

58. Landry ML. Parvovirus B19. *Microbiol Spectr* 2016;**4**(3). https://doi.org/10.1128/microbiolspec.DMIH2-0008-2015. 27337440.

59. Brennan ML, Schrijver I. Cystic fibrosis: a review of associated phenotypes, use of molecular diagnostic approaches, genetic characteristics, progress, and dilemmas. *J Mol Diagn* 2016;**18**(1):3–14. https://doi.org/10.1016/j.jmoldx.2015.06.010. Review, 26631874.

60. Granchelli AM, Adler FR, Keogh RH, Kartsonaki C, Cox DR, Liou TG. Microbial interactions in the cystic fibrosis airway. *J Clin Microbiol* 2018;**56**(8):e00354-18. https://doi.org/10.1128/JCM.00354-18. 29769279.

61. Wanders RJA, Vaz FM, Ferdinandusse S, van Kuilenburg ABP, Kemp S, et al. Translational metabolism: a multidisciplinary approach towards precision diagnosis of inborn errors of metabolism in the omics era. *J Inherit Metab Dis* 2019;**42**(2):197–208. https://doi.org/10.1002/jimd.12008. Review, 30723938.

62. Contrepois K, Mahmoudi S, Ubhi BK, Papsdorf K, Hornburg D. Cross-platform comparison of untargeted and targeted lipidomics approaches on aging mouse plasma. *Sci Rep* 2018;**8**(1):17747. https://doi.org/10.1038/s41598-018-35807-4.

63. Jaeken J. Congenital disorders of glycosylation. *Handb Clin Neurol* 2013;**113**:1737–43. https://doi.org/10.1016/B978-0-444-59565-2.00044-7. Review, 23622397.

64. Abu Bakar N, Lefeber DJ, van Scherpenzeel M. Clinical glycomics for the diagnosis of congenital disorders of glycosylation. *J Inherit Metab Dis* 2018;**41**(3):499–513. https://doi.org/10.1007/s10545-018-0144-9. 29497882.

65. Jaeken J, Péanne R. What is new in CDG? *J Inherit Metab Dis* 2017;**40**(4):569–86. https://doi.org/10.1007/s10545-017-0050-6. Review. Erratum in: J Inherit Metab Dis. 2017, 28484880.

66. Wevers RA, Blau N. Think big—think omics. *J Inherit Metab Dis* 2018;**41**(3):281–3. https://doi.org/10.1007/s10545-018-0165-4. 29541953.

67. Rodenburg RJ. The functional genomics laboratory: functional validation of genetic variants. *J Inherit Metab Dis* 2018;**41**(3):297–307. https://doi.org/10.1007/s10545-018-0146-7. 29445992.

68. Cummings BB, Marshall JL, Tukiainen T, Lek M, Donkervoort S, et al. Improving genetic diagnosis in Mendelian disease with transcriptome sequencing. *Sci Transl Med* 2017;**9**(386):eaal5209. https://doi.org/10.1126/scitranslmed.aal5209. 28424332.

69. Alston CL, Compton AG, Formosa LE, Strecker V, Oláhová M, et al. Biallelic mutations in TMEM126B cause severe complex I deficiency with a variable clinical phenotype. *Am J Hum Genet* 2016;**99**(1):217–27. https://doi.org/10.1016/j.ajhg.2016.05.021. 27374774.

70. Guerrero-Castillo S, Baertling F, Kownatzki D, Wessels HJ, Arnold S, et al. The assembly pathway of mitochondrial respiratory chain complex I. *Cell Metab* 2017;**25**(1):128–39. https://doi.org/10.1016/j.cmet.2016.09.002. 27720676.

71. Maldonado EM, Taha F, Rahman J, Rahman S. Systems biology approaches toward understanding primary mitochondrial diseases. *Front Genet* 2019;**10**:19. https://doi.org/10.3389/fgene.2019.00019. eCollection 2019, 30774647.

72. Simon MT, Eftekharian SS, Stover AE, Osborne AF, Braffman BH, et al. Novel mutations in the mitochondrial complex I assembly gene NDUFAF5 reveal heterogeneous phenotypes. *Mol Genet Metab* 2019;**126**(1):53–63. https://doi.org/10.1016/j.ymgme.2018.11.001. 30473481.

73. Argmann CA, Houten SM, Zhu J, Schadt EE. A next generation multiscale view of inborn errors of metabolism. *Cell Metab* 2016;**23**(1):13–26. https://doi.org/10.1016/j.cmet.2015.11.012. Review, 26712461.

74. Scriver CR, Waters PJ. Monogenic traits are not simple: lessons from phenylketonuria. *Trends Genet* 1999;**15**(7):267–72. Review, 10390625.

75. El-Brolosy MA, DYR S. Genetic compensation: a phenomenon in search of mechanisms. *PLoS Genet* 2017;**13**(7):e1006780. https://doi.org/10.1371/journal.pgen.1006780. eCollection 2017 July, 28704371.

76. Chen R, Shi L, Hakenberg J, Naughton B, Sklar P, et al. Analysis of 589,306 genomes identifies individuals resilient to severe Mendelian childhood diseases. *Nat Biotechnol* 2016;**34**(5):531–8. https://doi.org/10.1038/nbt.3514. 27065010.

77. Narasimhan VM, Hunt KA, Mason D, Baker CL, Karczewski KJ, et al. Health and population effects of rare gene knockouts in adult humans with related parents. *Science* 2016;**352**(6284):474–7. https://doi.org/10.1126/science.aac8624. 26940866.

78. Giordano C, Iommarini L, Giordano L, Maresca A, Pisano A, et al. Efficient mitochondrial biogenesis drives incomplete penetrance in Leber's hereditary optic neuropathy. *Brain* 2014;**137**(Pt 2):335–53. https://doi.org/10.1093/brain/awt343. 24369379.

79. Kuszak AJ, Espey MG, Falk MJ, Holmbeck MA, Manfredi G, et al. Nutritional interventions for mitochondrial OXPHOS deficiencies: mechanisms and model systems. *Annu Rev Pathol* 2018;**13**:163–91. https://doi.org/10.1146/annurev-pathol-020117-043644. 3. 29099651.

80. Therrell Jr BL, Lloyd-Puryear MA, Camp KM, Mann MY. Inborn errors of metabolism identified via newborn screening: ten-year incidence data and costs of nutritional interventions for research agenda planning. *Mol Genet Metab* 2014;**113**(1–2):14–26. https://doi.org/10.1016/j.ymgme.2014.07.009. 25085281.

81. Camp KM, Lloyd-Puryear MA, Yao L, Groft SC, Parisi MA, et al. Expanding research to provide an evidence base for nutritional interventions for the management of inborn errors of metabolism. *Mol Genet Metab* 2013;**109**(4):319–28. https://doi.org/10.1016/j.ymgme.2013.05.008. 23806236.

82. Haro D, Marrero PF, Relat J. Nutritional regulation of gene expression: carbohydrate-, fat- and amino acid-dependent modulation of transcriptional activity. *Int J Mol Sci* 2019;**20**(6):E1386. https://doi.org/10.3390/ijms20061386. 30893897.

83. Forlino A, Marini JC. Osteogenesis imperfecta. *Lancet* 2016;**387**(10028):1657–71. https://doi.org/10.1016/S0140-6736(15)00728-X. Epub 2015 Nov 3. Review, 26542481.

84. Ishikawa Y, Bächinger HP. A molecular ensemble in the rER for procollagen maturation. *Biochim Biophys Acta* 2013;**1833**(11):2479–91. https://doi.org/10.1016/j.bbamcr.2013.04.008. Review, 23602968.

85. Lindert U, Cabral WA, Ausavarat S, Tongkobpetch S, Ludin K, et al. MBTPS2 mutations cause defective regulated intramembrane proteolysis in X-linked osteogenesis imperfecta. *Nat Commun* 2016;**7**:11920. https://doi.org/10.1038/ncomms11920. 27380894.

86. Herst PM, Rowe MR, Carson GM, Berridge MV. Functional mitochondria in health and disease. *Front Endocrinol (Lausanne)* 2017;**8**:296. https://doi.org/10.3389/fendo.2017.00296. eCollection 2017. Review, 29163365.

87. Rabhi N, Hannou SA, Froguel P, Annicotte JS. Cofactors as metabolic sensors driving cell adaptation in physiology and disease. *Front Endocrinol (Lausanne)* 2017;**8**:304. https://doi.org/10.3389/fendo.2017.00304. eCollection 2017, 29163371.

88. Wacker M, Holick MF. Sunlight and vitamin D: a global perspective for health. *Dermatoendocrinology* 2013;**5**(1):51–108. https://doi.org/10.4161/derm.24494. 24494042.

89. Holick MF. Biological effects of sunlight, ultraviolet radiation, visible light, infrared radiation and vitamin D for health. *Anticancer Res* 2016;**36**(3):1345–56. Review, 26977036.

90. Giordano CN, Yew YW, Spivak G, Lim HW. Understanding photodermatoses associated with defective DNA repair: syndromes with cancer predisposition. *J Am Acad Dermatol* 2016;**75**(5):855–70. https://doi.org/10.1016/j.jaad.2016.03.045. Review, 27745641.

91. Yasuda M, Chen B, Desnick RJ. Recent advances on porphyria genetics: inheritance, penetrance & molecular heterogeneity, including new modifying/causative genes. *Mol Genet Metab* 2018. https://doi.org/10.1016/j.ymgme.2018.11.012. Review, 30594473.

92. Ahn AC, Tewari M, Poon CS, Phillips RS. The clinical applications of a systems approach. *PLoS Med* 2006;**3**(7):e209. 16683861.

93. Loscalzo J, Barabasi AL. Systems biology and the future of medicine. *Wiley Interdiscip Rev Syst Biol Med* 2011;**3**(6):619–27. https://doi.org/10.1002/wsbm.144. 21928407.

94. Osler W. *The principles and practice of medicine.* New York: Appleton; 1982.

95. Greene JA, Loscalzo J. Putting the patient Back together—social medicine, network medicine, and the limits of reductionism. *N Engl J Med* 2017;**377**(25):2493–9. https://doi.org/10.1056/NEJMms1706744. 29262277.

96. Sturmberg JP, Picard M, Aron DC, Bennett JM, Bircher J, et al. Health and disease-emergent states resulting from adaptive social and biological network interactions. *Front Med (Lausanne)* 2019;**6**:59. https://doi.org/10.3389/fmed.2019.00059. eCollection 2019, 30984762.

11

ENVIRONMENTS, RESOURCES, AND HEALTH

Planetary health is the health of human civilization and the state of the natural systems on which it depends.

Rockefeller Foundation Lancet Commission report.[1]

Green-house gas production and climate change problems and potential solutions

A 2017 report from the Environmental Protection agency in the USA[2] indicated that total world-wide emissions into the atmosphere on 2017 total 6547 million metric tons of CO_2 equivalent. The EPA report documented that 28.9% of the emissions were generated from transportation, 27.5% from electricity generations; 22.2% from industry, 12% from environmental and residential use and 9% from agriculture.

It is slightly encouraging to note that in 2017 emissions were lower than rates reported between 1997 and 2007. The decrease was due to slight decreases in use of fossil fuels and to slight increases in renewable energy sources.

Carbon Data from NASA (https://climate.Nasa.gov accessed 3 July 2019)[3] indicate that carbon dioxide levels in the atmosphere rose from approximately 310 ppm in 1950 to 400 ppm today. The capacity of carbon dioxide to trap heat has been proven several times since the 1860's, Tyndall,[4] Arrhenius.[5]

The NASA report[3] also noted that the warmest years on planet earth have occurred since 2010. Warming temperatures impact not only ground surfaces but also oceans, high temperatures lead to ice sheet melting and to rises in sea levels. High CO_2 levels also lead to ocean acidification.

There is also evidence that the rise in earth temperature is associated with widespread changes in weather patterns. These include higher than normal precipitation in some areas and drought in other areas. There are also increases in intense whether events including tropical storms, hurricanes and cyclones.

Min et al.[6] reported on the role of high green-house gas concentration in leading to intensification of heavy precipitation events on the

Gene Environment Interactions. https://doi.org/10.1016/B978-0-12-819613-7.00011-6

333

Northern Hemisphere in the late twentieth century. Perlwitz[7] reported studies that linked green-house gases and global warming with displacement of the North Atlantic air jet stream. This displacement was linked to heavy rains and flooding in England and Wales in 2000. Pall et al.[8] noted that rainfall changes had to do with warming and higher levels of water vapor in the high atmosphere leading to increased rainfall in some areas.

Mohtadi et al.[9] described changes in tropical zonal atmospheric circulation often referred to as Walker circulation and the effects of shifts in these circulation patterns that lead to droughts in some areas and extreme rainfall and flooding in other areas. The Walker circulation refers to surface westerly flow along the equatorial Indian ocean and eastern flow along the equatorial Pacific Ocean. They documented the sensitivity of tropical circulation patterns to earth and atmospheric temperature changes.

Ozone

The ozone layer in the atmosphere absorbs damaging ultra-violet radiation from the sun. The composition of the ozone layer is determined by specific processes both physical and biological that take place in the oceans and on land. These chemicals are then transported via atmospheric circulation patterns. To the upper atmosphere (stratosphere). In January 2018 NASA[10] reported that satellite observations revealed the first evidence that ozone depletion was declining due to the international ban on chlorfluorocarbons.

Countries with highest energy use and percentage of energy generated from renewable resources

Energy use as reported in the REN 2016[11] report is listed below.

Country	Total energy used gigawatts per hour	% of energy from renewables
China	1,522,585	24.5
USA	637,076	14.7
Brazil	465,579	80
Canada	433,597	65
Germany	188,342	29
Russia	184,171	16.9
Norway	145,378	92.2
Spain	104,639	38.1
United Kingdom	99,330	27.9
Sweden	89,127	57

Renewable energy resources

It is important to note that these can include wood, plant biomass, wind, solar and water sources, hydro and tidal energy and geothermal energy.

Rockefeller Foundation Lancet Commission on planetary health report 2015 (RFLC)

Recommendations included design of a framework to identify key biological and physical process related to Earth system function that require careful monitoring (see Ref. 12). The RFLC report emphasized monitoring of climate changes, stratospheric ozone depletion, atmospheric aerosol loading, ocean acidification, biogeochemical flow, fresh water use, land system changes, biosphere integrity.

Measurements in 2015 indicated that ocean acidification has increased by 26% since the beginning of the Industrial Revolution. This acidification reduces calcium carbonate saturation that is required for corrals mussels and clams to grow shells. The RFLC referred to reports of impaired larval development, impaired growth of ocean animals.

Freshwater includes water in rivers and streams, surface water, and ground water including water in deep aquifers. The RFLC noted that a 55% increase in water demand is projected to occur between 2000 and 2050. In addition to projected impacts on water supplies due to climate change with severe droughts in some areas and flooding on other areas a growing problem that emerged is pollution of surface water through increased run-off from agriculture and other human activities of water with high levels of nitrogen and phosphorus. Excessive nitrogen and phosphorus then accumulate in wetlands and rivers and even in the ocean.

Land clearance and intensive farming have contributed to soil degradation. The RFLC report noted that chemical and pesticides used in agriculture, domestic settings, and in medical facilities impact the ecosystem and human health. There is increasing evidence of endocrine disruptions in wild life, in marine organisms and possibly also in humans. The RFLC report noted that approximately 800 different chemicals have been reported to be potentially involved in exerting endocrine disruptions.

The RFLC report noted that key drivers of environmental change in the current era include population growth, over-consumption and technology.

The International Energy Agency monitors coal use. In the USA and Europe coal use declined between 2000 and 2007. However, in China coal use greatly increased between 2000 and 2007 https://www.iea.org/topics/coal/.[13]

Documenting the scope and depth of problems

Clearly it is important to document the scope and depth of problems and to find solutions. The scientific literature is replete with problem documentation and perhaps relatively deficient in identifying potential solutions.

Effects of climate change on waterways and marine ecosystems

Extreme heat led to death of freshwater fish in the Murray Darling river system in Australia in January 2019. Tonkin et al.[14] also reported that thousands of dead fish were found in the Rhine river system following a heatwave in 2017. They also noted that management of waterflow and river bank ecology is particularly important in areas where rivers have been damned.

Jonkers et al.[15] noted that climate change has significantly impacted the distribution of marine zoo-plankton.

Loss of protection in coastal regions

In an article published in National Geographic magazine in July 2019 Schwartzstein[16] documented evidence that rising ocean waters and logging have led to loss of Mangrove forests that previously flourished in the Sundarbans region along the coastline of the Bay of Bengal. The Mangrove loss, and higher seas increased flooding hazards that have led to migration of more than one million people from this region.

Loss of Behring sea ice and Arctic temperatures

In May 2019 Cornwall[17] reported that during 2018 and in early 2019 ice cover was absent in most of the Behring sea. In addition, changes in zooplankton algae and fish populations were documented.

In 2018 a report from the US National Oceanic and Atmospheric Administration noted that in the year between October 2017 and September 2018 temperatures in the Arctic reached record highs. The numbers of wild reindeer and caribou declined by 50%.[18]

Retreat of the Antarctic ice sheet

In June 2019 Steig[19] reported clear evidence of rise in atmospheric CO_2 leading to loss of ice in Antarctica and the retreat there of glaciers and increased release of ice from glaciers. Release of ice from the Thwaite Glacier was reported to have doubled in recent years The Amundsen glacier was also shown to have diminished.

Invasive insects

In a 2018 report Wild et al.[20] noted that world-wide invasive species are spreading and constitute threats to biodiversity. This spreading is likely to be due in part to climate change that facilitates movements of species into new ecosystems. An example of this is invasion of a new ant species *Linepithema humile* in South Africa. This ant species disrupts seed dispersal from certain indigenous South African plants.

Downgrading of biodiversity protection areas

Golden Kroner et al.[21] reported that between 1892 and 2018 governments of 73 countries in the world enacted legislation to establish areas that were protected from development. These areas constituted 15% of global lands and 7.3% of oceans. The goal of these endeavors was to protect biodiversity and promote conservation.

Golden Kroner et al. documented legal changes in recent years that have led to down-grading and down-sizing of protected areas. In recent years, 269 areas were downgraded in the USA and 440 areas were downgraded in Amazonian countries. In the USA 24 protective area downgrades were associated with industrial resource extraction and development. Examples of such events included approval for oil and gas extraction in the Arctic National Wildlife Refuge and approval for mining in Joshua Tree National Park in California.

Climate change, questions relating to sustainability and human health

In 2006 McMichael et al.[22] reviewed aspects of climate change and human health. They noted the connection of climate change to anthropogenic increases in greenhouse gases and noted the longevity of green-house gases. Climate change will then continue for periods even after the generation of additional greenhouse gases discontinued.

McMichael et al. noted that models of climate change must consider impacts on land surfaces, oceans and seas and ice. Important physical parameters measured included carbon dioxide, sulfate and non-sulfate aerosols, the carbon cycle, atmospheric chemistry and vegetation dynamics.

Sea and ocean surface temperatures are relevant to the generation of storms and cyclones and parameters referred to as thermoclines that are related to rainfall.

McMichael et al. summarized main pathways by which climate change affects population. Among these are:

Increased frequency of extreme weather events and their effects on ecosystems.

Sea-level rise with salination of coastal water, and fresh-water streams.

Storm surges and related environmental degradation and impact on fisheries.

Major health effects could result from:

Thermal stress, death from flood, cyclones or bush fires,

Microbial proliferation, changes in host vector relationship.

Impaired yields from life-stock, crops and fisheries.

Loss of employment, displacement, poverty, malnutrition, adverse physical and mental health.

McMichael et al. noted that extreme weather events can stress populations beyond their adaptation capabilities. There is evidence of increased mortality in severe weather conditions, particularly among the elderly and in individuals with pre-existing cardio-vascular or respiratory conditions. Excess heat leading to dehydration may be hazardous in children and in individuals at any age. In inner cities with low ventilation extreme heat is more hazardous than in rural environments.

As climate change escalates, floods and storm surges intensify. In addition to direct damage there are other hazards. These include sewage and animal wastes accumulating in drinking water. Another important consideration is that excessive rainfall can cause increased flow of toxic pollutants, e.g., mercury into streams rivers, oceans and fish.

McMichael et al. noted that the El Nino South Oscillation (ENSO) has led to increased frequencies of Malaria in South Eastern Asia and to increased incidence of Dengue fever in South America.

Indirect effects of climate change on human health include alterations in food supplies, conflicts, migrations and refugee situations.

Climate change and human well-being

Inka Weissbesker[23] edited a book that was based on topics presented at a United Nations convened conference for non-governmental organizations. The title of the conference was "Climate Change: How it impacts us all". The title of the book was "Climate Change and Human Well-being," Contributors to chapters in this book particularly emphasized the importance of including concerns about mental health and psychological well-being, and social justice issues in the climate change agenda.

Attendees at this conference and many other have emphasized that climate change most adversely affects individuals and populations who have not contributed to the causes of climate change.

Weissbecker detailed physical aspects of the parameters of climate change. The key cause of climate changes is known to be due to the rise in levels of greenhouse gasses generated through burning of fossil fuels and industrialization. Key components of greenhouse gasses include carbon-dioxide CO_2, methane, nitrous oxide and aerosols. Concentrations of greenhouse gasses have increased to be 31% higher than concentrations measured in the late eighteenth century. Along with the increase in greenhouse gases the average temperatures have steadily risen since 1906.

Concepts of tipping points have also been introduced into climate change assessment. The tipping point has been defined as a critical threshold that can dramatically alter the state of a system.

Weissbecker noted that a number of studies have determined that climate change will particularly impact individuals with lower economic resources and lower adaptive capabilities. Furthermore, it will exacerbate social and political inequities. Weissbecker noted that increasing questions of social justice and ethical issues will need to be addressed. Psychologists and mental health professionals will increasingly be required to promote adaptations to change. Displacement and refugee status bring about significant stresses and the skills and services of psychologists and mental health professionals will be required.

Weissbecker stressed that discussion and collaboration among people of different backgrounds and cultures will be necessary to address problems related to climate change.

In a chapter in the "Climate Change and Human Well-being" Simpson et al.[24] examined biobehavioral aspects of disaster preparedness and responses. In a chapter that dealt in part with capacity building, Weissbecker and Czincz[25] emphasized the importance of availability of mental health intervention during humanitarian crises. Reser et al.[26] addressed the impact of perceptions of threat in addition to stresses induced by the physical environmental impacts of climate change.

Questions of environmental state, sustainability human interactions and human health

In a comprehensive assessment published in 2015 the Rockefeller Foundation Lancet Commission on planetary health (see Ref. 12) detailed how changes in the function of earth's natural systems represent a growing threat to human health. Examples of key changes in the earth's natural include climate change, fresh-water depletion, land degradation, over-fishing, ocean acidification, biodiversity loss.

The committee also documented that global health of humans has apparently been improving even during the decades during which the changes in earth's natural systems documented above have occurred. Writers of the report concluded that in order to bring about these changes the current generation "have been mortgaging the health of future generations."

Key challenges to be addressed that the Commission report emphasized included over reliance on gross-domestic product as a measure of human progress. They wrote, "Policies should balance social progress, environmental sustainability and the economy."

The Commission also noted the need for more transdisciplinary research and funding and also more rapid responses to problems on the part of governments and institutions. They stressed the importance of promoting sustainable levels of consumption, more equitable distribution of resources and harnessing innovative technological approaches to problems.

Important data regarding human status from 2015, as presented by the Rockefeller foundation Lancet Commission (RFLC) report were the fact that overall life expectancy worldwide increased from 47 years between 1950 and 1955 to 69 years between 2005 and 2010. Furthermore, the death rate in childhood had fallen dramatically and extreme deprivation from poverty was greatly decreased.

The RFLC report documented information from the 2005 Millennium Ecosystem assessment that included estimations that 20% of the earth's ecosystem services showed evidence of degradation or unsustainable use.

Specific examples of natural environment degradation included degradation of mangroves, coral reefs and wetlands that provide protection from storm surges.

In addition to considering human health it is important that there be growing emphasis placed on planetary health. The RFLC developed a clear definition of planetary health. "Planetary health is the health of human civilization and the state of the natural systems on which it depends."

With respect to human health issues the report noted that food insecurity could increase as a result of impaired water resources and through temperature changes due to climate change. Food insecurity could also result from unsustainable levels of fishing, also from loss of biodiversity including reduction of pollinators of plants. The impact of increasing air pollution on human health must also be considered. Air pollution derives from combustion of fossil fuels. In some countries it also derives from burning of wood. With rising temperatures landscape fires have increased in frequency. Such fires lead to emission of air with high levels of particulate matter.

The RFLC report also noted the impacts of climate and environmental change on human mental health and behaviors. A new term, sostalgia, was coined to describe the mental distress caused by environmental changes. Drought periods were reported to be associated with increased anxiety, depression and suicidality.

Increased frequencies of certain vector borne disease are considered to be related to loss of forests and to rising temperatures leading to increases in the range of certain vectors. In 2015 Watts et al.[27] published a summary of tracking of health impacts of climate change. They illustrated the collective effects of extreme temperature, altered rainfall patterns and air pollution in leading to loss of habitable land space, and agricultural productivity and the impact of these on human physical and mental health. They noted that climate change led to increased poverty, to increased mass migration and to violent conflict.

Additional information on health effects related to climate change were reported by Hunter.[28] They noted that warmer temperatures were associated with increased growth of pathogenic waterborne organisms, including coliform and vibrio species. Also, as the range of vectors increased there were increases in the incidence of vector borne diseases, including West Nile Virus disease, Rocky mountain spotted fever and Lyme disease in the USA. In other countries the incidence of Zika virus infections, Dengue fever and Chikungunya increased.

In addition, there were outbreaks of cholera particularly in areas of destruction of support systems due to armed conflicts (e.g., in Yemen).

Spread of disease across borders

African swine fever virus (ASF virus) is reported to be endemic in Africa. In a report in May 2019 Normile et al.[29] noted that in 2007 ASF virus that causes a fatal disease in pigs was reported in Georgia, Russia. In 2018 the virus was reported to have spread to China. In China the epidemic was reported to lead to death or culling of more than 1 million pigs. There are reports that in 2019 the disease is present in Vietnam, Cambodia, Mongolia and Hong-Kong. Normile noted that ASF infections in Asia lead not only to economic hardships but could potentially compromise food security.

Specific measures including restructuring of the pig industry were reported to have eliminated African Swine fever from Spain and Portugal where it appeared in the 1960s.

Population aging

The UN report on world population in 2019 https://population.un.org/wpp/ revealed that in 1990 there were 54 million people aged 80 years and older and in 2019 there were 143 million and that this number is expected to triple to 436 million in 2150.

This report noted that this increase will significantly impact the support ratio calculated as the number of people still working (25–64 years) compared to the number of people older than 65 years. The report also noted that this data indicated that many countries will face problems providing pensions, healthcare and social protection for older persons.

Gerontology

Arai et al.[30] emphasized that in addition to geriatrics that focuses on the health care studies in the elderly, gerontology is important. Gerontology as they defined it is the discipline that focuses on the study of social welfare, psychology, environments and social systems for elderly persons.

Arai et al. noted that in 2008 in Japan 10% of the population were older than 75 years. They emphasized the importance of integration and interaction among administration, community, industry and educational institutions in approaching issues related to aging and the elderly.

They also noted that attention should be given to encourage possibilities for the elderly to continue to participate and to make contributions in society. They proposed that attention be paid to situations where elderly persons could participate in the workforce.

Arai also emphasized the benefits of home-based care for the aged.

In 2018 de Wind et al.[31] reported that in the Netherlands increasing numbers of individuals continue to work beyond the years when they are eligible to retire, and that work beyond retirement age was particularly feasible in individuals who were free of chronic diseases. They noted that being physically active and "feeling full of life" were positive predictors of working successfully beyond retirement.

Extending life and growing numbers of older individuals

This can be a privilege if learning from life and living successfully are considered as valuable assets and part of the wisdom that older individuals can share with younger members of society.

Downsides of aging come about when aged individuals, particularly through ill health, take up economic resources that could be devoted to other purposes.

Perhaps the economic drains brought about by illness in the aged can be better understood and perhaps prevented by understanding the determinants of aging.

Networks and determinants of aging

Riera et al.[32] reviewed signaling networks the determine the life span. They noted that the key defect in aging is failure of homeostasis of the system. They also noted that there is evidence of heterogeneity in aging and that through determining the factors that play roles in causing differences in aging rates, we can obtain insights into mechanisms. It is important to note that many of the factors and mechanisms that play roles in aging have been identified primarily through studies in model organisms, including *C. elegans* and mice. Riera et al. noted that key mechanisms in aging included the following:

Nutrient sensing pathways

Mitochondrial function

Oxidative stress

DNA damage

Telomere attrition

Epigenetic modifications

Alterations in proteostasis

Nutrient sensing pathways

Key elements in nutrient sensing include insulin, insulin receptor, insulin-like growth factor that can also be stimulated by other growth factors including fibroblast growth factor FGF21. Another important factor that interacts with growth factor receptors is Klotho a cellular membrane protein. Signal transmission pathways downstream of receptor activations is also key factors in nutrient sensing. These pathways include phospholipids such as phosphatidyl inositol, AKT serine threonine kinases and the MTOR pathway.

Riera et al. noted that MTOR plays a particularly important role in aging and inhibition of MTORC1 has been reported to extend life in model organisms. AMPK adenosine monophosphate kinase acts to block the activity of the MTOR containing complex MTORC1. AMPK adenosine monophosphate kinase acts to block MTORC1 activity and was shown to increase life span in some model organisms. Under conditions of nutrient scarcity AMPK activity is turned on.

Riera et al. also noted that there is evidence that dietary restriction increases life span in model organisms and delays emergence of age-related diseases.

Mitochondria

Aging has been reported to be significantly advanced when mitochondrial function is impaired. Wallace[33] reported that alterations in mitochondrial DNA and genetic and epigenetic changes in nuclear DNA can significantly impact mitochondrial bioenergetic processes. Decline in mitochondrial functions can lead to immunological dysfunction, metabolic diseases and degenerative diseases.

Jang et al.[34] emphasized that optimal mitochondrial functions are also dependent on mitophagy processes that remove damaged mitochondrial. In addition, stress signals transmitted to nuclei can lead to activation of gene expression designed to enhance mitochondrial biogenesis.

Oxidative stress and DNA damage

Riera et al.[32] noted that many studies have revealed that increased levels of reactive oxygen species lead to DNA damage and have long been considered to be important factors in aging. Specific nucleotide changes induced by reactive oxygen species include oxidation of guanosine leading to generations of 8-oxo-2-deoxyguanosine. This modification if unrepaired leads to conversion of guanosine to thymine that constitutes a mutation.

Specific enzymes exist to reduce reactive oxygen species, these include superoxide dismutases, peroxidases and thioredoxins. In certain model organisms increased expression of some of these factors were shown to increase life span.

Telomeres

There is evidence that maintenance of telomeres that occur at the ends of chromosomes is important in delaying aging. Telomere maintenance involves adequate generation of DNA repeat sequence TTAGGG and this maintenance requires adequate function of the enzyme telomerase. Riera noted however that it is important that telomerase activity be regulated since cancer cells have been shown to express high levels of telomerase.

Epigenetics and aging

Epigenetic studies focus on DNA and chromatin. Earlier studies focused on patterns of DNA nucleic acid modifications and histone modifications. Riera et al. reported that specific modifications of histones that were reported to occur in aging involved increased acetylation of histone H4 lysine 16 (H4K16ac) and increased frequency of trimethylation of histone H4 lysine 20 (H4K20me3).

In aging alterations in patterns of heterochromatin distribution were detected in micro histology. These changes involved loss of heterochromatin in certain genomic regions and development of heterochromatin foci in other regions.

Morris et al.[35] reported that epigenetic changes in aging dysregulated transcription networks and chromatin topological domains. They noted that these epigenetic changes led to inappropriate gene expression.

Kane and Sinclair[36] reported that senescent associated heterochromatin foci had increased levels of specific proteins including HP1 (also known as CBX5) a non-histone chromatin protein and increased levels of histone MacroH2A. They also noted that nucleosome positioning was altered in aging and that there was evidence for loss of core histones that build nucleosomes.

Sirtuins, NAD and aging

Imai and Guarente[37] reviewed Sirtuins. They noted that sirtuins are enzymes that are dependent on nicotinamide adenine dinucleotide (NAD) as coenzymes. Sirtuins can remove certain molecular groups from proteins, they can act as deacylases, desuccinylases, deamlonylases; in addition, they can act as ribosyltransferases.

Sirtuins function at various sites within the cell including in the mitochondria and nucleus. When NAD levels are low, activities of sirtuins are compromised. Decreased activity of SIRT1 results in decreased mitochondrial biogenesis, impairments in oxidative metabolism and impaired function of anti-oxidant pathways. Imai and Guarente noted that SIRT1 and SIRT3 are involved in mitophagy, the removal of damaged mitochondria.

NAD sources and availability

Johnson and Imai[38] reviewed NAD sources. They noted that NAD may be biosynthesized from ingested niacin; it may also be derived in a salvage pathway whereby nicotinamide is converted to nicotinamide adenine mononucleotide and then to nicotinamide adenine dinucleotide (NAD).

NAD deficit may result from decreased biosynthesis or increased utilization of NAD.

DNA methylation changes in aging

Kane and Sinclair[36] noted that there are numerous reports of decreased methylation of repetitive sequences in DNA in aging. In addition, there is evidence increased methylation in CpG islands located in gene rich regions.

Horvath[39] reported that DNA methylation patterns could be used as indicators of chronological age.

Quach et al.[40] reported that epigenetic biomarkers of age can be influenced by exercise, diet and lifestyle patterns.

Proteins and aging

Riera et al.[32] noted that removal of damaged proteins is delayed during aging. Efficiency of the proteolytic systems, including the autophagic-lysosomal systems and the ubiquitin proteasome system declines during aging.

Stimulation of processes that increase removal of damaged proteins during aging could potentially reduce the impacts of age-related diseases.

Kane and Sinclair[36] noted that the localization and abundance of transcription changed with age. The major transcription factors studied in connection with aging included NRF2 and FOXO.

Translation of information

Kane and Sinclair noted that aging was once considered immutable but that concept was now changing. Key interventions include caloric restriction to modulate nutrient sensing pathways. Other interventions being considered relate to the MTOR and sirtuin pathways.

The Millennium ecosystem assessment

This was a comprehensive study carried out between 2001 and 2005[41] that examined ecosystem data and was compiled from world-wide data gathered by more than 1360 experts. Key assessments involved how ecosystems had changed over a 50-year period. The report also focused on ecosystem services and linkages between ecosystems and well-being.

In documenting ecosystem services, the report considered provision of food, fresh water, fiber and fuel. The ecosystem was also considered in the context of aesthetic, spiritual, educational and recreational services.

Constituents of well-being analyzed included:

Security, personal safety, secure resources, security from disasters.

Basic life materials, livelihood, nutritious food, shelter.

Health categories, included access to clean air, clean water, concepts of feeling well and strength.

Good social relations, social cohesion, mutual respect and ability to help others.

Freedom of choice defined as opportunity to be able to achieve what is considered of value in being and doing.

Ecosystem changes

The Millennium report documented that over 50 years ecosystems had changed extensively. Changes were in part due to increasing demands for food, water and fuel. The changes had improved human well-being for some, had exacerbated poverty for others and had led to ecosystem degradation. Specific ecosystem changes documented included:

Conversion of more land to cropland.

Increased impounding of water into dams and its withdrawal from rivers.

Degradation of coral reefs and mangrove areas.

Increased generation of nitrogen and phosphorus in terrestrial ecosystems.

Significant changes in forest cover.

Loss of terrestrial biomass including woodland scrub, grassland and savannahs.

Attention was also focused on extinction of species, including mammals, birds, amphibians and marine species. And decline in pollinators.

Ecosystems being degraded or used in unsustainable measures included fisheries, water supplies.

Increased problems were documented regarding waste-water treatment and detoxification.

The report also documented economic and public-health costs of specific ecosystem losses. These included collapse of fishing due to overfishing and decrease in agricultural yield due to soil damage and water pollution.

Categories of food resources documented as increasing, included livestock.

Climate change

Abrupt changes in climate occurred in some areas and these were referred to in the Millennium report as non-linear ecosystem changes. These changes can include excess flooding, drought dust storms and desertification. Climate change has increased the frequency of weather-related disasters, severe storms, cyclones, drought and wild-fires.

Climate changes can increase population migrations and conflicts. These can result in physical injury, malnutrition displacements and psychological trauma.

Rapid population growth can result in over-crowding, unemployment, political unrest.

World population growth

Data published by the United Nations in 2019, https://population. un.org/wpp/ reported that the world population size continues to increase and in 2019 the world population is 7.7 billion. It is estimated to increase to 8.5 billion by 2030.

The report noted that many of the countries with the fastest growing population are also the poorest countries. This then adds to the challenges to projected millennial aims of combating poverty and malnutrition and efforts to improve health and education.

However, the report noted that in some there has been a decline in the number of childbirths. Fertility rates in Sub-Saharan Africa were estimated to be 4/6 births per woman. The global fertility rate has fallen from 3.2 births per woman in 1990 and was projected to decline to 2.2. births per woman in 2050.

World-wide data revealed that life expectancy in 1990 was 64.2 years in 1990 and 72.6 years in 2019. The fastest growing age group in the world are individuals older than 65 years. The fastest growing age group included persons over 65 and data revealed that the number of persons older than 65 outnumbered the numbers of children under 5 years. In addition, projection indicated that by 2050 the number of individuals in the 15–24 years age range will be outnumbered by persons older than 65 years.

The greatest population growth rates between 2019 and 2050 are projected to occur in the Democratic Republic of the Congo, in Egypt, Ethiopia, India, Nigeria, Pakistan, Tanzania and in the USA. Some countries, e.g., China project a decrease in population size between 2019 and 2050.

The most populous regions in the world in the 2019 report includes Eastern and South Eastern Asia with 30% of the world's population and South- Central Asia with 26% of the world's population.

In some countries population size increases can be attributed to increased migrations. These countries include, Belarus, Germany, Hungary and Italy.

Refugees and migrants

The United Nations defines refugees as follows: "Refugees are people who have fled wars, violence conflicts or persecution and crossed international borders to find safety in another country." Refugees are protected under international law.

A 2019 WHO report estimated that there were 258 million international migrants and that 71 million were reported to have fled war. The UN report emphasized that international human rights standards and conventions exist to protect the rights of migrants and refugees, and that these rights included access to health services. UN work on a new refugee and migrant global action plan began in 2019.

It is also interesting to note that the in the WHO mandate a human rights approach to healthcare was adopted. The rights proclamation emphasized non-discrimination and equality and noted: "the onus on states to redress discriminatory law, practice or policy."

Achievements in world-health

In the millennial project the WHO established sustainable development goals related to Healthcare. In 2018 a progress report was published on 36 health-related goals https://www.who.int/gho/publications/world_health_statistics/2018/en/.[42]

With respect to maternal, newborn health the report noted that in 2015 approximately 303,00 women died due to maternal related causes. Two thirds of these deaths occurred in the Africa region. It was estimated that 40% of the women who died did not receive ante-natal care. In addition, in low and low middle-income countries less than half of all births were assisted by trained mid-wives or physicians. The report also emphasized the risks of early child-bearing and noted that 12.8 million births occurred in females 15–19 years of age.

A key achievement noted included decreased mortality of children under 5 years of age. Predominant causes of death in newborn and young infants included prematurity, birth trauma neonatal sepsis. Causes of death before age 5 years included particularly respiratory illnesses, diarrhea and malaria. Impaired growth was reported to occur in 22% of children younger than 5 years of age. Highest incidences of growth impairments occurred in the WHO assessed regions in Africa and South-East Asia.

HIV incidence

The incidence of HIV was reported to have decreased. The Africa region had the highest incidence, 1.24 per 100 individuals. The WHO report noted that the introduction of antiretroviral therapy led to a 48% decline in deaths from HIV. Sadly, anti-retroviral therapy was reported to reach only 53% of people with HIV infection.

Malaria

The 2018 WHO report indicated that progress in eliminating malarial infections had stalled and malaria was reported to have led to death of 445,000 people in 2016.

Tuberculosis

Progress in achieving effective and lasting treatment of Tuberculosis was less than optimal particularly since the appearance of drug resistant forms of the *Mycobacterium tuberculosis*.

Hepatitis

The highest incidences of hepatitis in the world were reported in Somalia and South Sudan.

Trachoma that leads to blindness

Trachoma caused by the organism Chlamydia trachomatis was report by the WHO in 2019 to have decreased in incidence from 1.5 billion affected individuals in 2002, to 142 million in 2019. This was achieved largely through use of the antibiotic azithromycin. Eight countries were reported to have eliminated trachoma. However, trachoma infections remain prevelant in 44 countries; highest documented incidences occur in Cambodia, Ghana, Islamic Republic of Iran, Lao Peoples Democratic republic, Mexico, Morocco, Nepal and Oman.

Vaccines

In 2016 Barrett[43] reviewed vaccinology in the twenty-first century. He noted that vaccinology included basic science related to immunogens, and information on host responses. It also included information on manufacturing and delivery and regulatory, ethical and economic factors. Barrett emphasized the debt that the twenty-first century owed to vaccinologists in the twentieth century for their achievements.

With respect to development of vaccines in the twenty-first century Barrett noted that there was significant increase put in place regarding the quantity and quality of data needed before vaccines could be licensed. He emphasized that these changes had both positive and negative effects in that there is not an increase in the number of vaccines licensed.

In his 2016 report Barrett noted that only 50 human vaccines were available and that of these only between 40 and 45 were licensed in Europe and in the USA. This in contrast to the fact that worldwide more than 300 vaccines for veterinary use are available.

Questions that remain or are inadequately addressed are who will pay for vaccine research and vaccine development. In the USA the manufacturers count on primarily consumers costs to cover vaccines and associated costs.

Emerging diseases and vaccines

Barrett noted that in the twenty-first century molecular techniques can facilitate vaccine development in a short time to meet crisis demands. This was evident in dealing with Ebola and Zika virus outbreaks. In these circumstances, academia, governmental, non-governmental scientists, pharmaceutical companies, regulators and funders worked together to produce vaccines and make them available.

Vaccine availability

WHO information (accessed 28 June 2019)[44] listed 26 vaccines that were available for treatment of different infectious disease. They noted that in different countries health authorities provide guidelines for which vaccines should be used and at the ages when they should be administered. The WHO also noted that 24 vaccines were in the development pipeline.

GAVI the global alliance for vaccines and immunization

This is a public private global health partnership with the goal to increase vaccine availability in poor countries and to assist in immunization. GAVI also partners, through financing mechanisms, with companies that produce vaccines. A 2017 report from GAVI indicated that their mission was to have immunized 365 million children by 2020, www.gavi.org/progress-report.[45]

The report noted that shortages of vaccines were experience particularly in inactivated polio vaccine and in human papilloma virus vaccine. The 2017 GAVI report also noted difficulties in accomplishing vaccination in areas defined as "suffering from fragility" these include zones of conflict, economic decline or climate related pressures.

Vaccine shortages

The Center for Disease Control (CDC) in the USA noted that reasons for vaccine shortages were multi-factorial. They included production difficulties, insufficient stockpiles and companies leaving the market.

A shortage of polio vaccine occurred following withdrawal of the oral polio vaccine that contained attenuated poliovirus. This withdrawal was necessary following reports of studies that revealed that attenuated poliovirus could undergo genetic change and result in vaccine derived polioviruses.

Non-Communicable cases of death

These accounted for 71% of the total number of deaths reported in 2016 WHO statistics. In the non-communicable cause categories that together led to 57 million deaths, 44% of deaths were attributed to cardiac and vascular causes, 22% were due to cancer, chronic respiratory diseases accounted for 9% of deaths and diabetes accounted for 4%.

Key risk factors for non-communicable disease include tobacco use, unhealthful diet, air pollution excess alcohol use, and insufficient physical activity. Highest alcohol use was reported in European countries, Suicide deaths were also monitored and highest rates were reported to occur in males between the ages of 25 and 34 years. WHO statistics on homicides revealed that in the period 2012 to 2016 the death rate due to conflicts rose substantially from rates reported in 2007–2011.

Access to clean cooking fuel was particularly problematic in Sub-Saharan Africa. It was also problematic in India and South-East Asia. Household pollution was a causative factor in respiratory and cardio-vascular diseases in some regions.

Unsafe drinking water and poor sanitation were reported to occur in significant number of areas within Africa.

Obesity

WHO data in 2018[42] revealed that obesity rates in the world had tripled since 1975. Data gathered in 2016 indicated that 1.9 billion adults were overweight and this number included 650 million adults who were obese. Obesity is linked to cardio-vascular diseases and diabetes and may add to risk of certain cancers, e.g., breast and colon cancer.

The metric used to gauge weight is body mass index (BMI) calculated as weight in kilograms divided by the square of height in meters kg/m^2. For adults BMI equal to or greater than 25 is considered overweight and BMI greater than 30 is considered obese.

Obesity also increased in children under 5 years of age to reach 41 million in 2016 and 340 million children and adolescents (5–19 years) were determined to be overweight.

For children overweight is determine as weight for height greater than 2 standard deviations above median for age. Obesity is defined as weight for height greater than 3 standard deviations above the median.

Key factors impacting obesity include increased intake of energy dense foods and decreased physical activity.

Malnutrition

The 2018 WHO report revealed that worldwide 462 million adults were underweight. Malnutrition status in children is sub-divided into different categories. These include wasting with low weight relative to height; stunting with low height for age, and under-weight with low weight for age, collectively these conditions were reported to impact 224 million children under the age of 5 years.

Family planning

A number of different organizations, national and international public and private, endeavor to provide information on family planning. Studies carried out under auspices of the United Nations organization, https://www.unfpa.org/family-planning[46] determined that access to family planning empowers women, enabling them to complete their education and to strengthen the economic security of their families.

Contraceptive access also prevents the hazardous situations that can sometimes result from unintended pregnancies, including dangerous abortion practices.

Analyses reported by Wulifan et al.[47] revealed that family planning was an unmet need in many low- income and middle- income countries. Results of a study in Chiapas Mexico, were reported by Dansereau et al.[48] Their study revealed that there were intergenerational and cultural gaps in acceptability of family planning and also religious objective to contraception. In house-holds men strongly influenced choices. In addition, men did not participate in educational programs related to family planning. The report also noted that adolescent women requested more information on family planning.

Ecosystem services

Summers et al.[49] reviewed the interwoven relationship of human well-being and ecosystem services. They noted that ecosystems have contributed to the well-being of billions of peoples who have led to substantial strains on ecosystems to provide services and have led to loss of biodiversity. All of these factors impact human well- being. Summers and many others re-echo the Millennium Ecosystem assessment that emphasized that human well- being does not only comprise the absence of health and incapacity and having basic living needs met. Human well-being required that individuals have a sense of purpose, inclusiveness in society and supportive relationships, www.millenniumassessment.org.[41]

The one health concept

In recent years multiple organizations, including the World Health Organization and the Center for Disease Control (CDC) in the USA, have recognized the One Health Concept that emphasizes that the health of people is connected with the health of animals and with the health of environments.

This concept emphasizes local, regional, national and global collaborative efforts to analyze problems and to reach for and share solutions to problems.

Additional approaches to health: The one health concept

This concept was reviewed by Destoumieux-Garzón et al.[50] They emphasized that health security should consider human health, animal health, plant and ecosystems health and biodiversity.

They noted that the one-health concept had relevance to both communicable and non-communicable disease. Successful implementation of the one health concept requires dissolution of barriers that compartmentalize human medicine, veterinary medicine, ecology, evolutionary science and environmental science.

Destoumieux-Garzón defined three main categories in the one health concept and interactions between these categories. Main concepts in human health include non-communicable diseases, communicable diseases, antimicrobial resistance, multi-factorial disease, social networks and human medicine.

Components of animal health include veterinary medicine, domestication of animals and human animal relations. Human and animal health intersected in zoonoses and in epizootics.

Environmental health includes ecology cycles, reservoirs, ecotoxicology. Human and environmental health intersected in urbanization.

Human, animal and environmental health intersected in human animal relations, cultural practices and legal frameworks.

Ecosystem dynamics

In considering emergence and re-emergence of infectious disease Destoumieux-Garzón et al. emphasized that genomes of parasitic organisms evolve continuously through mutations, horizontal transfer and hybridization. In some environments specific genotypes are selected. Organisms with altered genotypes may colonize new hosts and become pathogenic. They noted that it is important to establish if specific environmental modifications favor the emergence and multiplication of specific organisms. Specific etiologic factors may also favor transmission of infectious agents from one species to another.

Destoumieux-Garzón noted that there is evidence that disruption of habitat, pollution and climate change have altered the geographic distribution of certain potentially pathogenic organisms. In addition, globalization, trade, travel, agribusiness and fish farming, alter distribution of human, plants animals and their accompanying microorganisms.

Destoumieux-Garzón et al. noted that increases in population sizes, industrialization, ecosystem deterioration, migrations of humans and other species were leading to emergence of new diseases and re-emergence of older diseases. Zoonoses such as bird-flu, Ebola and Zika virus infections demonstrated the connections between animal and human health.

Ecotoxicology

Toxic risk has been found to be particularly high in regions of high population density close to coastal regions. Emerging pollutants include toxin produced by microalgae that flourish in nitrogen rich effluents, Other hazardous pollutants include microplastics and nano plastics.

Destoumieux-Garzón et al. noted reports that document increased frequencies of diseases and impact of multifactorial diseases in diminishing populations of bees, amphibians, small vertebrates and shell- fish.

Child development and nurture

A 2017 report by Black et al.[51] documented processes and factors that begin before conception, and pre-natal and post-natal processes that can disrupt normal brain development. Black et al., emphasize that it is important to consider child development as the "ordered progression of perceptual, motor, cognitive, language, socio-emotional and self-regulation skills." They emphasized the key importance of nurturing care, opportunities for play and exploration and safety of the environment.

Nurturing includes food nutrition adequacy of micronutrients for prospective mothers, pregnant women, infants and children. Black et al. cited evidence that the key period when adequate nutrition was necessary for brain development included the 1000 days between conception and the age of 2 years.

UNICEF has recommended policies and procedures that support caregivers and that implement child protection. The United Nations High Commission on refugees reported in 2015 that 59 million people were displaced from their homes and that more than 50% of these people were children.

In a Lancet report in 2017 by Britto et al.[52] emphasized that advancement required application of emerging scientific knowledge through multi-sector approaches that included community-based strategies and social safety nets.

Britto et al. stressed women's health even in the preconceptual stage, this included attention to nutrition and micronutrients and in certain places iodine supplementation is necessary. They noted the importance of education on the deleterious effects of alcohol and tobacco use particularly during pregnancy. In addition, they noted that maternal stress and depression have been shown to co-exist with preterm births and low birthweight infants. They emphasized that living in poverty is associated with a high degree of stress.

Family violence was reported to be a key public health problem in low and middle-income countries. Britto et al. noted evidence that maltreatment in childhood impacts the development of the child's brain and leads to reduced development particularly of brain regions involved in learning and memory, including the hippocampus and mid-sagittal brain regions. Maltreatment prevention programs include professional home visits, nurse-family partnerships. Child-care facilities and facilitation of multigenerational care.

The Lancet report noted that early childhood education in pre- primary schools has been shown to be of value on intellectual and emotional development.

Environmental agent exposures and child development

Analyses of structural and functional birth and growth defects and possible links to environmental agent exposures have been carried out by several groups of investigators. Results of different studies have yielded conflicting results for certain environmental toxins. However, consensus has emerged regarding the hazardous effects of particular exposures.

Time periods of exposures have differed in different studies. In studies reported by Mattison[53] exposure period prior to conceptions were included in the study. In other studies exposures during pregnancy and in early post-natal periods were primarily studied. Methods of exposure assessment also varied in different studies.

Exposure assessment methods included verbal history of workplace type of work, residential location. In some studies, environmental monitoring and/or personal monitoring were carried out. Personal measurements of potential exposures included measurement of specific component in blood, hair or urine. Llop et al.[54] used assessment of urinary excretion of 1-hydroxypyrine to monito exposure to polycyclic aromatic hydrocarbons (PAH). PAH concentrations are high in tobacco smoke and in urban environments.

There is clear evidence of the harmful effects on fetuses, infants and children of specific heavy metals, including mercury and lead. Mercury exposure originaltes from various sources, including ingestion of fish that may be contaminated from mercury released into water sources. Exposure to mercury in the prenatal period or during early childhood has been reported to lead to cognitive, behavioral and motor defects, Axelrad et al.[55]

Mattison[53] noted that there is evidence from reports from different countries, including the USA, France and Poland that herbicides including triazine, metolachlor and cyanazine that accumulate in ground water and then in drinking water, lead to growth retardation and to infant being born small for gestational age.

Lyall et al.[56] reported that 11 studies in the USA during the prior decade had reported evidence that prenatal air pollution constituted a risk factor for autism. Hazardous air pollutants implicated included nitrogen dioxide (NO_2), ozone, particulate matter Pm 2.5 μm and 10 μm in size, and traffic fumes. They noted that the USA studies had included a socio-economic confounder. Comparing data on impact of air pollution in different countries is difficult given that methods used to measure specific pollutants vary.

Lyall et al. noted that early life exposure to endocrine disruptors are of concern. Endocrine disruptors have been shown to impact hormones involved in neurodevelopment and factors involved in development of the nervous system.

Child development, physical environment and green space

Much has been written about potential toxins, including heavy metals, inorganic solvent and pesticides and their impact on child development.

More recently researchers have begun to track evidence of the importance of access green space and natural elements in child development. McCormick[57] reported evidence that children's access to green space improved their mental well- being and cognitive development. Green space access was also shown to lead to behavioral improvement in children with attention deficit hyper-activity disorders.

Engemann et al.[58] reported evidence that higher levels of exposure to green space during childhood were associated with decreased frequency of later psychiatric illness. Benefits of green space exposure remained even after data were adjusted to include socio-economic factors. They emphasized the benefits of integration of green space and natural environments into urban planning.

It is perhaps important to consider alternate possibilities for children in dry climates; perhaps exposure to open space with natural elements.

Social ecology and health

Manstead[59] noted that modern Western values tend to emphasize meritocracy, linking achievement to work and talent but not social and economic distinctions, however the later were becoming more evident.

Manstead cited 2012 data from the UK that indicated that 40% of the national income is earned by 20% of the population in the high-income category and 8% of the income was earned by the 20% in the lowest income category. More equitable levels were achieved between 1938 and 1979. He noted evidence from the USA that indicated that 20% of individuals have 84% of the wealth.

Manstead cited evidence that differences in income were also associated with differences in social capital, defined as friendship networks, and cultural capital.

There is also evidence that individuals in the population were more likely to be concerned about immigration if they perceived that their own earning abilities were likely to be threatened by immigrants.

Health and employment

Brydsten et al.[60] presented data from a study carried out in Sweden on the impact of unemployment on health. They reported that 7.9% of the populations were unemployed, the unemployment rate was higher for men than for women and individuals in the 15–24 years reported experience highest levels of unemployment.

There was evidence of a relationship between unemployment and health problems including physical and mental health. Important consequences of unemployment included economic deprivation, loss of social interaction, and sensing loss of control.

Based on their study and previous studies Brydsten concluded that employment is one of the fundamental determinants of health and that although economic considerations were important social support and connections also contribute to health.

A study on health inequalities and social dimensions of health in South Africa was reported by Omotso and Koch in 2018.[61] They reported that socio-economic inequalities constituted one of the greatest hazards to health in South Africa. The authors proposed interventions to address problems. These included strengthening individuals and communities, improving living and working conditions and improving access to essential services.

Doyle et al.[62] reported results of investigation related to hypertension incidence and management. They noted that the incidence of hypertension in the USA is highest among the non-Hispanic Black populations however analyses revealed that therapeutic control of hypertension is highest in whites. There is also evidence that hypertension prevalence is higher in lower income groups,

There was also evidence for an increased risk of myocardial infarction following job loss. Living in disadvantaged neighborhood was associated with an increased risk of coronary heart disease in whites and non-whites.

Doyle et al. noted that intervention to improve blood pressure included at home blood pressure monitoring, physical activity, monitoring sodium intake and medication adherence.

Sustainable development goals

It is worthwhile to reiterate the 17 Sustainable Development Goals set forth by the United Nations General Assembly in 2015 to be achieved by 2030.[63]

1. No poverty
2. Zero hunger
3. Good health and well-being
4. Quality education
5. Gender equality
6. Clean water and sanitation
7. Affordable clean energy
8. Decent work and economic growth
9. Industry, innovation, infra-structure
10. Reducing inequality
11. Sustainable cities and communities
12. Responsible consumption and production
13. Climate action
14. Life below water
15. Life on land
16. Peace, justice and strong initiatives
17. Partnerships and goals.

Key concepts of the UN Sustainable Development Goals emphasize human capabilities and Human Capital and the principle, "leave no one behind".

The WHO comprehensive Mental Health Action plan 2013–2020[64]

This plan defined 4 major objectives.

Strength effective leadership and governance for mental health.

Provide comprehensive integrated and responsive mental health and social care in a community-based setting.

Implement strategies to promote mental health and to prevent mental disorders.

Strengthen information systems evidence and resources for mental health.

Mental health and sustainable development

A comprehensive report on global mental health and sustainable medicine was published by the Lancet Commission in October 2018 (see Ref. 65). The report noted that major demographic, environmental and sociopolitical transitions have led to deterioration in mental health in all countries and that it was important to consider not only mental disorders but also population mental health.

Interactions of social and environmental influences together with genetic and biological factors have been shown to determine neurodevelopmental and psychological growth.

The concept that human rights be considered in approaches to individuals with poor mental health was embedded in the WHO comprehensive Mental Health Action plan and was stressed by the Lancet Commission who also noted that physical health of individuals who needed mental health care must also be considered.

The Lancet commission report stressed that care of mental health enhances possibilities for individuals to contribute to sustainable development.

In tracing the history of approaches to mental health the Lancet Commission noted progressive shifts from institutionalized care to community care. Transitions had also occurred regarding assignment of care providers for individuals needing mental health care, from earlier model where these individuals were primarily psychiatrists to broader concepts of personnel and care givers.

Regarding social determinants of mental health, the commission report documented evidence for roles of poverty, violence, adversity, poor educational attainment, reduced earning capacity as pathways that can potentially impair mental functioning.

There is evidence that the global burden of mental health deterioration has increased steadily since 1996 along with increased numbers of individuals needing care; there is evidence of poor access to care.

Reframing mental health

The Lancet Commission report particularly emphasized the importance of adoption of broader concepts of mental health not focused primarily on recognition of designated clinical diseases. Mental disorders are defined in inventories of diagnostic criteria, e.g., DSM (USA) and WHO classification of disorders as, "disturbances of thought emotion, behavior and behavior that impair function and life activities." Other concepts include impaired functioning resulting from social suffering due to life circumstances, psychosocial disability due to social barriers.

Important positive concepts in mental health include happiness, perception of living a meaningful life, well-being, quality of life in relation to common goals.

Limitations in diagnosing mental disorders

The Commission Report emphasized that genomic studies revealed that specific risk variants were found to be shared across different clinical defined diagnostic phenotypes, including autism spectrum disorders, schizophrenia, bipolar disorder and depression. Also noted was growing evidence for concepts that included interacting biologic and environmental factors that function in different domains including cognition affect and behavior.

Genomic studies of value in certain neurodevelopmental disorders in childhood

It is however important to assign relevance to specific genomic alterations in the causation of neurodevelopmental disorders.

In 2019 Chawner et al.[66] reported results of studies on children with genomic defects defined as neurodevelopmental copy number variants (ND-CNVs). These CNVs encompass specific regions and the genomic alterations include deletions (dels) or duplications (dups). Important ND-CNV genomic regions include 1q21.1 del/or dup; 2p16.3 del; 9q34.3 del; 15q11.2 del; 15q13,3 del/dup; 16p11.2 del/dup, 22q11.2 del./dup.

Their studies revealed that these ND-CNVs led to similar adverse neurodevelopmental defects with overlapping manifestations. The neurodevelopmental defects found included autism spectrum disorder traits, hyperactivity, impaired social function, impaired motor-co-ordinations.

Importance for human health of contact with nature

Hartig et al.[67] have written of the impact of urbanization and lifestyle changes in limiting human contact with nature They stressed the importance of contact with nature in improving health, both physical and mental.

Hartig et al. noted that it is important to define nature and different forms of contact with nature and to document specific examples of the benefits of nature contact. They focused particularly on the benefits of nature for urban populations and emphasized that contact with nature should be considered in planning and design and in policy measures.

Natural features to be considered include, indoor plants, street trees, urban parks, community gardens. Also discussed were possibilities of viewing nature in photographs or through virtual means. In addition to derived health benefits and for stress reduction they stressed the importance of environments conducive to physical activity. Increased physical activity is important to reduce the growing incidence of obesity and cardiovascular disease. There is evidence that increased outdoor physical activity in an urban setting can increase social contact with neighbors and generates a sense of community, and social cohesion within neighborhoods. it can also lead to increased subjective well-being and increased performance. Hartig et al. emphasized that particular natural features may promote walking and cycling. They emphasized the need for suitable infra-structure to promote safety, examples include sidewalks and bike paths.

Another important development in recent years is increasing acceptance of the concept of health promoted by the UN-WHO: "Health is a state of complete physical, mental and social well-being and not merely the absence of disease and infirmity." Hartig et al. noted that through acceptance of this multi-dimensional concept of health, individuals in many different fields contribute to activities in support of health. Healthcare professionals are not limited to those who concentrate on manifestations of disease.

Hartig et al. noted evidence that specific natural surroundings play important roles in stress reduction. Landscaping can reduce feeling of crowding and reduce noise. There is also evidence that green space and natural features enhance attention, reduce feelings of sadness and anxiety.

The Biophilia hypothesis

The term Biophilia was popularized by Edward O. Wilson.[68] It is defined as "the connection that human beings sub-consciously seek with the rest of life."

Partnerships and interconnections in dealing with challenges

A remarkable short viewpoint piece was published in the Journal of the American Medical Association in March 2018, authored by Monsignor Sorondo, H. Frumkin and V. Ramanathan. The piece was entitled. "Health Faith and Science in a Warming Planet".[69] They stressed the moral obligation to safeguard the earth for future generations and of the moral obligation to care for the most vulnerable.

They closed with a call to "share commitment to human dignity, to pursuit of justice and world peace and exercise charity".

Efforts to counteract climate change and associated devastations

Adversity has the effects of eliciting talents which in prosperous circumstances would have been dormant.

Horace Quintus Horatius Flaccus 65 BC

Striving toward solutions

In 1999 The National Academy of Sciences published a comprehensive report entitled, "Our common journey, a transition toward sustainability".[70] This report documented that the world population in 1999 was estimated to be 6 billion, in 2050 the world population was projected to be 9 billion. The report documented key challenges in meeting the needs of the expanded population for goods and services and also the need to address the continued pressures on environment and on life support systems. The report also documented that searches for progress toward sustainability would be complicated by the resource intensive consumptive lifestyles in industrialized countries.

The report noted that human needs included provision of food and nutrition, nurturing of children, finding shelter, providing education, finding employment, reducing hunger and poverty. With respect to earth's life support systems attention needed to be paid to ensuring quality and supply of fresh water, controlling atmospheric pollution, protecting oceans, maintaining ecosystems and biodiversity.

The report also emphasized the importance of progressing from knowledge generation to application of knowledge. Priorities for action included acceleration of human fertility reduction through access to family planning and education; knowledge generation and action to accommodate the projected doubling or tripling of urban populations; addressing and reversing decline in agricultural production particularly in sub-Saharan Africa; acceleration in improved use of energy; retardation of ecosystem degradation and biodiversity loss.

Dietary changes to benefit health and mitigate climate change

Springmann et al.[71] reported that more than 25% of all greenhouse gas emissions are generated by the food system and that 80% of these emissions are generated in livestock facilities. They emphasized further that high consumption of red meat and processed meat together with low consumption of fruit and vegetables contributes to increased weight and to ill-health. Springmann et al. calculated that decreased consumption of animal derived products could reduce greenhouse gas emissions, improve health and be economically beneficial.

Proposed transformative changes in energy generation

It is important to note that intense efforts are being applied to generations of low carbon energy supply systems. The European Electricity Energy Initiative is pursuing development and instillation of systems to make energy production completely decarbonized by 2050.[72] The U.S. Department of Energy has active programs to stimulate generations and use of renewable forms of energy.

The fuel cell and hydrogen joint technology initiative implements systems for stationary power supplies and systems for energy supplies for transport systems.[72,73]

The Solar Europe initiative is focused on the use of solar power and photovoltaics to be integrated into the urban electricity grid.[72]

The European Wind Initiative includes mapping of wind systems, development and analysis of at least 4 prototypes of different structures to harness wind energy and to integrate wind generated energy into power grids.[74]

In addition, the European Technology Plan includes development of CO_2 capture systems and systems to store CO_2.[75]

Physiological adaptations

López-Maury et al.[76] reviewed adaptations to changing environments. They emphasized that abilities to sense environmental changes and to adapt to them are essential to all living organisms. Sensing of changes involved detection of extra-cellular signals and this must be followed by activation of signal transduction pathways, that lead to changes in gene transcription and to post-transcriptional changes in gene products.

Studies on lower eukaryotes such as yeast, have provided information on mechanisms of sensing changes and response strategies. Studies have revealed that key mechanisms include those involved in maintaining balance between stress related responses and growth. Stress related gene expression patterns differ from expression patterns related to growth. López-Maury et al. noted that gene expression patterns under stress conditions are directed at maintenance of cellular physiology and metabolism and protection against cell damage and cell death.

Studies in yeast revealed that under stress condition the stress signaling pathway is activated leading to expression of stress induced genes while expression of the growth-related pathway genes and MTOR expression are inhibited. In the presence of adequate nutrients and/or growth factors the expression of MTOR pathway genes is activated. Regulators of these pathways include oxygen and nutrient levels and pH.

López-Maury et al. also noted evidence for alterations in translation of gene transcripts under stress conditions. This translational inhibition was found to be in part due to inhibitory phosphorylation of the eukaryotic translation initiation factor EIF2.

Gene expression is dependent on chromatin structure that facilitates binding of transcription factors and polymerase II to the 5' and promoter region of genes. López-Maury et al. reported that the core promoter sequence element TATA is enriched in stress related genes and there is evidence that the TATA box is probably important in genes that undergo variable expression. The TATA box is frequently absent from the 5' regions of housekeeping genes.

López-Maury et al. emphasized that genetic variation in regulatory DNA elements constitute mechanisms for variations in expression and in phenotypes. Changing environments and adaptation may, in the long run lead to evolutionary changes.

The heat shock response and heat-shock proteins

Expression and activity of these proteins are now known to be induced not only by heat but also by a variety of different stresses and they are known to have cytoprotective effects. Heat shock proteins also act as molecular chaperones that serve in part to facilitate the refolding of proteins.

Kregel et al.[77] reviewed heat shock proteins. By 2002 9 different families of heat shock proteins had been described in humans and were named according to their molecular weights. Heat shock proteins were found to be located in different cellular regions, cytosol, nucleus, mitochondria and endoplasmic reticulum. The principal families of heat shock proteins include HSP27, HSP60, HSP70, HSP72, HSP73, HSP75, HSP90, HSP100 and HSP104. Proteins in specific families, HSP, HSP72, HSP104 were found to play roles in protein folding and to restore appropriate conformations structure and function. Proteins in the HSP70 family were reported to protect cells against apoptosis and cell death under conditions of heat stress.

Barna et al.[78] reported that the heat shock response system is activated by cellular stress, that results from factors that include heat, exposure to oxidizing agents, toxins, heavy metals. They noted that an evolutionarily conserved transcription factor HSF1 (Heat Shock factor 1), plays a key role in activated expression of heat shock proteins Under specific stress conditions HSF1 undergoes structural changes trimerization, and it is phosphorylated and moves from the cytoplasm to the nucleus, Within the nucleus its binds to specific DNA elements heat shock responsive elements (HSE) that are located in the

5′ untranslated region of target genes. A stable association then forms between a binding region in HSF1 transcription factor and the HSE in the target gene. In addition, chromatin structure in the target gene region may be impacted by binding of HSF1 to the heat shock response element. Barna et al. noted that HSF1 has been reported to regulate expression of genes involved in physiological stress and stress induced cellular processes.

Barna et al. reported that the oxidizing agent hydrogen peroxide triggered HSF1 trimerization and movement to the nucleus leading to expression of a number of genes included NRF2 (nuclear regulator factor 2), a key regulator of the response to oxidative stress.

The HSF1 transcription factor can also undergo trimerization, movement to the nucleus and bind to specific DNA response elements to trigger expression of genes that play roles in autophagy. Barna et al. also described interaction between HSF1 and genes that encode proteins that function in the ubiquitin proteasome system that function to ensure that functional proteins are present in the cells and dysfunctional proteins and proteins that cannot be appropriately refolded are removed from the cells.

Under specific adverse condition it is important that the cell cycle and cell proliferation be blocked. HSF1 also plays a role in these processes. It impacts cell growth through modulation of the MAP Kinase-MTOR system.

Documenting problems and sometimes proposing solutions

It is important that the scope and depth of problems are documented. There are also articles that describe problems and also address potential solutions.

Loss of large carnivores and consequences

Atkins et al.[79] reported that the numbers of large predators in the wild is rapidly declining. This results in herbivores being less afraid and undertaking grazing in new habitats. They reported that the loss of leopards and wild dogs in the Gorongoza National Park in Mozambique led to more expansive bushbuck grazing and to suppression of growth of indigenous plants. They also produced evidence that re-introduction of predators reversed plant loss.

Rewilding

In a news report in Science magazine in October 2018 Pennisi[80] documented efforts related to rewilding and re-introduction of large mammals to protect grasslands and the tundra. Evidence from Hluhluwe National Park in South Africa revealed that when number of large mammals including wildebeest, rhino, buffalo and impala decreased the numbers of wildfires increased.

Grazers have also been shown to be important in protecting the tundra where grazing keeps down the growth of certain plants that increase soil warmth and promote ice melting.

Perspectives on limitations

In an article published in 2018, Crist[81] wrote of "the planetwide sense of entitlement blind to the wisdom of limitations". She emphasized the importance of scaling down and reduction of consumptions of food energy and materials.

Gu et al.[82] wrote on aspects of food consumption and noted that this differs in different parts of the world. Consumption of meat and dairy products is much higher in Western countries than elsewhere in the world. However, meat consumption was reported to be increasing in non-western countries. Gu et al. noted that the shift to increased meat consumption has led to 15% of cropland in the world being devoted to production of animal feed.

Gu et al. proposed that governments set up campaigns to promote optimum diets. They noted that a report from the European Union recommended that meat and dairy consumption be halved by 2050. They noted that in addition it will be critical to reduce food waste. They also advocated improved methods of food storage and food sharing.

Other recommendations included sharing of knowledge between scientists and farmers regarding optimal choices of crop varieties, fertilization and irrigation practices.

A report on food wastage from the Food and Agriculture Organization C14 http://www.fao.org/food-loss-and-food-waste/en/[83] noted that approximately one third of the food produced in the world is wasted. Highest percentages of waste occur in North America, Europe and Industrialized Asia. The lowest degree of food wastage occurs in Sub-Saharan Africa. The report noted that in high income countries, wastage occurs primarily at late stages in the food chain. For fruit and vegetables 45% of produce was wasted; 20% of dairy products produced were wasted and 30% of cereal products produced were wasted.

Efforts to counteract climate change and associated devastations

Drawdown

The book "Drawdown" edited by Paul Hawken[84] and published in 2017 generated important information on plans to mitigate global warming and earth degradation and ideas and observations from 57 individuals.

Several articles present information on Farmland restoration; regenerative farming and multistrata farming practices were presented. Multistrata farming includes have segments of a land area devoted to crops and in between these segments natural grassland or trees can be grown. This strategy cuts down on soil erosion and soil degradation. In restorative farming segments of land can include trees interspersed with segments of cattle grazing. The cattle can help restore organic nitrogen to soil.

Other articles deal with soil managements use of organic fertilizer in place of inorganic fertilizer and interplanting or multistrata farming decreases runoff and contamination of streams and rivers.

Altered methods of rice planting have also been developed to replace constant flooding with periodic flooding.

Specific sections of Drawdown, present ideas on city planning, conducive to walking, transportation modes and enhancement of mass transport, building materials that would be advantageous in changing climates, dietary modifications and alternate forms of energy.

Celebrations of nature in human endeavors

In this closing section it seems important to consider examples of how writers, poets and musicians have endeavored, through their close observations and concentrated efforts, succeeded in capturing wonder, beauty and mystery of the natural world. Collectively their efforts deepen our perceptions, understanding, growth and well-being.

Perhaps, in the face of accounts of accounts of destruction of the natural world, we can gain strength to confront problems by revisiting explorations and creativity of those who have used their talents of convey wonders and observations great and small, and experiences they encountered in the natural world.

We each have our own collection of heroes of the human spirit whom we celebrate, I will take the opportunity to celebrate some of mine.

Rachel Carson[85] is perhaps best known for her book, *Silent Spring*, that began with an epitaph from John Keats, "The sedge is withered from the lake and no bird sings." In *Silent Spring* she reported damage to the natural world by toxic chemicals. In other works, Rachel Carson celebrated the wonders and mysteries of life on life in the seas. She drew attention to intricate connections and urged us to reflect on meanings and on nature's refrains with humility.

Aldo Leopold[86] expanded our visions of things small, and large and things commonplace. In *A Sand County Almanac,* he wrote with joy of the changes of seasons, of mists and winds, of sunsets of sunrises and birdsong. He delighted us with descriptions of the habits of creatures small and large, the meadow mouse, migrating cranes, and deer. He drew attention to the painful, and lasting effects of lack of appreciation of connections and of losses.

Leopold wrote of symbioses, of the evolution of modes of co-operation of communities in which soil plants and animals are members. He urged the need for changes in attitudes and convictions from concentrations on self-interest to include conceptions of diversity of biota.

Lewis Thomas[87] frequently wrote with humor about humans and about life. However, in a chapter in his book *The Fragile Species* his anger and dismay at human activity that exhausted earth's resources and destroyed species were forcefully expressed, "It is not simply wrong, it is a piece of stupidity..."

Lewis Thomas wrote too of the importance of co-operation, and of what was perhaps the most significant event in evolution, symbiosis.

E.O. Wilson[88] in his book *Consilience* noted with appreciation the works of Enlightenment thinkers who stressed the unity of knowledge. Wilson emphasized the need to link science and the humanities. He noted that the quests and achievements of science were to map and to understand reality and that the quests of those in the arts were to construct, images, narratives and rhythms.

Wilson concluded that problems that threaten humans and the environments could best be solved by integrating knowledge generated by science with information derived from the social sciences and the humanities.

It is clear that application of humanity's collective wisdom, innate creativity, scientific knowledge and innovations will be necessary to solve ecological problems.

Writers and Poets who capture in words events, and images in nature and changes in seasons and light stir souls, expand minds and surely inspire us to protect and conserve. Henry Beston[89] wrote in *The Outermost House* of a year spent on Cape Cod. He captured the changing of season, patterns of weather and ocean currents, bird migrations, movements of dunes and sands and lives of insects.

In *Gift from the Sea* Anne Morrow Lindbergh[90] wrote of the renewal of spirit she experienced spending days on a beach where there were opportunities for solitude and to pick up and reflect on shells, She drew analogies between structures of shells the creatures that made and housed in them, and life as a woman, a wife and a mother.

Peter Matthiessen[91] in *The Birds of Heaven; travels with Cranes* wrote of the value of capturing a habits and images of single species to understand evolution, adaptations and environments.

Terry Tempest Williams[92] in *Refuge* described bird species in the vicinity of the Great Salt Lake and Bear River Migratory Bird Refuge. She presented images and habits of birds and also documented losses that impacted creatures of the refuge and losses in her family.

Kathleen Dean Moore[93] writes of *The Solace of Nature* even as she mourns losses and grieves at the lack of appreciation of gifts of life and light. She writes and images and joys that unfold on hikes in the Pacific Northwest and while canoeing rivers or paddling on lakes. She welcomes the contributions of science to understanding and appreciation of life and its mysteries.

Musicians have at times sought to capture in notes and harmonies, the events of nature. Magnificent examples include Beethoven's *Pastoral symphony*[94] and Mendelsohn's[95] portrayal of rushing ocean waters and wind in *Fingal's Cave*. Ralph Vaughan Williams[96] memorably captured the calls notes, trills and joy of birdsong in *The Lark Ascending*.

References

1. Executive Summary Rockefeller Foundation Lancet Commission. *Safeguarding human health in the Anthropocene epoch: report.* www.thelancet.com/commissions/planetary-health.
2. Environmental Protection Agency. https://www.epa.gov/ghgemissions/sources-greenhouse-gas-emissions; 2017.
3. NASA. *Carbon data from NASA.* https://climate.Nasa.gov; 2018. [accessed 5 August 2018].
4. Tyndall J. On the absorption and radiation of heat by gases and vapours, and on the physical connexion of radiation, absorption, and conduction. *Philos Trans R Soc Lond* 1861;**151**:1–36.
5. Arrhenius S. On the influence of carbonic acid in the air upon the temperature of the ground. *London, Edinburgh, and Dublin: Philos Mag J Sci* fifth series. 1896;**41**:237–75.
6. Min SK, Zhang X, Zwiers FW, Hegerl GC. Human contribution to more-intense precipitation extremes. *Nature* 2011;**470**(7334):378–81. https://doi.org/10.1038/nature09763. PMID: 21331039.
7. Perlwitz J. Tug of war on the jet stream. *Nat Climate Change* 2011;**1**:29–31.
8. Pall P, Aina T, Stone DA, Stott PA, Nozawa T, et al. Anthropogenic greenhouse gas contribution to flood risk in England and Wales in autumn 2000. *Nature* 2011;**470**:382–5.
9. Mohtadi M, Prange M, Schefuß E, Jennerjahn TC. Late Holocene slowdown of the Indian Ocean Walker circulation. *Nat Commun* 2017;**8**(1):1015. https://doi.org/10.1038/s41467-017-00855-3. PMID: 29044105.
10. *NASA Ozone hole recovery.* https://www.jpl.nasa.gov/news/news.php?feature=7033.
11. REN21. *Renewables 2016 global status report.* www.ren21.net/reports/global-status-report.
12. Whitmee S, Haines A, Beyrer C, Boltz F, et al. Safeguarding human health in the Anthropocene epoch: report of The Rockefeller Foundation-Lancet Commission on planetary health. *Lancet* 2015;**386**(10007):1973–2028. https://doi.org/10.1016/S0140-6736(15)60901-1. PMID: 26188744.
13. *Coal demands in select countries/regions.* https://www.iea.org/topics/coal/.
14. Tonkin JD, Poff NL, Bond NR, Horne A, Merritt DM. Prepare river ecosystems for an uncertain future. *Nature* 2019;**570**(7761):301–3. https://doi.org/10.1038/d41586-019-01877-1. PMID: 31213691.

15. Jonkers L, Hillebrand H, Kucera M. Global change drives modern plankton communities away from the pre-industrial state. *Nature* 2019;**570**(7761):372–5. https://doi.org/10.1038/s41586-019-1230-3. PMID: 31118509.

16. Schwartzstein P. *As climate change shrinks the Sundarbans, lives are washed away.* www.nationalgeographic.com/magazine/2019/07/.

17. Cornwall W. Vanishing Bering Sea ice poses climate puzzle. *Science* 2019;**364**(6441):616–7. https://doi.org/10.1126/science.364.6441.616. PMID: 31097644.

18. National Oceanic and Atmospheric Administration US. https://arctic. noaa.gov/Report-Card/Report-Card-2018/ArtMID/7878/ArticleID/783/ Surface-Air-Temperature.

19. Steig EJ. How fast will the Antarctic ice sheet retreat? *Science* 2019;**364**(6444):936–7. https://doi.org/10.1126/science.aax2626. PMID: 31023892.

20. Wild S. South Africa's invasive species guzzle precious water and cost US$450 million a year. *Nature* 2018;**563**(7730):164–5. https://doi.org/10.1038/d41586-018-07286-0. PMID: 30401848.

21. Golden Kroner RE, Qin S, Cook CN, Krithivasan R, Pack SM, et al. The uncertain future of protected lands and waters. *Science* 2019;**364**(6443):881–6. https://doi. org/10.1126/science.aau5525.

22. McMichael AJ, Woodruff RE, Hales S. Climate change and human health: present and future risks. *Lancet* 2006;**367**(9513):859–69. PMID: 16530580, https://doi. org/10.1016/S0140-6736(06)68079-3.

23. Weissbecker I. *Climate change and human well being: global challenges and opportunities.* Springer; 2011. ISSN 1574-0455.

24. Simpson DM, Weissbecker I, Sephton SE. Extreme weather-related events: implications for mental health and well-being, Chapter 4. In: *Climate change and human well being: global challenges and opportunities.* Springer; 2011. ISSN 1574-0455.

25. Weissbecker I, Czincz J. Humanitarian crises: the need for cultural competence and local capacity building, Chapter 5. In: *Climate change and human well being: global challenges and opportunities.* Springer; 2011. ISSN 1574-0455.

26. Reser JP, Morrissey SA, Ellul M. The threat of climate change: psychological response, adaptation, impacts, Chapter 2. In: *Climate change and human well being: global challenges and opportunities.* Springer; 2011. ISSN 1574-0455.

27. Watts N, Adger WN, Agnolucci P, Blackstock J, et al. Health and climate change: policy responses to protect public health. *Lancet* 2015;**386**(10006):1861–914. https://doi.org/10.1016/S0140-6736(15)60854-6. PMID: 26111439.

28. Hunter DJ, Frumkin H, Jha A. Preventive medicine for the planet and its peoples. *N Engl J Med* 2017;**376**(17):1605–7. https://doi.org/10.1056/NEJMp1702378. PMID: 28249124.

29. Normile D. African swine fever marches across much of Asia. *Science* 2019;**364**(6441):617–8. https://doi.org/10.1126/science.364.6441.617. PMID: 31097645.

30. Arai H, Ouchi Y, Yokode M, Ito H, Uematsu H, et al. Toward the realization of a better aged society: messages from gerontology and geriatrics. *Geriatr Gerontol Int* 2012;**12**(1):16–22. https://doi.org/10.1111/j.1447-0594.2011.00776.x. PMID: 22188494.

31. de Wind A, van der Noordt M, Deeg DJH, Boot CRL. Working life expectancy in good and poor self-perceived health among Dutch workers aged 55–65 years with a chronic disease over the period 1992–2016. *Occup Environ Med* 2018;**75**(11):792–7. https://doi.org/10.1136/oemed-2018-105243. PMID: 30194272.

32. Riera CE, Merkwirth C, De Magalhaes Filho CD, Dillin A. Signaling networks determining life span. *Annu Rev Biochem* 2016;**85**:35–64. https://doi.org/10.1146/ annurev-biochem-060815-014451. Review. PMID: 27294438.

33. Wallace DC. A mitochondrial bioenergetic etiology of disease. *J Clin Invest* 2013;**123**(4):1405–12. https://doi.org/10.1172/JCI61398. PMID: 23543062.

34. Jang JY, Blum A, Liu J, Finkel T. The role of mitochondria in aging. *J Clin Invest* 2018;**128**(9):3662–70. https://doi.org/10.1172/JCI120842. PMID: 30059016.

35. Morris KV. Long antisense non-coding RNAs function to direct epigenetic complexes that regulate transcription in human cells. *Epigenetics* 2009;**4**(5):296–301. Review. PMID: 19633414.

36. Kane AE, Sinclair DA. Epigenetic changes during aging and their reprogramming potential. *Crit Rev Biochem Mol Biol* 2019;**54**(1):61–83. https://doi.org/10.1080/1 0409238.2019.1570075. PMID 30822165.

37. Imai SI, Guarente L. It takes two to tango: NAD^+ and sirtuins in aging/longevity control. *NPJ Aging Mech Dis* 2016;**2**:16017. https://doi.org/10.1038/np-jamd.2016.17. eCollection 2016. PMID: 28721271.

38. Johnson S, Imai SI. NAD^+ biosynthesis, aging, and disease. *F1000Res* 2018;**7**:132. https://doi.org/10.12688/f1000research.12120.1. eCollection 2018. Review. PMID: 29744033.

39. Horvath S. DNA methylation age of human tissues and cell types. *Genome Biol* 2013;**14**(10):R115. PMID: 24138928B21.

40. Quach A, Levine ME, Tanaka T, Lu AT, Chen BH, et al. Epigenetic clock analysis of diet, exercise, education, and lifestyle factors. *Aging (Albany NY)* 2017;**9**(2):419–46. https://doi.org/10.18632/aging.101168. PMID: 28198702.

41. *The Millennium ecosystem assessment*. www.millenniumassessment.org.

42. *WHO Goals and Health Statistics*. https://www.who.int/gho/publications/world_health_statistics/2018/en/.

43. Barrett AD. Vaccinology in the twenty-first century. *npj Vaccines* 2016;**1**:16009.

44. *Vaccines and availability*. https://www.who.int/immunization/diseases/en/.

45. *GAVI report on vaccines*. https://www.gavi.org/progress-report/.

46. *WHO family planning*. https://www.who.int/reproductivehealth/topics/family_planning/en/.

47. Wulifan JK, Brenner S, Jahn A, De Allegri M. A scoping review on determinants of unmet need for family planning among women of reproductive age in low and middle income countries. *BMC Womens Health* 2016;**16**:2. https://doi.org/10.1186/s12905-015-0281-3. Review. PMID: 26772591.

48. Dansereau E, Schaefer A, Hernández B, Nelson J, Palmisano E, et al. Perceptions of and barriers to family planning services in the poorest regions of Chiapas, Mexico: a qualitative study of men, women, and adolescents. B28. *Reprod Health* 2017;**14**(1):129. https://doi.org/10.1186/s12978-017-0392-4. PMID: 29041977.

49. Summers JK, Smith LM, Case JL, Linthurst RA. A review of the elements of human well-being with an emphasis on the contribution of ecosystem services. *Ambio* 2012;**41**(4):327–40. https://doi.org/10.1007/s13280-012-0256-7. Review. PMID: 22581385.

50. Destoumieux-Garzón D, Mavingui P, Boetsch G, Boissier J, Darriet F, et al. The one health concept: 10 years old and a long road ahead. *Front Vet Sci* 2018;**5**:14. https://doi.org/10.3389/fvets.2018.00014. eCollection 2018. Review. PMID: 29484301.

51. Black MM, Walker SP, Fernald LCH, Andersen CT, et alLancet Early Childhood Development Series Steering Committee. Early childhood development coming of age: science through the life course. *Lancet* 2017;**389**(10064):77–90. https://doi.org/10.1016/S0140-6736(16)31389-7. Review. PMID: 27717614.

52. Britto PR, Lye SJ, Proulx K, Yousafzai AK, et al. Early childhood development interventions review group, for the lancet early childhood development series steering committee. *Lancet* 2017;**389**(10064):91–102. https://doi.org/10.1016/S0140-6736(16)31390-3. Review. PMID: 27717615.

53. Mattison DR. Environmental exposures and development. *Curr Opin Pediatr* 2010;**22**(2):208–18. https://doi.org/10.1097/MOP.0b013e32833779bf. Review. PMID: 20216314.

54. Llop S, Ballester F, Estarlich M, Ibarluzea J, et al. Urinary 1-hydroxypyrene, air pollution exposure and associated life style factors in pregnant women. *Sci Total Environ* 2008;**407**(1):97–104. https://doi.org/10.1016/j.scitotenv.2008.07.070. PMID: 18804258.

55. Axelrad DA, Bellinger DC, Ryan LM, Woodruff TJ. Dose-response relationship of prenatal mercury exposure and IQ: an integrative analysis of epidemiologic data. *Environ Health Perspect* 2007;**115**(4):609–15. PMID: 17450232.

56. Lyall K, Croen L, Daniels J, Fallin MD, et al. The changing epidemiology of autism spectrum disorders. *Annu Rev Public Health* 2017;**38**:81–102. https://doi.org/10.1146/annurev-publhealth-031816-044318. PMID: 28068486.

57. McCormick R. Does access to green space impact the mental well-being of children: a systematic review. *J Pediatr Nurs* 2017;**37**:3–7. https://doi.org/10.1016/j.pedn.2017.08.027. Review. PMID: 28882650.

58. Engemann K, Pedersen CB, Arge L, Tsirogiannis C, Mortensen PB, Svenning JC. Residential green space in childhood is associated with lower risk of psychiatric disorders from adolescence into adulthood. *Proc Natl Acad Sci U S A* 2019;**116**(11):5188–93. https://doi.org/10.1073/pnas.1807504116. PMID: 30804178.

59. Manstead ASR. The psychology of social class: how socioeconomic status impacts thought, feelings, and behaviour. *Br J Soc Psychol* 2018;**57**(2):267–91. https://doi.org/10.1111/bjso.12251. PMID: 29492984.

60. Brydsten A, Hammarström A, San Sebastian M. Health inequalities between employed and unemployed in northern Sweden: a decomposition analysis of social determinants for mental health. *Int J Equity Health* 2018;**17**(1):59. https://doi.org/10.1186/s12939-018-0773-5.

61. Omotoso KO, Koch SF. Assessing changes in social determinants of health inequalities in South Africa: a decomposition analysis. *Int J Equity Health* 2018;**17**(1):181. https://doi.org/10.1186/s12939-018-0885-y. PMID: 30537976.

62. Doyle SK, Chang AM, Levy P, Rising KL. Achieving health equity in hypertension management through addressing the social determinants of health. *Curr Hypertens Rep* 2019;**21**(8):58. https://doi.org/10.1007/s11906-019-0962-7. Review. PMID: 31190099.

63. 17 Sustainable Development Goals set forth by the United Nations General Assembly in 2015 https://sustainabledevelopment.un.org/post2015/summit.

64. https://www.who.int/mental_health/action_plan_2013/en/.

65. Patel V, Saxena S, Lund C, Thornicroft G, Baingana F, et al. The Lancet Commission on global mental health and sustainable development. *Lancet* 2018;**392**(10157):1553–98. https://doi.org/10.1016/S0140-6736(18)31612-X. PMID: 30314863.

66. Chawner SJRA, Owen MJ, Holmans P, Raymond FL, Skuse D, et al. Genotype-phenotype associations in children with copy number variants associated with high neuropsychiatric risk in the UK (IMAGINE-ID): a case-control cohort study. *Lancet Psychiatry* 2019;**6**(6):493–505. https://doi.org/10.1016/S2215-0366(19)30123-3. PMID: 31056457.

67. Hartig T, Mitchell R, de Vries S, Frumkin H. Nature and health. *Annu Rev Public Health* 2014;**35**:207–28. https://doi.org/10.1146/annurev-publhealth-032013-182443. PMID: 24387090.

68. Wilson EO. *Biophilia*. Harvard University Press; 1984.

69. Sorondo MMS, Frumkin H, Ramanathan V. Health, faith, and science on a warming planet. *JAMA* 2018;**319**(16):1651–2. https://doi.org/10.1001/jama.2018.2779. PMID: 29543969.

70. National Academy of Sciences USA. *Our common journey, a transition toward sustainability.* https://www.nap.edu/catalog/9690/our-common-journey-a-transition-toward-sustainability.

71. Springmann H, Godfray CJ, Rayner M, Scarborough P. Analysis and valuation of the health and climate change co-benefits of dietary change. *PNAS* 2016;**113**(15):4146–51. https://doi.org/10.1073/pnas.1523119113.

72. SETIS Strategic Energy technologies Information Systems. https://setis.ec.europa.eu/european-industrial-initiative-solar-energy-photovoltaic-energy.

73. U.S. Department of Energy's Office of Energy Efficiency and Renewable Energy. https://www.energy.gov/eere/office-energy-efficiency-renewable-energy.

74. SETIS Strategic Energy Technologies Information Systems. European Industrial initiative on wind energy. https://setis.ec.europa.eu/accessibility-info.

75. The role of carbon capture utilization and storage in a low carbon economy, energy.gov https://www.energy.gov/articles/role-carbon-capture-utilization-and-storage-forming-low-carbon-economy.

76. López-Maury L, Marguerat S, Bähler J. Tuning gene expression to changing environments: from rapid responses to evolutionary adaptation. *Nat Rev Genet* 2008;**9**(8):583–93. https://doi.org/10.1038/nrg2398. PMID: 18591982.

77. Kregel KC. Heat shock proteins: modifying factors in physiological stress responses and acquired thermotolerance. *J Appl Physiol* 1985;**92**(5):2177–86. https://doi.org/10.1152/japplphysiol.01267.2001. PMID: 11960972.

78. Barna J, Csermely P, Vellai T. Roles of heat shock factor 1 beyond the heat shock response. *Cell Mol Life Sci* 2018;**75**(16):2897–916. https://doi.org/10.1007/s00018-018-2836-6. Review. PMID: 29774376.

79. Atkins JL, Long RA, Pansu J, Daskin JH, Potter AB, et al. Cascading impacts of large-carnivore extirpation in an African ecosystem. *Science* 2019;**364**(6436):173–7. https://doi.org/10.1126/science.aau3561. PMID: 30846612.

80. Pennisi E. Restoring lost grazers could help blunt climate change. *Science* 2018;**362**(6413):388. https://doi.org/10.1126/science.362.6413.388. PMID: 30361350.

81. Crist E. Reimagining the human. *Science* 2018;**362**(6420):1242–4. https://doi.org/10.1126/science.aau6026. 30545872.

82. Gu B, Zhang X, Bai X, Fu B, Chen D. Four steps to food security for swelling cities. *Nature* 2019;**566**(7742):31–3. https://doi.org/10.1038/d41586-019-00407-3. PMID: 30718889.

83. Food Loss and Wastage. Food and agriculture organization of the United Nations http://www.fao.org/food-loss-and-food-waste/en/.

84. Hawken P. *Drawdown: the most comprehensive plan ever proposed to reverse global warming.* Penguin Books; 2017. ISBN-10: 9780143130444.

85. Carson R. *Silent Spring.* Boston, NY: Houghton Mifflin Company; 1962.

86. Leopold A. *A Sand County almanac and sketches here and there.* New York, Oxford: Oxford University Press; 1949.

87. Lewis T. *The fragile species.* New York: Charles Scribner's Sons; 1992.

88. Wilson EO. *Consilience.* New York: Alfred A. Knopf; 1998.

89. Beston H. *The outermost house.* first published 1928, Henry Holt; 1949.

90. Morrow L. *Anne gift from the sea.* New York: Vintage Books: A Division of Random House; 1955.

91. Matthiessen P. *The birds of heaven; Travels with cranes.* New York: Northpoint Press: A Division of Farrar Straus and Giroux; 2001.

92. Williams TT. *Refuge: an unnatural history of family and place.* New York: Vintage Books: A Division of Random House; 1991.

93. Moore KD. *Wild comfort: the solace of nature.* Trumpeter Boston: An Imprint of Shambhala Publications, Inc.; 2010.

94. L. van Beethoven The Symphony No. 6 in F major, Op. 68, also known as the Pastoral Symphony, completed in 1808.
95. F. Mendelssohn The Hebrides (overture Fingal's cave) was composed in 1830, revised in 1832, and published the next year as his Op. 26.
96. Vaughan Williams Ralph in 1914 wrote music for the poem The Lark ascending a poem by George Meredith.

12

SUMMARY CHAPTER EPILOGUE

In this closing section, I will present what I consider to be some of the key conclusions presented in prior chapters; I will not include the references for information that led to these conclusions as they are presented in each of the earlier chapters.

Chapter 1 Interacting with the environment receiving and interpreting signals

A remarkable series of systems exist in the human body to facilitate interactions with the environment. These systems facilitate, vision, hearing, touch and peripheral pain sensation and components of the systems include not only specific anatomical structures but also single molecules, ions, channels and neural connections.

The early studies by anatomists, histologists followed by efforts of physiologists and biochemists who defined functional processes and geneticists who identified genes and their products have progressively led to our increased appreciation of the complexities and intricacies of sensory perception systems. In this regard it is perhaps important to draw attention to information gained through studies on individuals with compromised functioning of these sensory systems.

Studies on individuals and families with unusual peripheral pain sensitivity profiles have led to further discoveries on the structure and function of specific channels, ions and molecules involved in eliciting peripheral pain response and transmission. These include specific sodium ion channels, e.g. SCN10A, specific potassium channels and transient receptor potential (TRP) channels. Discovery of these channels has opened the way to new and different pharmacological approaches to treating pain.

Through defining the genetic basis of specific forms of blindness, deafness or of abnormal pain sensitivity, researchers characterized key proteins and molecules involved in the function of sensory organs. Through application of molecular histology techniques, and discoveries of the precise organization and position of gene products our insights into functional molecular mechanisms expanded. In some cases, studies of molecular mechanisms provided information that can be utilized in designing disease therapies.

Gene Environment Interactions. https://doi.org/10.1016/B978-0-12-819613-7.00012-8

Discoveries of the biochemical processes involved in the visual cycle and discovery of vitamin A and its conversion to retinal, essential in function of visual pigments led to the proposal that vitamin A deficiency may be responsible for certain forms of blindness. Application of these discoveries and action by the WHO and philanthropic organizations to provide vitamin A in resource poor settings has led to a significant achievement in our time, namely reduction in the incidence of a form of blindness Xeropthalmia keratomalacia.

A key advance in reducing the incidence of blindness in infants born prematurely, was the realization that exposure of premature infants to high levels of oxygen, administered clinically to compensate for impaired respiratory function in these infants, led to blindness. The mechanism involved in generating blindness was shown to be increased proliferation of retinal small blood vessels that were frequently leaky, leading to perivascular exudates. Lowering levels of oxygen administered and also insurance of administration of adequate levels of fatty acids in nutrients in infant feeding, decreased the incidence of retinopathies in premature infants.

Studies in individuals with age related macular dystrophy that can lead to severe visual impairment, led to determination of at least one environmental factor that plays a role in pathogenesis of the disorder in some patients, namely smoking.

Significant expansion has occurred in our understanding of mechanisms of sound passage and leading to movements of small bones in the middle ears. These movements trigger hair cell activity in the inner ear. Hair cell deflections and fluid movement, changes in ion movements through channels, lead to changes in membrane potentials and conversion of potential changes generate electrical signals that trigger neuronal signals. Analyses of hearing mechanisms and related structures in the ear and identification hearing-related nerves and brain regions, have greatly enhanced development of adequate devices to assist individuals with hearing deficits.

The realization that effective hearing was essential to neurodevelopment led several countries to initiate newborn screening programs to detect impaired hearing in young infants so that therapeutic measures could be applied early.

Chapter 2 Environment as provider

Over the past two centuries geologists, farmers, biologists and physicians have accumulated information on the essential ingredients for life that soil and earth systems provide. It seems that only more recently have we begun to face the reality that earth's resources may be limited.

In the case of soil analyses have revealed that certain practices that increased crop production, including use of inorganic fertilizers and pesticides and mechanical processes have impacted soil integrity and increased water runoff, have been found to impact soil fertility, to lead to soil erosion and to reduce productivity in the long run.

In recent years there has been increased focus on more ecologically sustainable processes including use of organic materials for fertilization and tilling methods that do not facilitate erosion. To address evidence that pesticide and chemical induce loss of pollinators including bees, ecologists and farmers have introduced mixed farming approaches. In mixed farming strips of a field planted with food crops are alternated with strips planted with wildflowers or native grasses. Alternate farming methods also include having cattle graze on grasses between rows of planted trees., thus also providing organic fertilizer for trees.

Intense focus on components important in human nutrition began in the early 1900s when vitamins were identified as essential components of co-factors in metabolic reactions. Studies on severe impairments of growth in children in particular focused attention on the need for adequate sources of energy and protein.

Deficiencies of protein and energy sources still present significant problems in many parts of the world, despite the tremendous growth in agriculture. Statistics from the World health Organization have revealed that the incidences of nutritional deficiencies have decreased in many countries in the world in recent decades; sadly however, the incidence of nutritional deficiency has increased in Sub-Saharan Africa during this same period.

However currently in many western countries in particular, many health problems arise from excess nutrient consumption in combination with decreased physical activity.

During the past century information was gathered on the roles of specific vitamin and micronutrient and mineral deficiencies in the role of specific diseases, e.g. vitamin C deficiency leading to scurvy and deficiency of vitamin D and calcium in leading to poor bone health. Physicians also began to clearly demonstrate that iodine deficiency led to impaired growth and development in children.

Our understanding of roles of specific vitamins and micronutrients in biochemical and physiological processes also expanded as a result of increased research on specific conditions associated with genetic alterations that impacted specific gene products. Examples include diseases due to gene defects that impacted metabolism of vitamins folate and B12 and, defects in systems that convert iodine into the biologically active form, defects that impacted normal transport of iodine into thyroid cells and its linkage to proteins to form active thyroid hormone. These defects led to hypothyroidism. A great medical advance was the initiation of screening programs in infancy to defect thyroid hormone deficiency. Thyroid deficiency in infants and young children can lead to impaired growth and to impaired physical and cognitive development and a condition known in prior centuries as cretinism.

Our knowledge of nutritional factors and the role of sunlight in generation of active vitamin D grew in parallel with observations on the increased frequency of rickets in children who lived in crowded closely packed housing communities in industrialized north European cities into which sunlight seldom penetrated. These conditions in combination with nutritional inadequacies of calcium and vitamin D led to rickets associated with poor growth and bone abnormalities.

Insights into processes involved in bone generation and bone maintenance have also been gathered through studies in osteogenesis imperfecta a genetically determined condition. Osteogenesis imperfecta can arise due to defects in any one of a number of different genes involved in osteoblast function and stability, defects that impair proteins involved in formation of bone matrix and genes involved in bone mineralization. Bone mineralization involves calcium and phosphorus and activity of a specific form of the enzyme alkaline phosphatase. Deficiency and abnormal function of the bone isoform of alkaline phosphatase can lead to hypophosphatasia and bone abnormalities.

In prior decades, particularly in the nineteenth century and early twentieth century iron deficiency was a problem that particularly affected women of child bearing age and young children. Iron deficiency anemia occurs much less frequently these days given frequent supplementation. However specific instances of excess iron levels in plasma and certain body organs have emerged. Excess iron levels may in some cases be due to presence of specific genetic variants. Excess iron intake has also emerged as a problem. In some populations this has been attributed to brewing of alcoholic beverages in iron containers.

Population growth, increased requirement for agriculturally produced food components and increased industrial activities also progressively add strain on resources of clean, unpolluted fresh water another resource provided by earth systems and required by all life forms.

Humans, animals, plants and marine creatures also require components provided in earth's atmosphere, including oxygen and nitrogen and water vapor. Human and other life forms are deleteriously impacted by increasing levels of air pollution particularly in environments where energy generation is dependent on combustion of fossil fuels.

Chapter 3 Evolution

Intense research by archeologists, geologists and anthropologists especially in the early and mid-twentieth century provided evidence that homo lineages emerged particularly in Africa between 7 and 9 million years before the present. In the late twentieth century development of techniques to extract DNA from ancient fossil bones and teeth and advances in DNA sequencing techniques in combination with efforts of bioinformaticians and geneticists greatly expanded our knowledge of evolution of Homo species.

Specific genomic studies have been carried out to attempt to obtain insight into factors that led to divergence of homo species from their primate ancestors. Specific genes have been identified that show differences between primates and Homo species. These include genes that encode proteins and protein complexes that are associated with increased density of neuronal dendrites in humans and that expand neuronal connectivity and genes that encode products that play roles in human language capabilities.

It is important to note that there is not only evidence of changes in the genome leading to Homo but also evidence of the continuous impact of climate, environmental changes and migrations.

There is also evidence that evolutionary changes involve not only changes in specific nucleotide sequences but also involve duplication of genomic segments. The generation of segmental duplications also increased the susceptibility to structural chromosome rearrangements both within and between chromosomes. Another important aspect of genome evolution involves increases in copy number of specific genes and the capacity of the newly developed gene copies to partly diverge and assume different functions.

In recent years significant information has been obtained on nucleotide sequences in DNA of Homo neanderthalensis and Denisova hominins who are directly related to modern humans. We have learned that these species also migrated not only within Europe but also into Asia. We are also beginning to gather information on other ancient Homo species that arose in Asia and seemed to remain confined to specific islands.

Perhaps among the most surprising sequence findings in recent years is that that some specific nucleotide changes found in Neandertals and/or Denisovans remain conserved in modern humans. These sequence elements, referred to as introgressed elements, have therefore been conserved throughout extended periods and may have functional relevance.

Detailed studies of crania from Neanderthals and comparison with crania of modern human indicate that Neanderthals had larger cranial sizes that average humans. However, it appears that in Neandertals specific brain regions likely differed in size from related regions in humans.

Evidence has been obtained for connections between environmental and climate changes and adaptation to changing conditions.

Phenotypic changes in response to environmental conditions can also arise due to changes in levels of expression of genes. These phenotypic changes have been shown in some instances to manifest plasticity, and phenotypes reverted to earlier patterns when environmental conditions reverted to previous patterns.

Many of the studies on DNA variants in evolution concentrated on studies of nucleotide sequences of specific genes, e.g. globin genes. More recently attention has also been directed to polygenic effects on environmental adaptation. In addition, researchers are also focusing attention on sequences in the non-proteins coding regions of the genomes that can involve sequences that impact levels of gene expression.

Studies on evolutionary changes and impact of migrations leading to population differences have involved not only nuclear DNA but also mitochondrial DNA.

Chapter 4 Gene and environment interactions and phenotypes

Insights into the composition and deposition of pigment in skin were greatly expanded through analyses of genes and gene products altered in cases of albinism. In this condition there is significant reduction of dark pigment in skin, hair and eyes.

Key genes involved in pigmentation include tyrosinase that converts tyrosine to melanin and genes that encode products involved in transfer of melanin into melanosomes.

It is interesting to note that specific cellular organelles, lysosome endosome related organelles are important in formation of melanosomes. Lysosome endosome related organelles also occur in other cells and in blood platelets. Albinism can therefore also be a feature of syndromic conditions in which abnormal blood clotting and immune function aberrations occur.

Congenital patches of hypomelanosis occur in the genetic condition Tuberous sclerosis that is associated with impaired inhibition of MTORC1. Acquired hypopigmentation, usually of a patchy nature occurs in a skin condition referred to as vitiligo. Vitiligo is reported to likely represent an immune function disorder.

Increased pigmentation in discrete lesions occurs in several condition due to mutations in genes that encode products involved in signaling pathways.

Specific population differences in pigmentation often involve variants in genes. There is evidence that skin pigmentation is an oligogenic trait that is due to the combined effects of variants in several genes and likely involves genes that have also been shown to be involved in tanning response. These include the melanocortin receptors, transporters of melanin and structural proteins within melanosomes.

A genetic trait that has demonstrates differences in population frequencies is retention of lactase function beyond childhood. Connections have been made with retention of lactase production into adult life in populations where herding is common and nutrition is derived particularly from milk. The primary nucleotide sequence variants that are involved in continued lactase production are located in a non-protein coding genomic segment upstream of the lactase gene and this region likely represents a regulatory element that impacts the lactase gene transcription.

Intense studies have been carried out on how immune system functions were shaped over time by exposure to micro-organism components. Analyses have been carried on both innate immune systems and adaptive immune system components.

Generation of genomic diversity in humans has played key roles in adaptation to malaria infection. Heterozygotes for certain globin gene mutations and for specific mutations in the glucose 6-phosphate dehydrogenase encoding gene have increased levels of protection against certain forms of malarial parasites. However, homozygotes for these same mutations have increased frequency of certain hematologic diseases.

Interesting examples have been described of sequence variants, both in organisms and human genomes, that play roles in susceptibility to trypanosome induced diseases that include Sleeping sickness. Human gene mutations including altered gene copy numbers led humans in certain populations, to develop resistance to prevalent Trypanosome organism. However, some trypanosome organisms underwent mutations that led them to overcome the human resistance mutations.

Adaptations to high altitude and reduced atmospheric oxygen concentrations have been identified in different human populations who live in mountainous regions. It is interesting to note that the specific genes that have undergone such mutations differ to some degree in different world regions.

Chapter 5 Signals, epigenetics, regulation of gene expression

Key steps in signaling include binding of activated proteins, and other molecules to specific receptors attached to the cell surfaces and subsequent triggering of down-stream signaling. These processes were elucidated primarily in the twentieth century. Principal processes involved included phosphorylation reactions, activity of kinase enzymes, conversion of nucleotide mono or diphosphates to nucleotide triphosphates. Specific cytoplasmic proteins were also found to anchor key kinases and nucleotide phosphate converting enzymes. A number of signaling proteins were also found to have abnormal functions in certain tumors.

In parallel with activation of transmembrane receptors, ion channel activity was shown to occur and to play roles signaling processes. Passage of ions across cell membranes via ion channels, led to changes in intracellular concentrations of ions, particularly calcium. Changes in intracellular concentration of specific ions were shown to impact different intracellular signaling pathways. Key intracellular signaling pathways were shown to include the phosphoinositide signaling pathway and the MAP kinase pathways and activity in these pathways ultimately led to transcription factor activity and altered gene expression.

Transcription factors are key to gene transcription and there is evidence that specific transcription factors in the cytoplasm, (e.g. NFKB) may be activated by phosphorylation and then pass into the nucleus to activate gene expression. Transcription factors are highly abundant in cells and there is evidence that 8% of the human genome is involved in encoding transcription factors. Transcription factors bind to specific site in genomic DNA within promoter regions and outside of genes in non-protein coding DNA.

An additional important signaling mechanism involved the activation of a specific kinase AKT that serves to activate MTORC1 the key metabolism related complex. Activated MTORC1 promotes processes that facilitate cell proliferation and growth. One of the functions of the MTORC1 complex is facilitate translation of mRNA to protein.

Epigenetic processes play important roles in regulation of gene expression. These processes include secondary modification of nucleotides in DNA and modifications of specific amino acids in DNA associated histones. Histones surround DNA strands and also occur in nucleosomes. Secondary modifications occur primarily on lysine residues in histone and modifications primarily involve acetylation, methylation and phosphorylation. One mechanism through which epigenetic mechanisms alter gene expression, is through changes in transcription factor binding following binding of methyl groups to nucleotides in DNA. Another important epigenetic process includes chromatin modification and movement of nucleosomes. Specific multiprotein complexes play roles in nucleosome movement and movements also depend on supplies of ATP as energy source. Efficient gene expression requires that nucleosomes move, leading to open stretches of DNA to which transcription factors can bind.

Studies on the passage of proteins, transcription factors and other molecules into the nucleus from the cytoplasm and passage of mRNA out of the nucleus into the cytoplasm, occurs through activities of nuclear pore complexes. The intricate structures of the nuclear pore complexes were recently revealed.

Epigenetic changes in DNA nucleotides primarily involve methylation of cytosine residues. There is some evidence that degrees of DNA methylation increase with age and with increasing numbers of cell divisions. Given the important roles of DNA methylation in impacting gene expression it is important that supplies of methyl groups in the body are adequate. Methyl groups are derived from specific nutrition resources and are dependent on effective metabolic processes. Key nutrition sources include vitamins, folate (vitamin B9) and cyanocobalamin (vitamin B12), choline and the amino acid methionine.

A number of studies have revealed that nutritional deprivation in early childhood and social stress situations impact epigenetic mechanisms.

More recent studies have revealed that particularly in brain, modification of cytosine residues occur not only when those residues are adjacent to guanine nucleotides (CpG) but may also occur at sites where cytosine is adjacent to adenine, thymine or a second cytosine residue, therefore at CC, CA or CT sites. This specific form of epigenetic modification was reported to be play important roles in synaptic activity and memory consolidation.

In recent decades increased information has been gathered on factors that influence regulation of gene expression. These include localization and characterization of regulator sequences scattered throughout the genome including enhancers and inhibitors. Information has also been gathered on factors that influence movements of chromatin and generation of chromatin loops that bring specific distant regulatory elements in contact with gene promoter regions.

There is also evidence that factors that impact the levels and precise locations of mRNA splicing events in particular gene transcripts impact MRNA sequence and subsequently protein sequence and function of products derived from a particular gene. Much information on cell type and tissue regulation has been obtained through efforts in collaborative projects, e.g. the ENCODE project.

There is also evidence that secondary modifications of nucleotides in mRNA influence levels of expression. These modification on cytosine in MRNA and also conversion of cytosine to uridine and conversion of adenosine to inosine.

Chapter 6 Maintaining homeostasis and mitigating effects of harmful factors in the intrinsic or extrinsic environment

Metabolic processes play key roles processing nutrients, using derived molecules to provide energy and resources for physiological functions, and body maintenance. The MTORC1 complex plays a central role in correlating process of growth and maintenance with available levels of nutrients. The MTORC1 complex is composed of a number of different proteins. MTORC1 is activate when nutrient and energy levels are adequate, and when levels of proteins lipids and nucleotides are adequate to promote growth. MTORC1 activity stimulates synthesis of nucleotides and of coenzymes. MTORC1 activity also suppresses autophagy and cellular breakdown.

Under conditions when nutrients are in short supply, MTOTC1 activity is decreased. This is achieved in part secondary to increased levels and activation of AMPK. These result when the cellular concentration of adenosine monophosphate (AMP) is higher than the concentration of adenosine triphosphate ATP. The AMPK enzyme stimulates activity of the tuberous sclerosis complex (TSC1-TSC2) that suppresses conversion RHEBGDP to RHEBGTP. Decreased availability of RHEBGTP inhibits MTORC1 activity and this in turn lead to increased expression of the kinase ULK1 that stimulates autophagy and cell breakdown.

A number of molecules and metabolites have been found to act as sensors for the metabolic state. The relative levels in cells of AMP to ATP constitute one signaling mechanism. Another important sensor involves the levels of NADH to NAD. The NAD form is required as an essential cofactor for many enzymes.

Studies over the past 21 years have expanded insights into mechanisms involved in maintaining cholesterol homeostasis. These factors include effective function of receptors on the outer cell membrane that bind cholesterol present in plasma and body fluid and promote its transfer it into the cell. Intracellular levels of cholesterol impact activity of the cellular cholesterol synthesis system.

Effective binding of cholesterol containing low density lipoproteins to the receptor on the outer cell membrane and effective cholesterol transfer into the cell are associated decreased cellular synthesis of cholesterol. This is achieved because high cellular cholesterol levels suppress the activity of the enzyme hydroxymethyl glutaryl-CoA synthase or HMG-CoA synthase, a key enzyme in the cholesterol synthesis pathway. Mutations in the receptors that impair the binding of low-density lipoproteins to the LDL receptor on the outer cell membrane lead to impaired transfer of cholesterol into the cell and continued cellular synthesis of cholesterol that makes its way into the plasma and forms lipoproteins.

Free fatty acid receptors also occur on cells and mutations in free fatty acid receptors have been found to play roles in diabetes and obesity.

Mitochondria have long been known to play key roles in energy generation. New research has revealed that the hormone insulin plays key roles in stimulating formation of new mitochondria and maintaining appropriate levels of proteins that function in mitochondria.

Mitochondrial function has more recently been shown to be impacted by proteins known as sirtuins. Increased levels of the sirtuin SIRT3 along with decreased nutrient consumption, have been shown to increase the efficiency of mitochondrial function and oxidative phosphorylation.

Recent studies have revealed the importance of connections between endoplasmic reticulum and mitochondria in maintaining energy homeostasis.

Mitochondrial metabolism can also lead to the generation of reactive oxygen species (ROS). Levels of ROS generation are impacted by nutrient levels and may be higher when nutrient intake exceeds levels of energy expenditure.

In recent years detailed studies have been carried out regarding the impact of physical exercise on metabolism, including glycolytic activity and mitochondrial metabolism. Attention has focused on the production of reactive oxygen species generated in mitochondria. ROS molecules can cause oxidative damage to proteins and tissue. Certain environmentally derived molecules can also accumulate in the body and also increase levels of oxidative stress since they potentially inhibit the activity of natural anti-oxidant molecules present in the body, including ascorbic acid, reduced glutathione and the superoxide dismutase enzymes. The superoxide dismutase enzyme SOD2 functions primarily in mitochondria.

Researchers have studied different forms of body fat, including white fat, brown fat and fat cells (adipocytes). A specific hormone adiponectin was found to be produced by adipocytes. Specific adiponectin mutations have been identified in some patients with obesity and diabetes.

Specific mechanisms exist in the body to maintain metal level homeostasis. Metals, including iron, copper, manganese, cobalt, zinc, magnesium play important roles in cofactors and in enzyme functions.

It is important to note that specific inborn errors of metabolism in humans arise as result of specific gene mutations can be associated with the generation of harmful metabolites. Examples include phenylketones that can occur at high levels in patients with phenylketonuria and have been reported to damage myelin of nerves.

A number of compounds generated in the internal environment or derived from the external environment have been shown to damage DNA. DNA can also be damaged by external sources including ultra-violet rays or other forms of radiation energy and by specific chemicals. During the course of evolution many different physiological processes have developed in humans and other life forms to detect DNA damage and to initiate repair mechanisms. The specific repair mechanism utilized depends on the type of DNA damage that must be corrected.

Significant insights into DNA repair mechanisms have been gained through studies in patients with specific disorders characterized by higher than normal levels of unrepaired DNA damage. These disorders include conditions such as Xeroderma pigmentosum, Ataxia telangiectasia, and Fanconi syndrome. In addition, defects in repair of nucleotide mismatches in DNA have been found to occur in patients with increased risk of certain cancers, particularly colon cancer. In some forms of cancer, e.g. breast cancer, there is also evidence of defects in repair of double stranded DNA breaks.

It is important that cells with damaged DNA do not continue to divide and specific enzymes and proteins are activated by DNA damage to inhibit cell division, these factors include the checkpoint kinase enzymes.

Specific physiological mechanisms have also been established during evolution to repair shortened telomeres that may arise on chromosomes.

Enzyme systems exist in the body to modify toxic chemical and consumed medications; these enzymes include cytochrome p450 enzymes. Toxic chemicals and other foreign substances can also be removed from cells through activity of specific transporters that remove chemicals from cells.

Certain chemicals have been shown to bind to specific gene products referred to as the arylhydrocarbon receptors. These proteins functions as nuclear receptors and bind to certain sites on DNA. Binding of a chemical to the arylhydrocarbon receptor and linkage of this combination to a specific site in DNA, can lead to increased transcription of genes that encode products that can detoxify or destroy harmful chemicals.

Chapter 7 Microorganisms and microbiome

In this chapter information on microorganisms in soils was discussed in the context of their contributions to human well-being. Information on the microbiome that exists at different sites in the body was presented. Interactions between pathogenic organisms and human systems was briefly presented.

In recent years researchers have paid increased attention to the types of organisms that exist is soil in different environments and to interactions between soil microorganisms and their functions. Microorganisms in soil include fungi, bacteria, archaea, viruses, and protists unicellular organism that are neither plants or animals. In addition to organisms in soil there is also growing information on organism present in plant root systems, the rhizobiome. There is evidence that a diverse microbiome in soil promotes plant growth.

Increasing attention is being focused on microorganisms in the ocean particularly on plankton protists. Plankton species provide important resources as fish nutrients. There is evidence that climate change effects have reduced plankton populations in the oceans.

Microorganisms play important roles in degradation of wastes, primarily organic wastes including agricultural and human wastes. Interactions between microorganisms at waste sites and the presence of pesticides and antibiotic residues in effluents introduced into those waste sites, is gaining increased attention.

In past decades soil organisms, especially fungi, have served as important resources for isolation of antibiotic substances. However, isolation of new antibiotics has slowed in recent years. This slowing in discovery of new antibiotics occurs at a time when there are growing numbers of reports of resistance of organisms to available antibiotics.

Antibiotic resistant organisms have been shown to commonly arise in wastes and waste effluents due to gene transfer between organisms and due to mutations in organisms. Microfilms that develop on devices inserted into the body, e.g. catheters and prostheses, are also sites of development of antibiotic resistant organisms.

Antibiotic use stewardship and recommended measures to reduce the frequency of emergence of antibiotic resistant organisms have been formulated by the World Health Organization and by governmental organizations, such as the Center for Disease Control (CDC) in the USA.

Scientists have also developed new methods to facilitate isolation of new antibiotics from soil sources. These methods include the use of diffusion chambers where organisms can be grown in wells and arrays and have indirect contact with soil and nutrient factors it contains, but remain separated from direct contact with soil by membranes.

The development of nucleic acid sequencing techniques has expanded possibilities for classification of distinct microorganism species based on sequence. Nucleic sequence studies have also facilitated identification of specific mutations that render the organism resistant to antibiotics.

Other sources of antimicrobial factors are being explored, these include plants and bacteriophage, specific viruses that can infiltrate and destroy bacteria.

The human microbiome in different body parts has been studied in great detail in recent years. It has become clear that the indigenous microbiome provides essential services to the host. Comprehensive approaches to characterization of the component organisms present in the microbiome in different body regions have developed. These include nucleic acid sequencing, and proteomic analytics and single cell isolation and characterization. Sequencing of the microsome small ribosomal subunit 16SRNA has proved to provide important information on which to classify types and subtypes of organisms. Studies are carried out to identify the different organisms present at different sites, including gut respiratory tract and skin, at different life stages.

Changes in the range of organisms present in the microbiome at a specific site have been documented in different human diseases. There is also evidence that damaged epithelia lead to changes in the types or microbiome derived molecules and substances that enter the circulation.

A significant change in the gut microbiome composition can occur as a consequence of antibiotic intake. Particularly problematic is the emergence and overgrowth in the gut of antibiotic resistant bacterium Clostridium difficile.

Children who are malnourished and have limited protein intake have been shown to develop an abnormal composition of organisms in the gut microbiome that amplifies the pathological features of malnutrition and is also associated with inadequate immune response.

There is evidence that altered gut microbiome composition exacerbates the manifestations of lactose intolerance.

The gut microbiome has also been shown to be altered in colon cancer.

Attention is also being paid to the microbiome composition in the respiratory tract and there the microbiome composition has been found to be altered in patients with chronic obstructive lung diseases and in patients with cystic fibrosis.

Skin microbiome changes have been found in certain skin conditions including Psoriasis.

Chapter 9 Precision medicine and personalized medicine

Precision medicine goals include the use of data on genome sequence and metabolism along with clinical information and entry of data into electronic medical records to improve patient care. In addition to collection of data, precision medicine promotes analysis of data and analysis of outcomes of therapeutic approaches. Inclusion of information on patient exposures, on environmental conditions and the patient's social environment are also part of personalized medicine.

Genome analyses studies that expanded significantly at the time of development of precision medicine concepts included genome wide association studies that aimed to identify common genetic variants associated with specific complex common disorders, e.g. cardiovascular disorders. Some emphasis was also placed on gathering more detailed information on phenotypic characteristics of specific genetic disorders, including rare genetic disorders. A further emphasis was placed on analyzing effects of variations in gene expression levels.

In the UK Biobanks were developed and the purpose of these was to recruit individuals who would provides health and social information and contribute blood and tissue samples for genetic testing.

Reports of studies from such the UK and other biobanks have contributed significantly to our understanding of genetic factors involved in complex common diseases. Biobank information and deep phenotype studies have also expanded our insights into the frequency of rare genetic disorders and on the phenotypic features of many of these disorders.

Availability and analysis of comprehensive genotype and phenotype data has provided information relative to variable penetrance, i.e. the fact that all individuals with exactly the same pathogenic mutation do not always have the same phenotypic features or the same degree of disease severity.

It has become increasingly clear that the effects of a specific deleterious gene mutation may be impacted by the effects of variants in other genes, or by environmental factors. Another factor that needs to be considered in variable expression is mosaicism that sometimes occurs, so that all the cells in a particular individual do not necessarily carry the same gene mutation.

Examples of pathogenic mutations in specific genes that vary in impact in different individuals and under different conditions include a specific mutation in coagulation factor V (FA5), referred to as the factor 5 Leiden mutation. This mutation ARG506GLN Snp rs6025. Other mutations with variable expression include the HFE mutation p. Cys282Tyr that may lead to clinical manifestations particularly when iron consumption is high.

The specific F5 Leiden mutation and the specific HFE mutation referred to above, are relatively common in some populations yet do not commonly lead to disease.

Combining DNA sequencing and transcriptome analyses have revealed that in autosomal dominant conditions where individuals are heterozygous, i.e. have a pathogenic mutation in a specific gene locus on one member of the chromosome pair and have a normal version of that gene on the other member of the chromosome pair, the degree of expression from the normal allele may impact the degree of clinical manifestations.

Following availability of comprehensive reference human sequence data and DNA probes precisely mapped to positions on human chromosomes, microarrays were developed that could be used to screen for deletions and duplications of specific chromosome regions, referred to as copy number changes.

Applications of microarray technologies have expanded information on the types and frequencies of structural chromosome changes and their roles in human genetic disease. However, it has become clear that not all copy number variants and not all structural chromosome changes have pathologic significance.

Availability of genomic data and its correlation with phenotype and clinical information, have clearly demonstrated that that individuals with the same genetic defects do not always have the same clinical manifestations. In some cases, the impact of a specific mutation also varied dependent on specific environmental factors.

It has also become clear that certain variants in a particular gene may lead to rare diseases, while other variants in that same gene are associated with common disease.

Personalized medicine in Mendelian disorders

Diagnoses in Mendelian Disorders have long been based on clinical findings, on results of biochemical studies on metabolites, enzymes and proteins. However, in a number of cases with metabolic disorders apparently of a genetic origin, biochemical studies can frequently not have diagnostic utility and cannot provide insights into pathogenetic mechanisms.

In recent years nucleic acid sequencing has been increasingly applied to attempt to precisely identify associated genetic variants. However precise determination of the pathologic significance of a particular DNA variant has not always been possible. Bioinformatic evaluation methods have often proved inadequate. Standards have been developed for classification of nucleotide variants of unknown significance. Databases have been developed listing information collected on assessment of mutations and their likely pathogenicity. These databases include information on phenotypes observed in individuals with a specific mutation classified as pathogenic. Databases have also been developed that provide information on the frequency of a specific variant in apparently healthy individuals in the general population. Efforts continue to be made to include information derived from individuals in different world populations.

It may be necessarily to study different cell types in a specific patient to rule out the presence of mosaicism.

In addition to database information it will continue to be important to undertake functional assessment of the products of specific genetic variants. Such studies could be carried out on accessible patient cells, or on developed stem cells or stem cell derived differentiated cells. In addition, studies on model organisms, e.g. yeast drosophila and mice, zebra fish, are sometimes undertaken. Such studies can involve introduction of specific gene variants into model organisms.

Efforts to combine sequencing analysis data and phenotype information on patients with apparent genetic disorders including congenital abnormalities, who are seen at different medical centers, and in different countries. Such studies have facilitated accurate diagnoses in some patients and have led to the discovery of previously unknown disease- causing genetic variants.

It remains important to keep in mind the possibility that specific disorders thought to be monogenic may be in fact be oligogenic or even polygenic in origin.

It is also important to note that clinical abnormalities due to known pathogenic founder gene mutations in specific families, may only present under certain environmental conditions. Examples include the cardiac condition long-Q-T syndrome, and dilated cardiomyopathy.

It has become clear that exact prediction of phenotype from genotype may often not be possible. As research proceeds, it will be necessary to obtain more information on modifier genes, regulatory factors and gene environmental interaction.

Identification of cancer predisposing mutations in healthy individuals can potentially lead to implementation of clinical screening procedures to identify malignancies in the very early stages. A number of different genes have been found to carry breast cancer predisposing mutations. A specific set of genes carry mutations that predispose to colon cancer.

In 2018 germline cancer predisposing mutations were reported to occur in 33 different forms of cancer. However, there is also evidence that specific environmental factors can influence the penetrance of cancer predisposing mutations.

Personalized precision medicine in multifactorial diseases

A number of later onset disorders have been found in some individuals to be due to rare mutations in a single gene, while in other patients the same disorder is apparently due to polygenic gene defects.

Autosomal dominant forms and rare autosomal recessive forms of Parkinson's disease occur. The most frequent forms of Parkinson's disease due to single gene mutations involved the synuclein gene SNCA or the LRRK2 kinase gene. There is evidence that heterozygous state for certain damaging mutations in genes that encode products involved in lysosomal function, e.g. the glucocerebrosidase gene (GBA) play roles in Parkinson's disease. However common variants in other genes have also been associated with increased Parkinson's disease risk More recently evidence had been put forward indicating that factors generated in the gut may reach the brain and impact brain nuclei involved with functions that are altered in Parkinson's disease.

Specific genetic variants have been identified that increase the risk of age- related macular degeneration (ARMD); smoking has been shown to increase risk.

Chapter 10 Integrating genetic, epigenetic and environmental information to improve health and well-being

It has become clear that certain genetic variants impact responses to medicinal chemicals and need to be taken into account when efficacy or side effects of medications are determined.

Much attention has been focused on factors involved in the etiology and progression of complex common diseases such as cardiovascular diseases.

In a number of common complex genetic disorders, it has become evident that variants at a number of different loci likely play roles even in individual patients. Given the ability to sequence genomes, and to use microarrays to assess nucleotide status at specific loci, investigators are analyzing the possible utility that can be provided by assessing in an individual polygenic risk scores.

Hypertension can arise from altered functions at in several sites in the body including endocrine, renal systems and altered neural functions. It can arise as a result of alterations in levels of the pituitary hormone ACTH, or abnormalities in productions of adrenocortical hormones, corticosterone and adrenal derived mineralocorticoids. Some forms of hypertension are referred to a neurogenic hypertension and are related to chronic sympathetic nervous system activity. Specific central nervous system regions have been reported to play roles in hypertension, particularly in response to psychological stress.

Epigenetic factors including altered methylation at specific genomic sites, have been reported to play roles in hypertension.

A comprehensive study of 277,005 individuals led to the identification of 314 loci that influenced hypertension risk. In addition, in this study life style factors were reported to play important roles in hypertension etiology. These factors included sedentary life style, high salt intake, increased body mass and smoking. Results of this study revealed that lifestyle factors most significantly influenced blood pressure levels.

Extensive genome-wide studies were carried out to identify genetic variants that play roles in stroke etiology and led to identification of 32 significant loci with variants the impacted stroke risk. Products connected with stroke risk loci functioned in several different physiological pathways including the coagulation pathway and nitric oxide pathway.

Analyses of polygenic risk scores revealed that high polygenic risk scores increased stroke risk by 35%. Unfavorable life style increased stroke risk by 66%.

Small vessel brain disease is now recognized as an important cause of stroke. Specific single gene mutations have been found to be causative in certain cases of this disorder. Cadasil is a specific autosomal dominant arteriopathy due to a mutation in the NOTCH gene on human chromosome 19 that impacts cerebral circulation.

Presence of the APOE4 allele has been shown to be associated with increased deposition of amyloid in cerebral vessels, a condition referred to amyloid angiopathy that increases risk for stroke.

Atherothrombotic events can impact blood vessels in the heart and in the brain and can lead to stroke. Specific variants in coagulation factors and variants in the thrombin receptor can increase the frequency of thrombotic events.

Increased environmental pollution has been shown to increase risk of stroke.

Detailed studies in patients with coronary heart disease have provided evidence that both rare gene variants and common variants increase the risk for coronary heart disease (CAD). In some case the functional effects of the identified risk variants have been determined. Key functions impacted include levels and properties of lipoproteins, altered levels of insulin resistance, alterations in coagulation factors, nitric oxide signaling and vascular tone have also been found.

Importantly 38% of the heritability of coronary heart disease was reported to be attributable to common genetic variants. The specific genes with variants that significantly impact risk of CAD were found to be different in different populations.

Studies reveal that clinically important factors to be considered in management of coronary heart disease risk include systolic blood pressure, lipid levels and possible presence of diabetes mellitus. Smoking is also known to increase risk of CAD.

Increasing evidence of the role of polygenic risk variants in complex common diseases points to the importance of considering gene-gene interactions. Increased efforts are also been focused on determination of the role of regulatory elements and on alterations of levels of gene expression in complex common diseases.

Genetic and metabolomic studies have been undertaken to identify gene variants associated with an increased risk of development of diabetes mellitus. Evidence indicates that 40% of the risk for this condition is due to genetic factors.

Comprehensive genomic analyses in individuals with psychiatric disorders have revealed that there is genetic variant overlap in the different clinically distinguished psychiatric disorders, e.g. bipolar disease, schizophrenia, attention deficit disorder, depression.

Rare genetic defects and approaches to management

Rare genetic variants may alter the impact of certain environmental factors and responses to infections. One example is that individuals with genetically determined red blood cell disorders including hemoglobinopathies, are at increased risk for severe complications of Parvo virus infection.

Individuals with cystic fibrosis due to pathogenic mutations in the CFTR gene, are at increased risk for severe respiratory tract infection due to *Staphylococcus aureus*, Pseudomonas species and certain other organisms including fungi.

Adjusting environments to compensate for inborn errors of metabolism

Accurate diagnosis is important for optimal management of inborn errors of metabolism. Accurate diagnosis of these disorders has been facilitated by the development of new technologies such as metabolomics and nucleic acid sequencing. Other important advances in diagnostic procedures have occurred in proteomics in enzyme assay methods and advances in molecular histopathology. Recently advances in glycomics have also occurred.

Analyses of mitochondrial function currently can include metabolite analysis, and analysis of individuals mitochondrial enzymes. In addition functions of individual respiratory complex systems can be analyzed. Histology and molecular histopathology studies can also be valuable.

Early diagnoses due to detection of inborn errors of metabolism in newborn screening programs has improved outlook.

Treatment of inborn metabolic disorders can include dietary manipulations to limit intake of nutrients that cannot be appropriately handled due to the genetic defect, e.g. low phenylalanine diet in cases of phenylketonuria. In cases of urea cycle defects for example protein intake must be monitored to be only adequate and specific amino acids may need to be added to the diet. In certain inborn errors enzyme function can be a boosted by administration of extra-supplies of coenzymes including vitamin related substances. In some inborn errors f metabolism therapeutic advances have been made to restore normal forms of gene products, e.g. enzyme replacement therapy in certain lysosomal storage diseases.

Specific inborn errors of metabolism may render individuals abnormally sensitive to sunlight and this must be taken into account. Abnormal sensitivity to sunlight may occur individuals with disorders in DNA damage repair. Increased sunlight sensitivity also occurs in individuals with some forms of porphyria.

Networks and gene interactions

Knowledge has expanded on the propensity of proteins that are individually produced by different genes, to form complexes in order to function appropriately. In addition, there is growing evidence that genes encode products that function in specific networks. It is therefore necessary to form broader concepts regarding the number of genes that can be implicated in a specific disorder. Since the products of different genes need to work together in specific functions it may not be sufficient to address defects in a single or the product of a single gene when designing therapy for a specific genetic disorder.

Increasingly attention will need to be focused on genetic variation and on interacting factors in internal and external environments. Health also depends on neuronal processes that can be influenced by societal factors. Health is also dependent on adaptive measures and therapeutic approaches that foster adaptions will continue to prove valuable.

Chapter 11 Environments, resources, and health

Greenhouse gas production climate change and potential solutions

There is clear evidence of increasing concentrations of carbon dioxide (CO_2) and other greenhouse gases, produced by human activities, that damage the health of the planet. Detailed scientific studies are being applied to study the mechanisms through which greenhouse gases alter climate and affect the planet. The anthropogenic generated gas increases now threaten life on the planet through initiating climate changes associated with increased levels of thermal stress and fire; altered wind and ocean current flow patterns increase risk of severe weather events, cyclones and floods.

Fortunately, increasing levels of human energy and ingenuity are being applied to the search for energy production systems that do not generate greenhouse gases.

Although humans have benefited greatly from expansion of agricultural activities over the past century, human stewardship of land, soil, and water resources have been inadequate. Extensive efforts are now being applied to analyses of soil conditions and agricultural practices that favor soil health and that do not waste water resources.

Climate change, societal unrest and violence lead to growing levels of human migrations from different regions of the world. In addition, patterns and frequencies of infectious disease outbreaks have emerged.

Substantial growth in human populations have occurred in recent decades. Concerns are raised about abilities to provide food sources for a continually growing human population. A startling disturbing fact however is that in a number of regions of the world there is over consumption of energy rich foods along with decreased physical activity, leading to health problems while in other regions of the world malnutrition and undernutrition remain significant problems.

The increasing size of the world human population and the increasing lifespans achieved by humans reflect the successes in agriculture and medical care over previous decades. However, societies have yet to successfully adapt to the larger number of older individuals and to appropriately incorporate their skills and remaining energies into beneficial activities.

The one health concept

A new concept of health has been developed. In this concept health security takes into account human, animal, plant and ecosystem health and biodiversity.

Measures to promote ecosystem health and human health are particularly important for children and young individuals, upon our future depends.

Through various activities in education, literature and the arts, there are individuals who tirelessly document and celebrate the ecosystem for its role in human well-being, beyond the supply of basic life materials, food and shelter.

Index

Note: Page numbers followed by *f* indicate figures and *t* indicate tables.